Communications
in Computer and Information Science 103

Prahlad Vadakkepat Jong-Hwan Kim
Norbert Jesse Abdullah Al Mamun
Tan Kok Kiong Jacky Baltes John Anderson
Igor Verner David Ahlgren (Eds.)

Trends in Intelligent Robotics

13th FIRA Robot World Congress, FIRA 2010
Bangalore, India, September 15-17, 2010
Proceedings

Main Editors

Prahlad Vadakkepat
National University of Singapore
E-mail: prahlad@nus.edu.sg

Jong-Hwan Kim
Korea Advanced Institute of Science and Technology, Daejeon, South Korea
E-mail: jhkim@ee.kaist.ac.kr

Associate Editors

Norbert Jesse
Technische Universität Dortmund, Germany
E-mail: norbert.jesse@udo.edu

Abdullah Al Mamun
National University of Singapore
E-mail: eleaam@nus.edu.sg

Tan Kok Kiong
National University of Singapore
E-mail: eletankk@nus.edu.sg

Jacky Baltes
University of Manitoba, Winnipeg, Manitoba, Canada
E-mail: jacky@cs.umanitoba.ca

John Anderson
University of Manitoba, Winnipeg, Manitoba, Canada
E-mail: andersj@cs.umanitoba.ca

Igor Verner
Technion - Israel Institute of Technology, Haifa, Israel
E-mail: ttrigor@tx.technion.ac.il

David Ahlgren
Trinity College, Hartford, CT, USA
E-mail: david.ahlgren@trincoll.edu

Library of Congress Control Number: 2010934021

CR Subject Classification (1998): I.2, H.4, H.3, C.2.4, I.2.11, F.1

ISSN 1865-0929
ISBN-10 3-642-15809-9 Springer Berlin Heidelberg New York
ISBN-13 978-3-642-15809-4 Springer Berlin Heidelberg New York

springer.com

Printed in Germany

Typesetting: Camera-ready by author, data conversion by Scientific Publishing Services, Chennai, India
Printed on acid-free paper 06/3180 5 4 3 2 1 0

Preface

This volume contains the papers selected for the 13th FIRA Robot World Congress, held at Amrita Vishwa Vidyapeetham Bangalore, India, September 15-17, 2010.

The Federation of International Robot-soccer Association (FIRA – www.fira.net) is a non-profit organization that annually organizes robotic competitions and meetings around the globe. The robot soccer competitions started in 1996, and FIRA was established on, June 5, 1997. The robot soccer competitions are aimed at promoting the spirit of science and technology to the younger generation. The congress is a forum to share ideas and future directions of technologies, and to enlarge the human networks in the robotics area.

The objectives of the FIRA Cup and Congress are to explore the technical developments and achievements in the field of robotics, and provide participants with a robot festival including technical presentations, robot soccer competitions, and exhibits under the theme "Where Theory and Practice Meet." FIRA India aims to propagate and popularize robotics and robotic competitions across India.

Under the umbrella of the 13th FIRA Robot World Congress 2010, four international conferences were held for greater impact and scientific exchange:

- 7th International Conference on Computational Intelligence, Robotics and Autonomous Systems (CIRAS)
- 2nd International Conference on Advanced Humanoid Robotics Research (ICAHRR)
- 2nd International Conference on Entertainment Robotics (ICER)
- 2nd International Robotics Education Forum (IREF)

This volume consists of selected quality papers from the four conferences. The volume is intended to inform readers of recent technical advances in autonomous underwater vehicles, humanoid robotics, cooperative robotics, and related fields.

The volume contains the 46 papers contributed to the FIRA Robot World Congress 2010. It has been organized into seven sections on the following topics: autonomous underwater vehicles, humanoid robotics and multi-legged systems, multi-agent/robot systems, path planning and navigation, robotics and education, pattern analysis and classification, and control and sensing.

The editors hope that this volume is informative to the readers. We thank Springer for undertaking its publication.

July 2010 Prahlad Vadakkepat

Organization

International Advisory Board

Jong Hwan Kim	KAIST, Republic of Korea
Mantha SS	AICTE, India
Lee Tong Heng	ECE, NUS, Singapore
Saurabh Srivastava	NASSCOM, India
Venkat Rangan	Amrita Viswa Vidyapeetham, India

General Chair

Prahlad Vadakkepat	National University of Singapore, Singapore

Conference Committees

CIRAS - International Conference on Computational Intelligence, Robotics and Autonomous Systems

General Chair	Prahlad Vadakkepat, National University of Singapore, Singapore
Program Chairs	Abdullah Al Mamun, National University of Singapore, Singapore
	Tan Kok Kiong, National University of Singapore, Singapore

ICAHRR - International Conference on Advanced Humanoid Robotics Research

General Chair	Jacky Baltes, University of Manitoba, Canada
Program Chair	John Anderson, University of Manitoba, Canada

ICER - International Conference on Entertainment Robotics

General Chair	Norbert Jesse, Technical University of Dortmund, Germany

IREF - International Robotics Education

General Chair	Igor Verner, Israel Institute of Technology, Israel
Program Chair	David Ahlgren, Trinity College, USA

Technical Program Committee

Secretariat

Local Organizing Committee

Table of Contents

Autonomous Underwater Vehicles

Humanoid Robotics and Multi Legged Systems

Multi-Agent / Robot Systems

Path Planning and Navigation

Robotics and Education

Pattern Analysis and Classification

Control and Sensing

Navigation of Autonomous Underwater Vehicle Using Extended Kalman Filter

Ranjan T.N[1], Arun Nherakkol[1], and Gajanan Navelkar[2]

[1] VIT University, Tamil Nadu, India
ranjanmsritster@gmail.com, arunnherakkol@gmail.com
[2] Marine Instrumentation Division,
National Institute of Oceanography, Dona Paula, Goa, India
navelkar@nio.org

Abstract. To navigate the Autonomous Underwater Vehicle (AUV) accurately is one of the most important aspects in its application. A truly autonomous vehicle must determine its position which requires the optimal integration of all available attitude and velocity signals. This paper investigates the extended Kalman Filtering (EKF) method to merge asynchronous heading, attitude, velocity and Global Positioning System (GPS) information to produce a single state vector. Dead reckoning determines the vehicle's position by calculating the distance travelled using its measured speed and time interval. The vehicle takes GPS fixes whenever available to reduce the position error and fuses the measurements for position estimation. The implementation of this algorithm with EKF provides better tracking of the trajectory for underwater missions of longer durations.

Keywords: Navigation, Extended Kalman Filter (EKF), Autonomous Underwater Vehicle (AUV), data fusion.

1 Introduction

Precise navigation remains a substantial challenge to all platforms moving underwater [1] due to unavailability of GPS signals. With the emergence of new applications and the growing acceptance of AUVs for both military and civilian use, the need for enhanced accuracy and robustness in under-water navigation is required. Also this leads to the emerging applications like fully autonomous naval operations for mine detection, polar deployments and under ice surveys for oceanographic research, pipeline inspection and other works related to oil exploration [2],[3]. The process and sensor noise levels present are responsible for errors in the position estimated which is important for navigating underwater autonomously. Surface navigation of AUV can be accomplished with the use of the GPS. However, the high frequency signals from GPS satellites cannot travel through the water to the AUV. Therefore, another method needs to be incorporated in to the system for under-water navigation.

Methods like geophysical, acoustic and image-based navigation can also be used for AUV trajectory tracking [4]. An accurate map of the environment combined with measurements of geophysical parameters, such as bathymetry, magnetic field, or

P. Vadakkepat et al. (Eds.): FIRA 2010, CCIS 103, pp. 1–9, 2010.

gravitational anomalies is required for geophysical and image-based navigation and is technically not viable for a survey application. Acoustic navigation requires deployment of array of acoustic transducers to track the vehicle trajectory. Dead reckoning is a more graceful solution in such applications.

2 Dead Reckoning

Dead reckoning is the process of measuring vehicle attitude (roll, pitch, and yaw) and velocity (forward, lateral, and water body currents) and integrating these measurements over time from an initial position to determine the current position. Attitude and velocity measurement errors will cause the integrated position estimate to deviate from the true position over time. Depending on the estimated position error built up over time, the AUV resurfaces to get the GPS-referenced position. This improves the accuracy, robustness and sustainability of the integrated INS/DVL/GPS navigation system presented in the study.

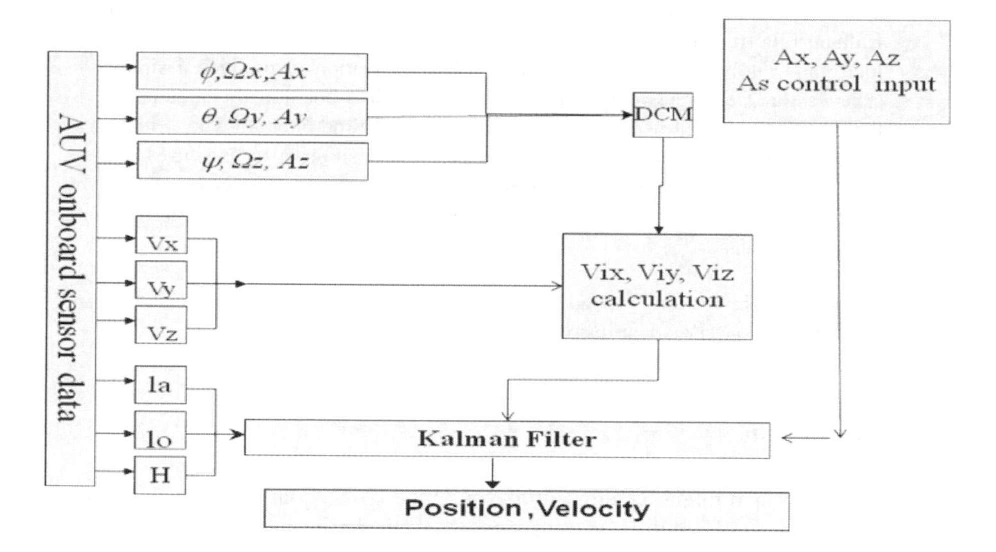

Fig. 1. Dead reckoning block diagram

Roll, pitch & yaw from INS are used to calculate the direction cosine matrix (DCM) [eqn.3]. DCM is used to transform velocity data (Vx,Vy & Vz) in body-reference frame to earth-reference frame(inertial) [eqn.4]. Latitude, Longitude information is supplied as measurement input to correct Z [eqn.5]. If this information is not available, the EKF takes velocity values to make the estimates with the help of dead reckoning (Fig.1).

On board sensors used for AUV navigation provide measurements in different co-ordinate frames and for using dead reckoning requires the acquired data to be re-framed in identical projections before integrating them.

2.1 Transformation of GPS Data

The GPS is a space-based global navigation satellite system that provides reliable location and time information in all weather and at all times and anywhere on the Earth where there is an unobstructed line of sight to four or more GPS satellites. In present case, the latitude and longitude measured from the GPS is converted to Universal Transverse Mercator (UTM) coordinate system [5], which is a grid-based method of specifying locations on the surface of the Earth as a practical application of a 2-dimensional Cartesian coordinate system.

The latitude and longitude measured from GPS is converted in to meters as shown below.

$$X = (\mathrm{lo} - \mathrm{lf}) * \cos(\mathrm{laf}/(180 * \pi)) * 60 * 1852 \quad (1)$$

$$Y = (\mathrm{la} - \mathrm{laf}) * 60 * 1852 \quad (2)$$

Where X - Northing in meters, Y - Easting in meters, Lo - longitude in decimal degrees, la - latitude in decimal degrees, lf - longitude reference, laf - latitude reference (i.e. Latitude & Longitude at starting point).

2.2 Transformation of DVL Velocity into Earth-Fixed –Frame

Components of velocity from Doppler Velocity Log (DVL) sensor are obtained in the body-frame (b-frame) of reference of AUV. For using these measurements with inertial navigation system (INS), they have to be transformed into a common coordinate system [6]. In the designed integrated positioning system, g-frame (inertial-frame) has been chosen as the common frame of reference for velocity calculations. The coordinate transformation requires knowledge of the orientation angles of AUV body with respect to the Navigation frame. The orientation angles (roll ϕ, pitch θ and heading φ) from Attitude Heading Reference System (AHRS – three axis sensor) have been used to calculate a coordinate transformation matrix (C_{body}^{nav}) from the body-frame to the inertial frame of reference. Transformation matrix for DVL velocities from body-frame to g-frame is given below.

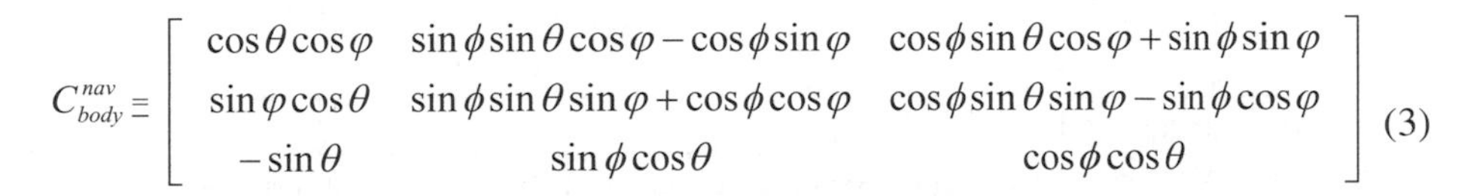

$$C_{body}^{nav} = \begin{bmatrix} \cos\theta\cos\varphi & \sin\phi\sin\theta\cos\varphi - \cos\phi\sin\varphi & \cos\phi\sin\theta\cos\varphi + \sin\phi\sin\varphi \\ \sin\varphi\cos\theta & \sin\phi\sin\theta\sin\varphi + \cos\phi\cos\varphi & \cos\phi\sin\theta\sin\varphi - \sin\phi\cos\varphi \\ -\sin\theta & \sin\phi\cos\theta & \cos\phi\cos\theta \end{bmatrix} \quad (3)$$

$$[\mathbf{V}_{ix}, \mathbf{V}_{iy}, \mathbf{V}_{iz}]^{T} = C_{body}^{nav} [\mathbf{V_x}, \mathbf{V_y}, \mathbf{V_z}]^{T} \quad (4)$$

V_{ix}, V_{iy}, V_{iz} → Components of AUV velocity in inertial frame (g-frame).

V_x, V_y, V_z → Components of AUV velocity in body-frame (b-frame),

ϕ, θ, φ → Orientation angles of b-frame with respect to g-frame (roll, pitch, yaw),

2.3 Methodology Used for Navigation

The received sensor data often contains partial and diminished information due to high update rates of the sensor. During the calculation of rotation matrix it is required to consider complete AHRS strings. The strings that contain erratic time tokens, incomplete angle tokens and acceleration values have been neglected. The same approach has been used for DVL and GPS measurements. The fusion code must take care of the GPS status (valid or in-valid), gap in UTM time, missing tokens etc. The erroneous velocity values are not considered for the fusion. The methodology adapted has been shown in flow chart below (fig.2).

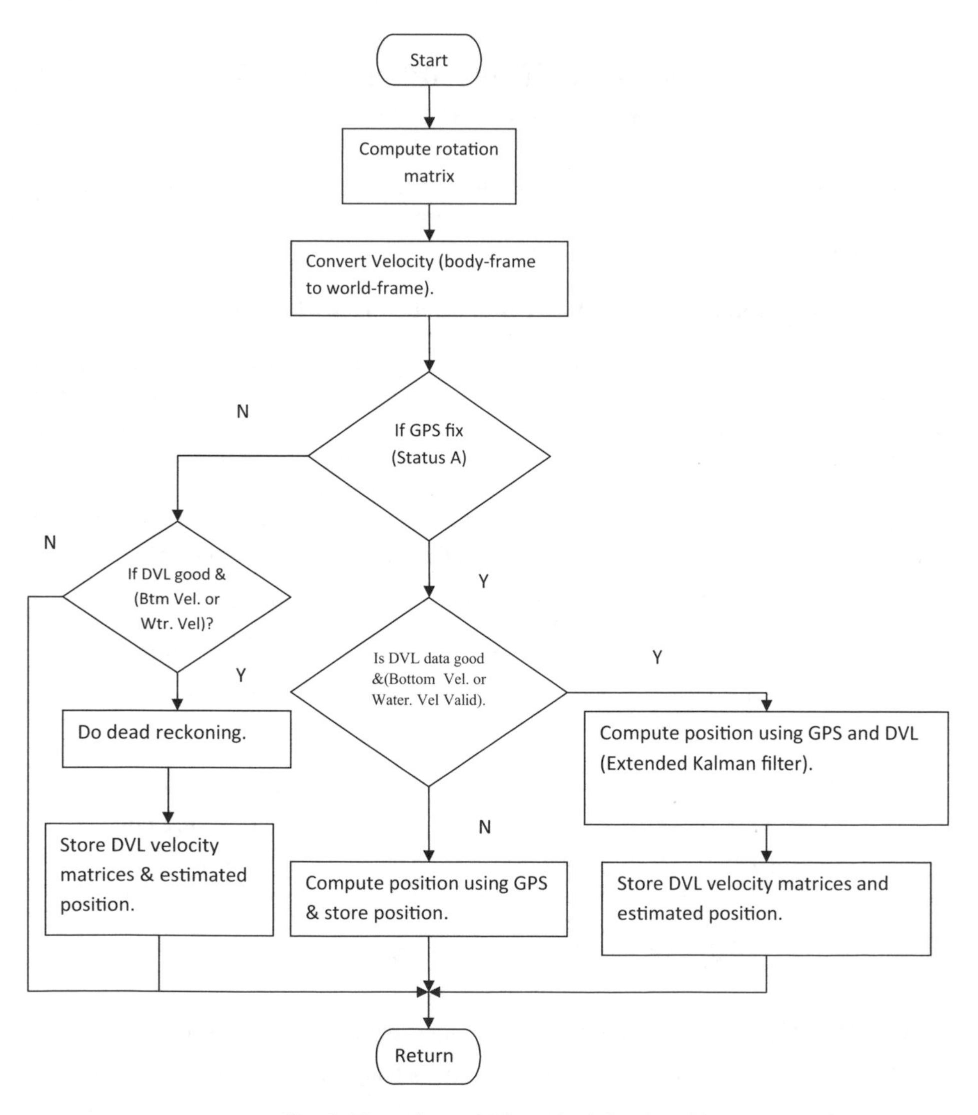

Fig. 2. Flow chart of the methodology used

3 Discrete Kalman Filter Algorithm

A Kalman filter is a programmed process that provides a best estimate of state variables with a least square error for a given a set of measurement data and model. The data fusion from the INS and the aiding sensors is typically accomplished using an Extended Kalman filter (EKF). The Kalman filter addresses the general problem of trying to estimate the state of a discrete-time controlled process that is governed by the linear stochastic difference equation [7].

$$\begin{aligned} x_k &= Ax_{k-1} + Bu_{k-1} + w_{k-1} \\ z_k &= Hx_k + (v_k) \end{aligned} \tag{5}$$

In these equations k is the time step, A is the state transition model which is applied to the previous state X_{k-1} , B is the control-input model which is applied to the control vector U_{k-1} , H is the observation model which maps the true state space into the observed space and Z is the measurement. The random variables w_{k-1} and v_k represent the process and measurement noise.

The EKF process is similar to a feedback control, i.e., the filter estimates the process state at some time and then obtains feedback in the form of (noisy) measurements. The current state estimate x_k is corrected by an amount that depends on the error between predicted and measured value. This is done with a gain that provides a set of influence coefficients on each state estimate.

The equations for the Kalman filter fall into two groups: *time update* equations and *measurement update* equations. The time update equations are responsible for projecting forward (in time) the current state and error covariance estimates to obtain the *a priori* estimates for the next time step.

The measurement update equations are responsible for the feedback, i.e. for incorporating a new measurement into the *a priori* estimate to obtain an improved *a posteriori* estimate.

Time update equations:

$$\begin{aligned} x_k^- &= Ax_{k-1} + Gw_{k-1} \\ P_k^- &= AP_{k-1}A^T + BQB^T + Q \end{aligned} \tag{6}$$

Measurement updates equations:

$$\begin{aligned} K_k &= P_k^- H^T (HP_k^- H^T + R)^{-1} \\ x_k &= x_k^- + K_k (z_k - Hx_k^-) \\ P_k &= (I - K_k H) P_k^- \end{aligned} \tag{7}$$

The time update equations can also be termed as *predictor* equations, while the measurement updates equations as *corrector* equations. P_k^- is Predicted covariance, x_k^- is predicted state, K_k is Kalman gain , P_k is corrected covariance and x_k is the corrected state at time step k. Modeling the bias or noise present in the process and sensors reduces the difference between predicted model and corrected model.

4 Experimental Setup

The experimental setup including vehicle particulars, employed navigation sensors, mission trajectories, and the processing of raw navigation data, is given in this section.

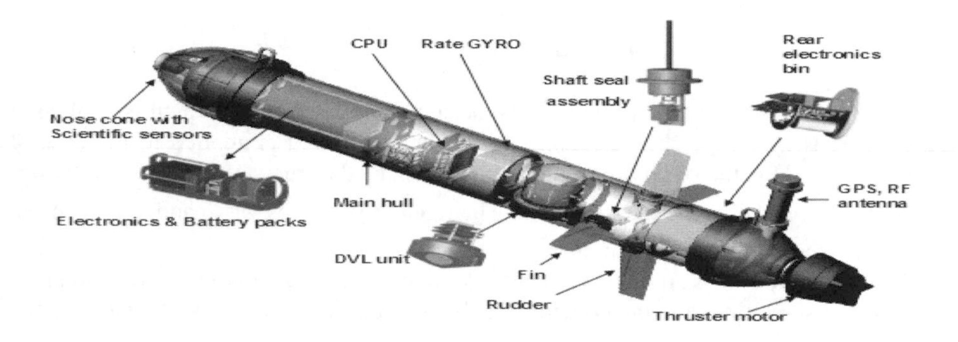

Fig. 3. AUV developed at NIO, Goa, India

A small AUV (Fig.3) has been developed at National Institute of Oceanography (NIO) for oceanographic applications [10]. Raw sensor data collected during deployments of this small AUV in the vicinity of 15° 28'N latitude and 74° 52'E longitude near Goa coast, India is used for evaluating the performance of Extended Kalman filter algorithm. The sensors used on AUV for navigation are AHRS [11], GPS [12], DVL [13] and a Pressure Sensor. AHRS measures the orientation (Euler angles), angular rates and linear accelerations of the vehicle. DVL measures velocity of the vehicle. The navigation sensors have different update rates and this data during AUV dives gets stored on a flash disk. Using this raw data stored, the performance testing of navigation algorithms and Kalman filter analysis has been carried out in MATLAB.

5 Results and Conclusions

The offset in the heading angle of the AUV is studied by comparing measured heading, GPS heading and true heading using the above algorithm for a surface run of AUV. The results obtained show that the the heading angle from the GPS path is different from the INS heading obtained (Fig.4). This difference is due to the currents in the sea and also errors that are not modelled in the system which affect the vehicle dynamics. Also the yaw rate and heading bias is shown in the Fig.5.

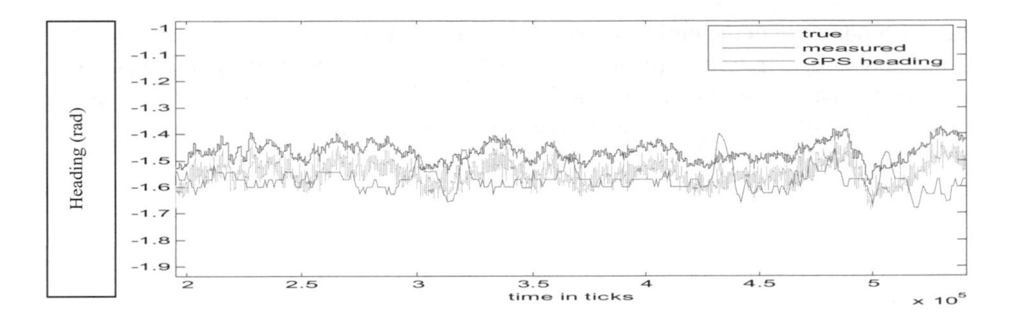

Fig. 4. Plot of Heading bias Vs Time

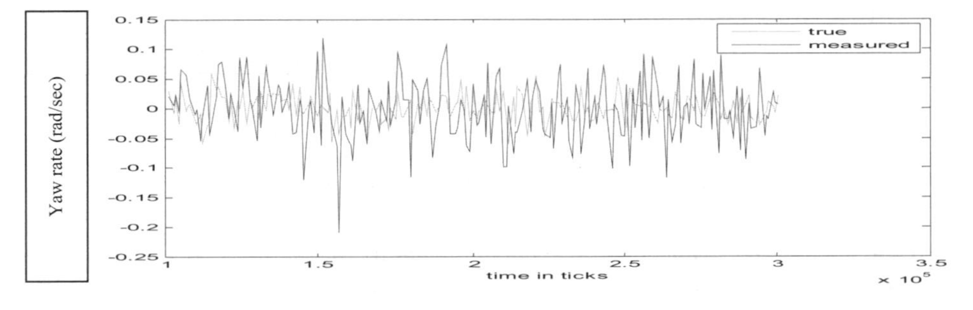

Fig. 5. Plot of Yaw rates bias Vs Time

The plots obtained for estimated AUV position using above algorithm with GPS available on surface [fig.6] and similar underwater mission [fig.7] of 500 meters are shown below.

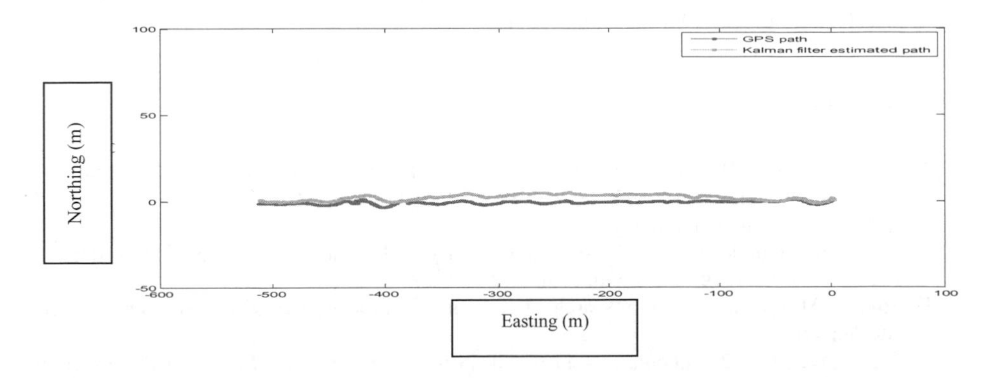

Fig. 6. Plot of predicted value Vs GPS

For surface mission, the GPS was available for the complete trajectory. In another experiment, GPS information was available for only for a short duration of about 20 meters at the start. After 20 m GPS signals were not available as AUV dived underwater. The Kalman filter predicted path is updated with the measurement information i.e. GPS positions in the initial 20. Now when the under-water mission starts, the EKF continues to update with estimated positions using above dead reckoning algorithm. It is noted that due to continuous integration of velocity information, the error in the estimated position increases. The trajectory obtained yielded a maximum deviation of about 10 meters for under-water mission.

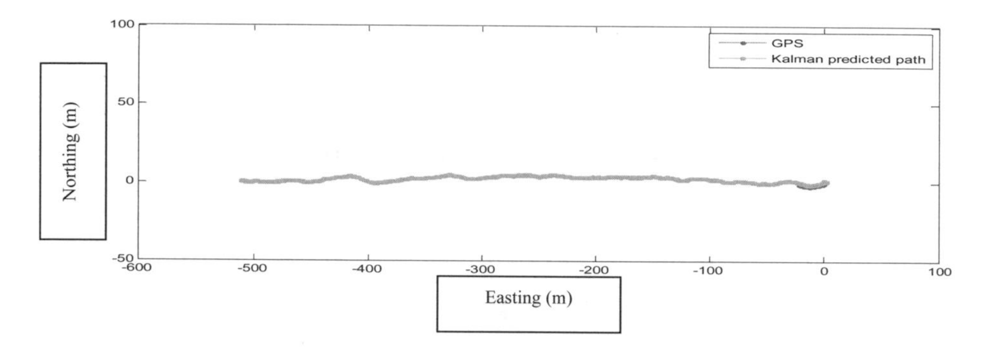

Fig. 7. plot of Kalman filter predicted path without GPS

Also it was observed in other runs (not presented here) that the AUV trajectory was tracked more accurately for linear missions compared to those with curved paths. Also the heading bias increases when the vehicle is following a curved trajectory.

Acknowledgements

We would like to thank Dr. S.R.Shetye, Director, NIO and Dr. Elgar Desa, Project leader of AUV project for allowing us to carry out this research work at National Institute of Oceanography, Dona Paula, Goa, India.

References

[1] Hegrenæs, Ø., Berglund, E.: Doppler Water-Track Aided Inertial Navigation for Autonomous Underwater Vehicles. In: Proceedings of the IEEE Oceans Conference and Exhibition, Bremen, Germany (2009)

[2] Autonomous undersea vehicle application center, Classification and application details, `http://www.ausi.org/auvs/auvs.html`

[3] Hugin AUVS in marine research-Today and Tomorrow-Per Espen Hagen & Nils Storkerson

[4] Barngrover, C.: Documents related to The Stingray Project. In: Sandiago Ibotics 2009 (2009), `http://cse.ucsd.edu/~cbarngrover/embedded.html`, `http://www.sdibotics.org/stingray.cfm`

[5] MAP Projections, A working Manual, US Geological Survey Paper 1395

[6] Datum Transformations of NAV420 Reference Frames - NAV420CA Application Note-Giri Baleri, Sr. Application Engineer,Crossbow Technology, Inc., `http://www.xbow.com/pdf/NAV420AppNote.pdf`

[7] An Introduction to the Kalman Filter Greg Welch and Gary Bishop, Department of Computer Science, University of North Carolina (2006)
[8] Grewal, M.S.: Application of Kalman filtering to the calibration and alignment of inertial navigation systems. In: Proceedings of the 29th Conference on Decision and Control, Hewall (December 1990)
[9] Integration of Inertial Navigation System and Global Positioning System Using Kalman Filtering -Theses Report- Vikas Kumar N. Dept of Aerospace Engineering, IIT Mumbai (July 2004)
[10] The small AUV Maya: Initial Field Results - Elgar Desa & Maya Team, National Institute of Oceanography, Goa, India, `http://drs.nio.org/drs/bitstream/2264/601/1/Int_Ocean_Syst_11_3.pdf`
[11] AHRS400 Series User's Manual for Model AHRS400CC- Release Revision A, Cross Bow Technology (June 2005)
[12] Lassen iQ GPS Receiver System Designer Reference Manual (February 2005)
[13] SonTek/YSI Argonaut-SL Application Notes (January 2007)

Safety Aspects for Underwater Vehicles

R. Madhan, G. Navelkar, Elgar Desa, Sanjeev Afzulpurkar, Shivanand Prabhudesai, Nitin Dabholkar, Antonio Mascarenhas, and Pramod Maurya

Marine Instrumentation, National Institute of Oceanography, Dona Paula, Goa, India 403004
madhan@nio.org

Abstract. Advances in technologies related to miniature sensors, memories, embedded controllers, power systems and materials has resulted in development of variety of autonomous underwater platforms for ocean exploration. The future in oceanographic instrumentation is intelligent small Autonomous Underwater Vehicles (AUV's), autonomous profilers, gliders [1], etc. The ultimate aim in all autonomous platforms research and development is to reach the stage of unescorted missions with minimum failures. This stresses for implementation of multiple safety measures of a high degree so that the platform operates continuously in a fail-safe mode. This paper discusses issues on safety measures implemented on the autonomous underwater platforms namely MAYA AUV and the Autonomous Vertical Profiler [AVP] developed at NIO, Goa. The safety aspects in Design & Construction, Operations, Monitoring and Emergency have been addressed.

Keywords: AUV, AVP, safety, collision.

1 Introduction

AUVs are free-swimming marine robots that require little or no human intervention. They are compact self-contained low drag profile crafts powered (in most cases but not all) by a single underwater DC thruster and are used for surveys and applications in the horizontal plane. In contrast, the AVPs profile the vertical water column. These vehicles use on-board computers, power packs and vehicle payloads for automatic control, navigation and guidance [2]. They are equipped with state-of –the-art scientific sensors to measure oceanic properties, or specialized biological and chemical payloads to detect marine life. The advent of autonomous operations with underwater vehicles, high cost and reliability of these platforms has resulted in incorporation of multiple safety features.

2 Safety

Multilayered safety needs to be incorporated from design stage in the hardware and software level of the vehicles. It is also essential that all safety measures are conceptualized at the initial design stage so as to incorporate the ideas in every sub system. These include:

P. Vadakkepat et al. (Eds.): FIRA 2010, CCIS 103, pp. 10–16, 2010.

- safety at design & construction level
- operational safety
- monitoring health
- emergency systems

These safety aspects are discussed in succeeding pages.

2.1 Design and Construction

The important issues in design& construction are the choice of materials and the sealing of static and dynamic ports [3]. These were addressed by:

(a) Selecting the right wall thickness of the hull by way of using standard pressure housing design tables followed by verification using FEM analysis.
(b) Confirming of material defects by performing non destructive tests before and after all machining
(c) Adding a small positive buoyancy by using divinylcell foam (~50 to 200g positive in case of smaller vehicles) so as to ensure that the system floats in case of any electronics failure. Right choice and testing on the foam is critical in the design process to verify the compressibility limits and minimum water absorption. Fig.1 shows the result of such a test wherein the super-high density (marked as SHD5) was seen to withstand the pressure of 10 bar while other varieties failed to do so.
(d) Using additional backup O-ring at critical openings to prevent water ingress. Special care needs to be taken to maintain machining accuracies at O-ring locations

In addition to the safety aspects it is necessary that the system design gives attention to a large separation between the Centre of Buoyancy (Cb) and the Centre of Gravity (Cg) which helps the vehicle to be stable in all conditions.

Fig. 1. Pressure test for Foam compressibility

2.2 Operational Safety

The major operational safety issues addressed are depth cutoff, bottom collision and obstacle avoidance. When the vehicle is rated to a particular depth, it is essential to prevent it from exceeding this limit. The AUV and the AVP has a pressure sensor to measure depth which is used to abort a mission in case of exceeding the design depth limit. Another aspect is the overshoot distance traveled by AVP, due to the momentum, after thruster power is switched off. The quantum of overshoot depends on the velocity for a given AVP buoyancy configuration. Fig.2 shows an overshoot of 2m for a programmed depth of 7m.

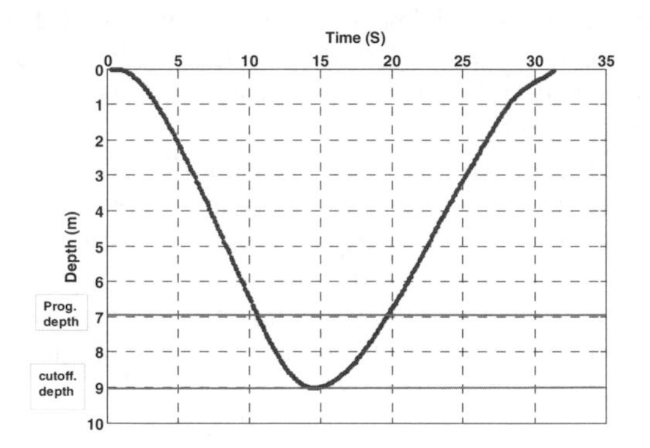

Fig. 2. Excess depth traveled

Overshoot can be minimized by ramping down thruster voltage such that AVP will just reach the programmed depth. The overshoot distance for different control voltages has been experimentally determined and the relationship between control voltage and overshoot distance has been incorporated in the software. Thus correct control voltage is applied at right time to the thruster to achieve the programmed depth.

A formula was derived from tests in field and incorporated in the system to control the overshoot distance. After implementation it was found that AVP traveled to 5.7m (Fig. 3), which is very close to the programmed depth of 6.0m. The experimental verification was carried out and the result is shown in Fig.3. The small deviation observed was because of the minor change in the overall buoyancy of the AVP which affects the velocity at different control voltages. This feature is particularly useful in shallow waters.

In case of profilers it is possible that the vehicle hits the bottom if the programmed depth is more than actual depth at the location of operation. To avoid bottom collision in such scenarios an echo sounder has been used at the AVP nose tip to detect the bottom and abort the dive when the echo range is less than programmed cut off range which is normally set to 5m.

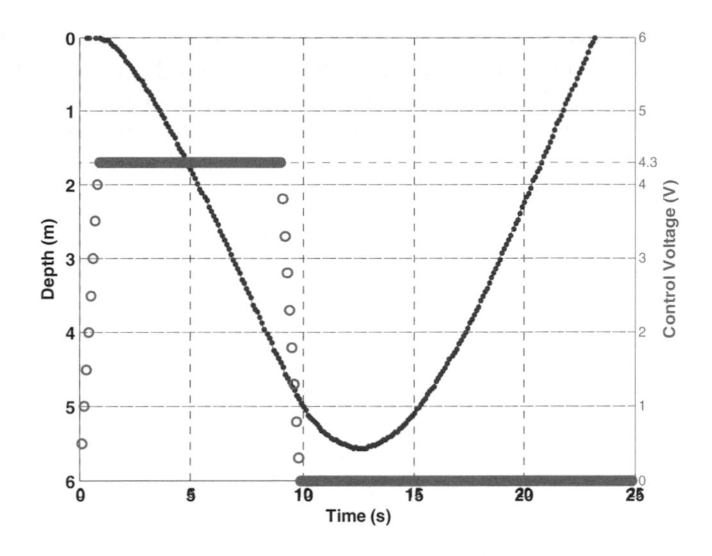

Fig. 3. Experimental verification of overshoot control

In the test the AVP was programmed to dive up to 7m, but performed an echo cut off by switching off thruster power at 5m from the bottom and floated up (Fig. 4).

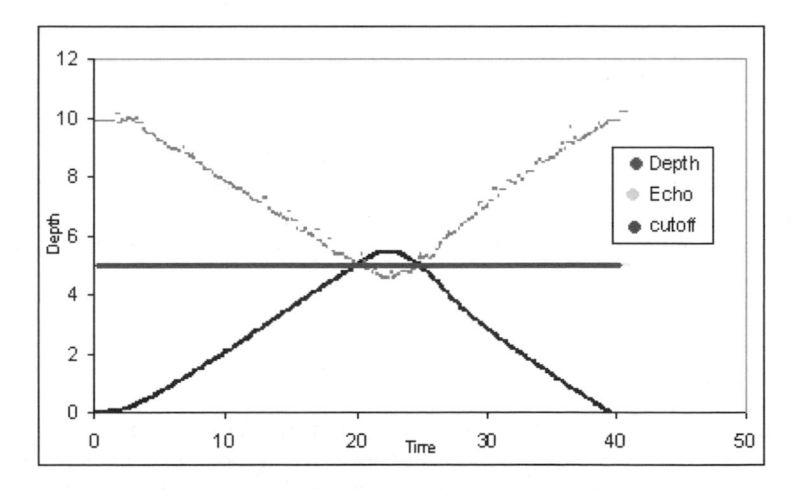

Fig. 4. Bottom collision avoidance

In comparison to the bottom collision safety feature of AVP, the AUV needs to avoid obstacles on its planned path. Thus Collision avoidance is an important aspect for AUVs while operating in confined spaces. Scanning sonar has been used for obtaining obstacle avoidance data ahead in the path of AUV [4]. The sonar has the capacity to determine a maximum range of 75 meters. With vehicle velocities around 1 m/s the obstacle avoidance has a maximum time of 75 seconds. The sonar, placed in the nose cone and controlled by a micro-controller node provides the range and angle data which is processed to determine the obstacles. The data obtained will be analyzed for any obstacle in the path, determine the range and the angle with respect to the AUV direction. This is conveyed to the navigation controller which executes the avoidance measure. Figure 5 below shows the overall interface for obstacle avoidance sonar.

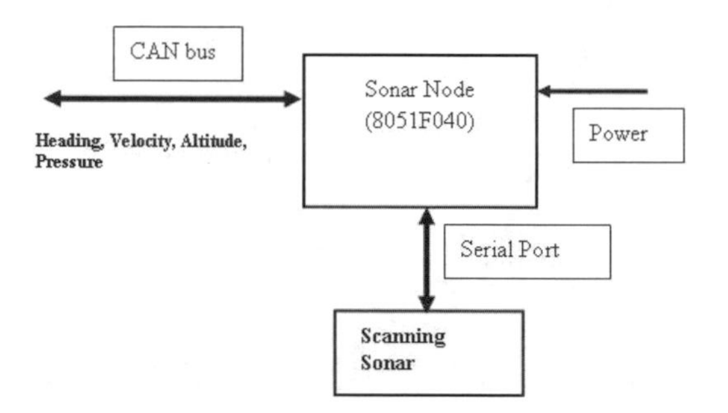

Fig. 5. Sonar control and data acquisition node

2.3 Monitoring

It is essential to take precautions in the form of monitoring several health parameters rather that taking action after a failure of a system. In this regard we monitor key parameters like battery level, temperature and humidity inside the housings which are good indicators of a failure.

In case of AVP a low battery level will lead to system shut down and it floats up. No further dives are performed. The AVP continues to communicate with the base station through RF modem. In case of AUV, the low battery level triggers a systematic shutdown of the controllers preventing loss of data. However the RF communication link is kept on to acquire AUV position for recovery.

An increase in temperature may indicate component failures or could be the result of a developing failure. An increase in humidity will indicate water entry into the hull. In both cases the mission is aborted and retrieval is carried out.

2.4 Emergency

Emergency will come into effect when the system (a) fails and the communication with the system is lost and (b) the system sinks or is trapped underwater. When on surface, in case of failure of the entire communication, a stand alone radio beacon cum flasher could be used to locate the system based on the last GPS position transmitted by the system within a range of 4 -5 kms.

A miniature acoustic pinger (tag) and receiver system is used to locate the autonomous platform. The complete acoustic tag receiver and transmitter system shown in Fig.6. The directional hydrophone provides greatest range and precision in locating acoustic tags mounted on the underwater autonomous platform. This has been successfully integrated with the AVP and tested over a range of 2km.

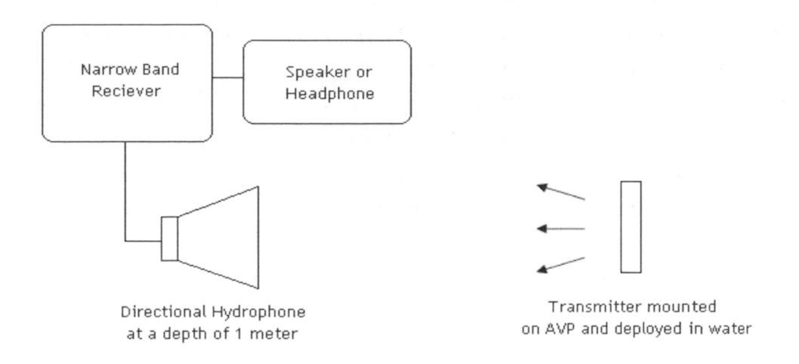

Fig. 6. Block Diagram of acoustic tag system

3 Conclusion

The safety features on depth cutoff, bottom collision, temperature, battery status and humidity monitoring have been incorporated in the AUV and AVP. These have improved reliability and increased confidence level in operations. Additional work is in progress on the obstacle avoidance feature for AUV and will be incorporated at a later stage.

The advantages in use of Marine Robots like AUVs, Profilers has resulted in an increase in their numbers worldwide. These vehicles now play an important role in costal zone monitoring, deep ocean profiling and climate research [5]. As the oceanographic community becomes more dependent on such vehicles, the issues of safety on these vehicles will be more challenging for the technologists. The aspects of safety discussed in this paper need to be used and improved for better reliability of underwater vehicles.

Acknowledgements

The authors are thankful to the Department of Information & Technology, New Delhi for supporting the AUV and AVP programs. Additionally, the authors would also like to thank all the project students who have contributed to the success of these projects.

References

1. Webb, D.C., et al.: SLOCUM: An Underwater Glider Propelled by Environmental Energy. IEEE Journal Of Oceanic Engineering 26(4), 447–452 (2001)
2. Dabholkar, N., et al.: Development of an Autonomous Vertical Profiler for Oceanographic Studies. In: International Symposium on Ocean Electronics, SYMPOL 2007 (2007)
3. Madhan, R., Desa, E., Navelkar, G., Prabhudesai, S., Afzulpurkar, S., Dabholkar, N., Mascarenhas, A., Maurya, P.: An Autonomous Zone Keeping Vertical Profiler for Coastal Oceanography - Concept and Challenges, USYS, Bali (2008)

4. Afzulpurkar, S., Desa, E., Navelkar, G.S., Mascarenhas, A.A.M.Q., Maurya, P.K., Martins, H., Madhan, R., Prabhudesai, S., Pinto, R., Marchon, N.: Miniature sonar for obstacle detection on small AUV Maya. In: Pillai, P.R.S., Supriya, M.H. (eds.) Proceedings of the International Symposium on Ocean Electronics (SYMPOL 2007), December 11-14, pp. 244–249. Cochin University of Science and Technology (Cusat), Cochin (2007)
5. Desa, E.S., Maurya, P., Madhan, R., Navelkar, G., Desa, E.: Environmental monitoring of the coastal zone by robot platforms. In: International Workshop on Underwater Robotics - Sustainable Management of Marine Ecosystems and Environmental Monitoring- IUWR 2005, Genoa, Italy, November 9 - 11 (2005)

Autonomous Vertical Profiler Data Management

Sanjeev Afzulpurkar*, Gajanan Navelkar, Elgar Desa, R. Madhan, Nitin Dabholkar, Shivanand Prabhudesai, and Antonio Mascarenhas

National Institute of Oceanography
Dona Paula, Goa
India 403002
sanjeev@nio.org

Abstract. The Autonomous Vertical Profiler (AVP), developed at NIO [1] [2], collects position and water column data over a period of 3 days and transmits through a satellite modem which is collated and stored on a PC.

Data includes GPS positions, water column properties related to Conductivity-Temperature - Depth (CTD), Dissolved Oxygen (DO), Chlorophyll–a along with profiler status. The receiving satellite modem feeds data to the profiler GUI and is also deposited into the server. Software has been developed, using freeware Java development tools to acquire the data from the serial port, accumulate in the desired formats, archive and then is expected to be made available to users. The following paper describes the technique and resources used to implement the data management.

It is expected that there would be multiple profilers operating at various locations, such as coastal seas, dams and other water bodies. Data would be relayed for archival, processing and be made available to the communities who are interested in research, commercial activities etc., in oceans.

Keywords: Autonomous Vertical Profiler (AVP) , satellite modem, Graphical User Interface (GUI), Conductivity-Temperature - Depth (CTD), Dissolved Oxygen (DO).

1 Introduction

The autonomous vertical profiler, a powered profiler [3] [4], is developed for applications in fresh water bodies, coastal waters, deep seas etc. Operable to depths of 200 M maximum, it allows users to effectively sample the water column in the euphotic zone and beyond. The AVP is programmed before deployment and the users can configure it as per their data requirements. The AVP is shown in Fig. 1 while Table 1, outlines the specifications. The unique features of the AVP are

- decoupled from external perturbation of the ship/boat providing a true vertical profile
- repetitive dives offer adequate statistics of the profile shape variability
- use of an echo-sounder and pressure sensor to avoid crashing into the seabed.

* Corresponding author.

P. Vadakkepat et al. (Eds.): FIRA 2010, CCIS 103, pp. 17–24, 2010.

- capability to hover at any set depth
- autonomous profiling drifter in coastal and open ocean waters by reporting its GPS coordinates and reduced data
- Ease in operation and one man deployable.

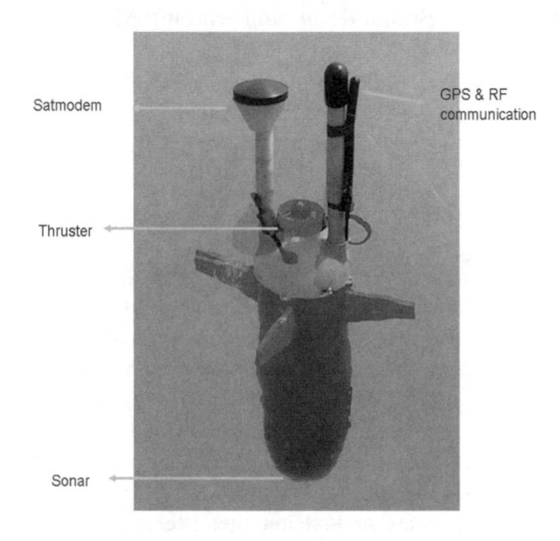

Fig. 1. Autonomous Vertical Profiler

Table 1. Autonomous Vertical Profiler Specifications

Length	1.17 M
Diameter	0.18 M
Weight	~13 to 16 Kgs based on the sensors
Rated Depth	200 M (max)
Hull	Aluminum alloy, Acetal nose & tail cones
Propulsion	Single DC thruster
Electronics	8051 and ARM7 microcontroller based
Communication	Radio modem (2.4 GHz) & Satellite Transmission (Iridium)
GUI	Labview based
Energy Source	Lithium Ion Polymer batteries (324 Whr)
Battery banks	Electronics & payloads - 12 V, 18Ah, Propulsion – 24 V, 4.5 Ah
Speed	0.4 – 1 m/s
Endurance	~ 3 days with 12 dives/day to 100 m

2 Communication

Communication with the AVP is through the satellite modem or through the RF modem. This is established using a GUI developed [5] for programming and field operation of the AVP. The GUI displays complete status and operational information related to the AVP. This includes the power, sensors, mission and other related parameters. The GUI is built with capability to process the data and show plots of parameters with depth. Fig. 2 shows the GUI front page.

The GUI is used to load the mission through the modem and then command it to execute the same. The AVP will continue to operate with programmed mission parameters till the end of mission. After every dive the AVP will transmit its position and the data is downloaded through the satellite modem/RF modem.

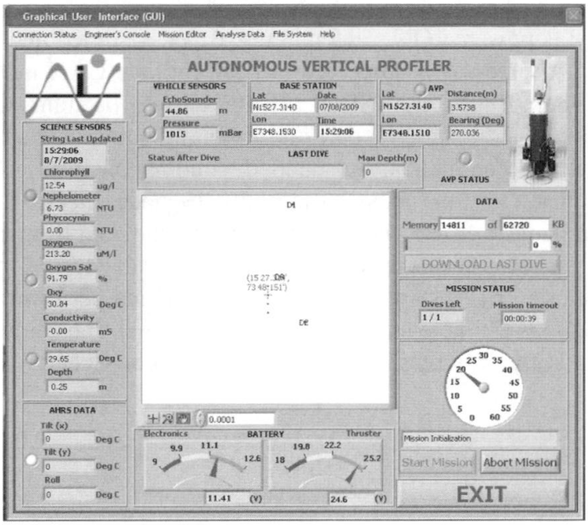

Fig. 2. AVP Graphical User Interface

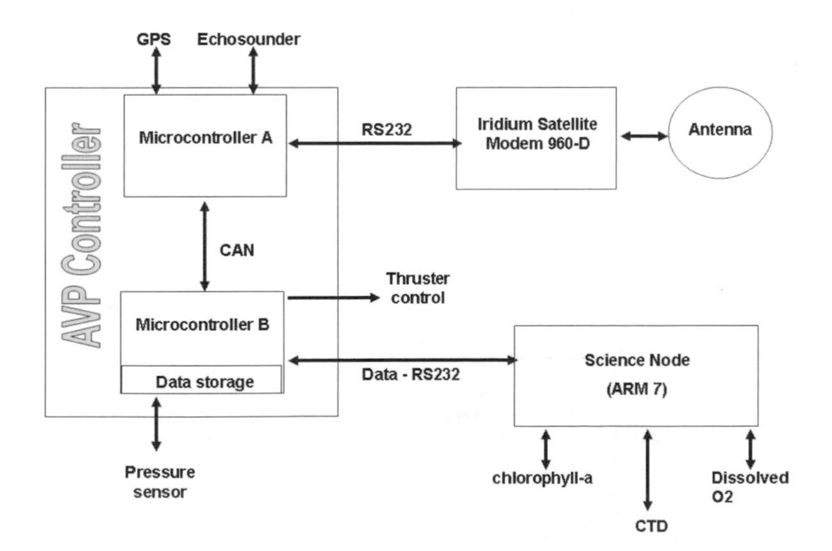

Fig. 3. Satellite modem interface to AVP electronics

The Iridium modem 9601-D [6] from NAL Research, USA is used for satellite communication. This allows only short-burst data (SBD) connectivity through the Iridium satellite network. Two modems will be used – one mounted inside the AVP and other on land communicating globally. Current AVP electronics uses one ARM 7 microcontroller as science node for data acquisition and two 8051 microcontrollers for control of AVP. One of the 8051 micro-controller is used for communicating with the satellite modem and interfaces to the main controller using CAN protocol for data transfer as shown in Fig 3.

When the AVP is powered ON, a 15 minutes timer is started and the 8051 modem node checks for CAN messages. On receipt of appropriate CAN message the modem node builds the Short Burst Data (SBD) [7] message to be transmitted via satellite. This includes latitude and longitude, date and time and battery status of the AVP. The SBD message will be continuously updated depending on the CAN posts received from main controller. Once the message is built, the node checks for required signal strength and if available, a SBDI session is performed and data from satellite modem is send to Iridium gateway. If successful then modem waits for 15 minutes and then sends the next data set or else starts checking the signal and performs SBDI again until it is successful. SBD is a simple and efficient bi-directional transport capability used to transfer messages with sizes ranging from zero (a mailbox check) to 1960 bytes for

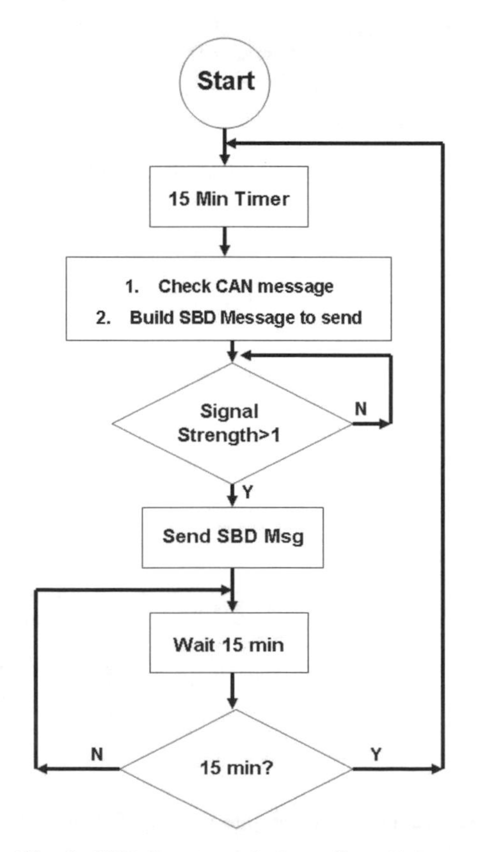

Fig. 4. AVP data transmission – flow diagram

Mobile Originated (MO-SBD) and zero to 1890 bytes for Mobile Terminated (MT-SBD). Whenever AVP sends the message, these are transmitted across the Iridium satellite network utilizing inter-satellite links to reach the gateway. From there, messages are disseminated to the receiving satellite modem. Global network transmit latency for delivery of messages ranges from ~5 seconds for short messages to ~20 seconds for maximum length messages. The ground system (PC) can receive these messages any time from any place through the receiving modem. More than one message will be queued like FIFO and are received sequentially. The data transfer between the AVP and the modem node is carried out as per the flow diagram in Fig. 4.

3 Data Management

The AVP generates lot of data during it's programmed mission period. Being autonomous and deployed at locations unattended, it is mandatory that the data be obtained on a continuous basis and archived so that users can access the same through a focal point. The objective is to develop a website that displays the acquired data graphically and stores raw as well as processed data.

Part of the development is done as per the flowcharts and has been divided as

a) Acquiring data from AVP through SAT Modem.
b) Tokenizing & Archiving of received data (raw).
c) Graphical Display of data on web page.
d) Archiving of the data (processed) & the charts.

NThis has been done using the JAVA freeware available over the net and includes Java Development kit (JDK 1.6), Netbeans IDE 6.0, Apache Tomcat Server 6.0, Servlet, JSP, JFree Charts and Java Comm API. A web portal needs to be developed as a central facility for storage and access of the data.

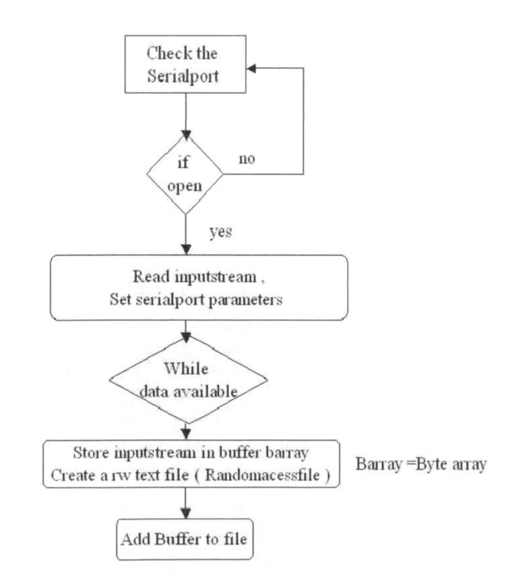

Fig. 5. Flow chart - acquire data

The data acquisition through satellite modem is done using Java Comm API by reading the serial port as per the flow diagram Fig. 5. Whenever data is received it is saved as a text file on the system (raw data file). The 'SimpleRead' program opens a serial port and creates a thread for asynchronously reading data through an event callback technique. The "inputstream" data is stored in a byte buffer (24 bits at a time). A text file with read and write (R/W) access is created. The buffer data is then added to the file. All subsequent data strings are read and appended to the same file.

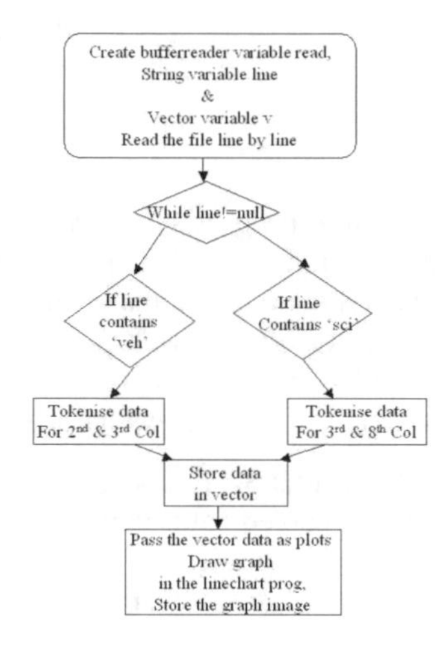

Fig. 6. Flow chart - tokenize data

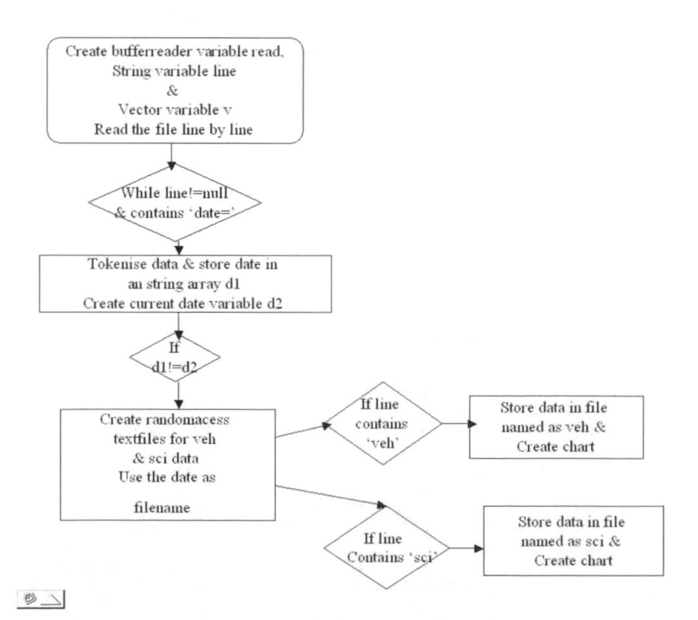

Fig. 7. Flow chart - Data archiving

Data is tokenized and archived as per the date on which it is received. In the 'CsvReader' program the data string from the text file is read line by line and is tokenized using "," delimiter as shown in Fig. 6. The text file consists of two types of strings: 'veh' & 'sci' . From the 'veh' string the 2nd & 3rd column data are saved as two different vectors. Similarly from the 'sci' string the 3rd & 6th column data are saved as two different vectors. These vectors are then called in the linechart program.

Archiving of the data is done by the program, as shown in Fig. 7. At first the text file is tokenized for the date string. If the date string does not match then two different text files namely date_veh.txt and date_sci.txt, are created and data is stored accordingly.

4 Conclusions

The data from AVP is transmitted to the base facility and acquired by the software developed for data management. The data is stored as a raw file as backup and processed so that users can access the same as per their requirements.

Further work involves the data to be made available on a website with appropriate security features that address the concerns of groups deploying the AVP's and those of general users. Live graphical displays need to be embedded on this website. Facility to download the data has to be incorporated on this site.

Future operations of multiple AVP's, their data acquisition and archiving needs to be integrated into this website. Capability to program/modify the AVP missions may also be added to the site. Stability of the entire AVP network to be operational over the lifetime of AVP's needs to be established.

Acknowledgements

The authors wish to acknowledge the grant received from Department of Information and Technology, New Delhi, India in supporting this development work. We thank Director NIO for the kind support. Finally we wish to acknowledge all the students who contributed towards this work.

References

1. Dabholkar, N., Desa, E., Afzulpurkar, S., Madhan, R., Mascarenhas, A.A.M.Q., Navelkar, G.S., Maurya, P.K., Prabhudesai, S., Nagvekar, S., Martins, H., Sawkar, G., Fernandes, P., Manoj, K.K.: Development of an autonomous vertical profiler for oceanographic studies. In: Pillai, P.R.S., Supriya, M.H. (eds.) Proceedings of the International Symposium on Ocean Electronics (SYMPOL 2007), December11-14, pp. 250–256. Cochin University of Science and Technology (CUSAT), Cochin (2007)
2. Madhan[1], R., Desa[2], E., Navelkar, G., Prabhudesai, S., Afzulpurkar, S., Dabholkar, N., Mascarenhas, A., Maurya, P.: An Autonomous Zone Keeping Vertical Profiler for Coastal Oceanography - Concept and Challenges. USYS, Bali (2008)
3. Maclane Moored Profiler, http://www.mclanelabs.com/mmp.html

4. Toole, J.M., Doherty, K.W., Frye, D.E., Millard, R.C.: A Wire-Guided Free Fall System to Facilitate Shipborne Hydrographic Profiling. American Meteorological Society, 667–675 (1997)
5. Sawkar, G., Maurya, P.K., Afzulpurkar, S., Desa, E., Navelkar, G., Prabhudesai, S., Madhan, R., Mascarenhas, A., Dabholkar, N., Martins, H., Fernandes, P., Manoj, K.K.: A Lab View based Graphical User Interface for Autonomous Vertical Profiler. NIO Internet Report
6. 9601-D manual, ftp://ftp.nalresearch.com/SatelliteProducts/StandardModems/9601-D
7. Network Reference - SBD, http://www.nalresearch.com/NetRef_SBD.html

Navigation and Position Control of Underwater Vehicle ROSUB 6000

Manecius Selvakumar Joseph, Atmanand M.A., Ramadass G.A., Ramesh Raju, and Jayakumar V.K.

Submersibles & Gas Hydrates, National Institute of Ocean technology, Chennai, Tamil Nadu, India
{manecius,atma,ramadass,rramesh,v_k_jayakumar}@niot.res.in

Abstract. Remotely operated vehicles are the main tools in exploring and exploiting subsea resources. The design, development and implementation of navigation control algorithms for deepwater work class Remotely Operated Vehicle – ROSUB 6000 is described in detail in this paper. The objective of this paper is to present the mathematical formulation and dynamics of ROV along with simulation results towards perfect tuning of ROV motion control algorithms for all three angular motions heading, pitch and roll. In this paper, the derivation of algorithm and automatic control of nonlinear systems using reverse dynamic task theory is stated. The basic essential input and output parameters of algorithm are also described. The paper presents simulation results, tank test results and results of some initial sea trials.

Keywords: Remotely Operated Vehicle; Navigation Control; Mathematical Modeling; Automatic Control Algorithms; Reverse Dynamic Task Theory.

1 Introduction

Remotely operated, underwater robotic vehicles have become the important tool to explore the secrete life of undersea. They are used for various purposes: inspection, recovery, construction, etc. Engineering problems associated with the high density, non uniform and unstructured seawater environment, and the nonlinear response of the vehicle make a high degree of autonomy difficult to achieve enhanced operator efficiency.

For a rigid body motion on the earth's surface, the vector of dynamic parameters is constituted by mass, inertia matrix and the first moment of inertia. For a rigid body motion in a fluid, the dynamic parameters include the first moment of buoyancy and the hydrodynamic effects. Linearity in the parameters for ground-fixed robotic structures can be easily extended to underwater vehicles in order to include the buoyancy effect.

This paper is organized as follows. In section 2, we describe the mathematical modelling of ROV. Section 3 presents the navigation control system of ROV along with detailed theory of developed control algorithms. Section 4 describes the control simulation results using the derived ROV model and algorithms. Section 5 illustrates the field testing results. Finally, section 5 has final remarks.

P. Vadakkepat et al. (Eds.): FIRA 2010, CCIS 103, pp. 25–32, 2010.

2 Mathematical Modeling of ROV

2.1 The ROV Model

2.2.1 Kinematics of ROV

The kinematic equations of angular movement of the underwater vehicle can be received using a vector of angular speed $\vec{\Omega}$ projected onto the axis of the connected coordinate system. These equations establish analytical dependence's of projections of a vector $\vec{\Omega}=\{\omega_x \quad \omega_y \quad \omega_z\}^T$ with values of rate of change of angle of a roll, a yaw and a pitch, which will be designated hereinafter through $\omega_x, \omega_y, \omega_z$ - the projections of a vector $\vec{\Omega}$ of angular speed of the underwater vehicle about its center of mass. In the above equations T is a symbol of transposing.

Relationship of Euler Angles with other kinematic parameters of rotary motion is established by projections of a vector of angular speed to axes of the connected coordinate system. Time derivatives of heading $\psi(t)$, pitch $\theta(t)$ and roll $\varphi(t)$ is equal to rate of change of corresponding angle and is directed on an individual axis of rotation of the underwater vehicle about its center of mass.

Kinematic parameters of rotary motion of the underwater vehicle are defined by equations (1) – (3):

$$\omega_x = \dot{\varphi} - \dot{\psi} \cdot \sin\theta \tag{1}$$

$$\omega_y = \dot{\psi} \cdot \cos\theta \cdot \sin\varphi + \dot{\theta} \cdot \cos\varphi \tag{2}$$

$$\omega_z = \dot{\psi} \cdot \cos\theta \cdot \cos\varphi - \dot{\theta} \cdot \sin\varphi \tag{3}$$

Using co-ordinate transformation matrix, ROV mass centre coordinates are determined using the differential equations as given in the following equations (4)-(6).

$$\dot{x}_{g0} = V_x \cdot \cos\theta \cdot \cos\psi + V_y (\sin\varphi \cdot \sin\theta \cdot \cos\psi - \cos\varphi \cdot \sin\psi) + V_z \cdot (\cos\varphi \cdot \sin\theta \cdot \cos\psi + \sin\varphi \cdot \sin\psi) \tag{4}$$

$$\dot{y}_{g0} = V_x \cdot \cos\theta \cdot \sin\psi + V_y (\sin\varphi \cdot \sin\theta \cdot \sin\psi + \cos\varphi \cdot \cos\psi) + V_z \cdot (\cos\varphi \cdot \sin\theta \cdot \sin\psi - \sin\varphi \cdot \cos\psi) \tag{5}$$

$$\dot{z}_{g0} = -V_x \cdot \sin\theta + V_y \cdot \sin\varphi \cdot \cos\theta + V_z \cdot \cos\varphi \cdot \cos\theta \tag{6}$$

2.2.2 Dynamics of ROV

Using engineering calculations on dynamics of movement of the underwater vehicle and the analysis and synthesis of control systems, it is possible to simplify assumptions. But it is always necessary to take into account that the added mass and the moments of inertia can render appreciable influence on dynamics of the vehicle.

By assuming further:

1 The origin of the connected coordinate system is in the center of mass of the underwater vehicle.
2 Coordinate axes of the connected coordinate system are the main central axes of symmetry.

3 The case of the underwater vehicle is symmetric about an axis OX of the connected coordinate system.

In view of the above dynamic equations of the underwater vehicle, it is possible to present the equations of spatial motions as follows

$$\dot{V}_x = b_{Vx}\left[a1_{Vx}\cdot V_y\cdot\omega_z - a2_{Vx}\cdot V_z\cdot\omega_y - a3_{Vx}\cdot\omega_y^2 + a4_{Vx}\cdot\omega_z^2 + R_{x\Sigma}\right] \quad (7)$$

$$\dot{V}_y = b_{Vy}\cdot\{a1_{Vy}\cdot V_x(t)\cdot V_y(t) - a2_{Vy}\cdot V_x(t)\cdot\omega_z(t) + a3_{Vy}\cdot V_z(t)\cdot\omega_x(t) + \\ + a4_{Vy}\cdot\omega_x(t)\cdot\omega_y(t) + a5_{Vy}\cdot R_{y\Sigma} - a6_{Vy}\cdot M_{z\Sigma}\} \quad (8)$$

$$\dot{V}_z = b_{Vz}\cdot\{-a1_{Vz}\cdot V_x(t)\cdot V_z(t) + a2_{Vz}\cdot V_x(t)\cdot\omega_y(t) - a3_{Vz}\cdot V_y(t)\cdot\omega_x(t) - \\ - a4_{Vz}\cdot\omega_x(t)\cdot\omega_z(t) + a5_{Vz}\cdot R_{z\Sigma} - a6_{Vz}\cdot M_{y\Sigma}\} \quad (9)$$

$$\dot{\omega}_x = b_{\omega x}\left[a1_{\omega x}\cdot V_y\cdot V_z - a2_{\omega x}\cdot V_y\cdot\omega_y + a3_{\omega x}\cdot V_z\cdot\omega_z + \\ + a4_{Vx}\cdot\omega_y\cdot\omega_z + M_{x\Sigma}\right] \quad (10)$$

$$\dot{\omega}_y(t) = b_{\omega y}\cdot\{a1_{\omega y}\cdot V_x(t)\cdot V_z(t) + a2_{\omega y}\cdot V_x(t)\cdot\omega_y(t) + a3_{\omega y}\cdot V_y(t)\cdot\omega_x(t) + \\ + a4_{\omega y}\cdot\omega_x(t)\cdot\omega_z(t) - a5_{\omega y}\cdot R_{z\Sigma} + a6_{\omega y}\cdot M_{y\Sigma}\} \quad (11)$$

$$\dot{\omega}_z(t) = b_{\omega z}\cdot\{a1_{\omega z}\cdot V_x(t)\cdot V_y(t) - a2_{\omega z}\cdot V_x(t)\cdot\omega_z(t) - a3_{\omega z}\cdot V_z(t)\cdot\omega_x(t) + \\ + a4_{\omega z}\cdot\omega_x(t)\cdot\omega_y(t) - a5_{\omega z}\cdot R_{y\Sigma} + a6_{\omega z}\cdot M_{z\Sigma}\} \quad (12)$$

Where,

$R_{x\Sigma} = R_{x\Gamma} + R_{xB} + R_{xT}$ - Resultant force acting on ROV – (13)

$M_{z\Sigma} = M_{z\Gamma} + M_{zB} + M_{zT}$ - Resultant moments acting on the ROV - (14)

3 Navigation Control System of ROV

3.1 Propulsion System Configuration

The ROSUB 6000 vehicle has seven DC thrusters (Make: Technadyne, USA). The propulsion system of the vehicle is shown in fig. 1.

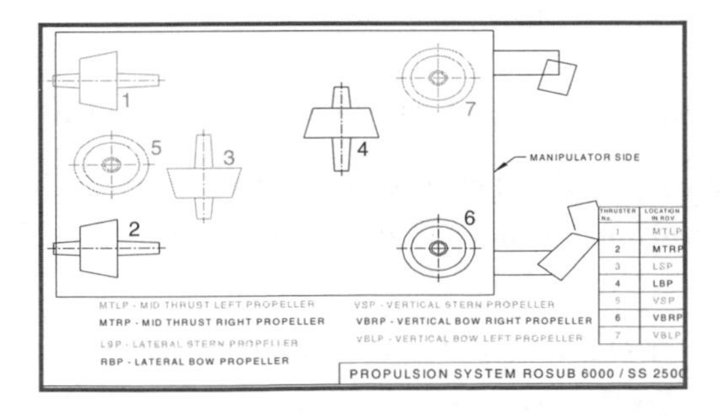

Fig. 1. Propulsion System Configuration – ROSUB 6000

3.2 Navigation Control Scheme

Fig. 2 shows the navigation control scheme of ROSUB 6000. The control drives of all the propellers are assembled in dedicated pressure cases and connected with the main data telemetry and control system pressure case for getting control signal and sending feedback information to and from each of the propeller respectively. The control signals (± 5 V) are the signal from the navigational joysticks of the console, send to the main data acquisition and control system pressure case through the telemetry cable. The power for the drives is fed from sub sea power converter through underwater cable from ship.

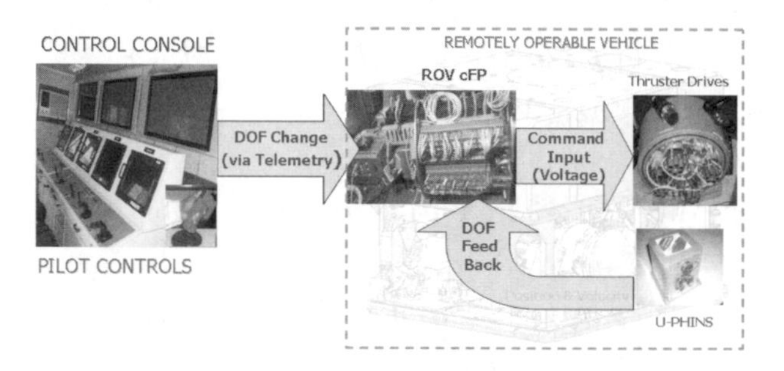

Fig. 2. Navigation Scheme- ROSUB 6000

The main control software was developed using National Instruments LabVIEW™ (Version7.1).

3.3 Navigation Control Algorithm

The control algorithms for semi and full automatic modes were derived after successful study of the mathematical modeling of the ROV followed by simulation to fine tune the performance of the control algorithms using MATLAB and Lab VIEW software. The main task of automatic control is to obtain the control force algorithm from reference equation and equations of system motion using reverse dynamics task theory.

By considering the following equation of motion of system:

$$\ddot{x}(t) = f(x, \dot{x}, u) \quad - (15)$$

Where *x(t) – position* coordinate of the system, u – control function, $f(x, x, u)$ - unknown force which normalized by mass. Let $x(0) = 0, \dot{x}(0) = 0$.

After resolving and simplification of the motion equation, the following equations were received.

$$\dot{u}(t) = k(f^*(x, \dot{x}) - f(x, \dot{x}, u)) \quad - (16)$$

$$\ddot{x}(t) = f(x, \dot{x}, u) \Rightarrow \dot{u}(t) = k(f^*(x, \dot{x}) - \ddot{x}) \quad - (17)$$

$$f^*(x,\dot{x}) = \lambda_1\lambda_2(x^0 - x) + (\lambda_1 + \lambda_2)\dot{x} \qquad -(18)$$

Taking into account of $k_1 = -(\lambda_1 + \lambda_2), k_0 = \lambda_1\lambda_2$, the following final algorithm is received:

$$f^*(x,\dot{x}) = k_0(x^0 - x) - k_1\dot{x} \qquad -(19)$$

$$u(t) = k\int_0^t \left[f^*(x,\dot{x}) - \ddot{x} \right] dt \qquad -(20)$$

Equations (19) and (20) are base control algorithm for all coordinates.

So, equation (20) can also be represented in another form as mentioned below:

$$\dot{u}(t) = k(f^*(x,\dot{x}) - \ddot{x}) \qquad -(21)$$

This features the control algorithm to control both derivative of control signal $\dot{u}(t)$ and the control signal $u(t)$.

4 Simulation Study

As a result of high cost of off shore tests, computer simulation has gained absolute priority in the process of designing control architectures for unmanned underwater vehicles

The results of control simulation for the movement of the underwater vehicle from a zero initial condition to 5 meters forward along an axis OX_{g0} are presented. During this movement, the other channels of control of the underwater vehicle work in a mode of stabilization of their zero initial conditions.

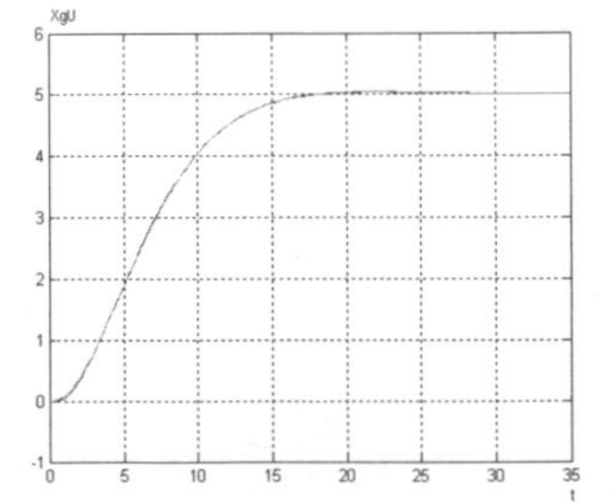

Graph 1: Simulation results for motion of ROV (5 m in forward and all other linear and angular co-ordinates in holding mode)

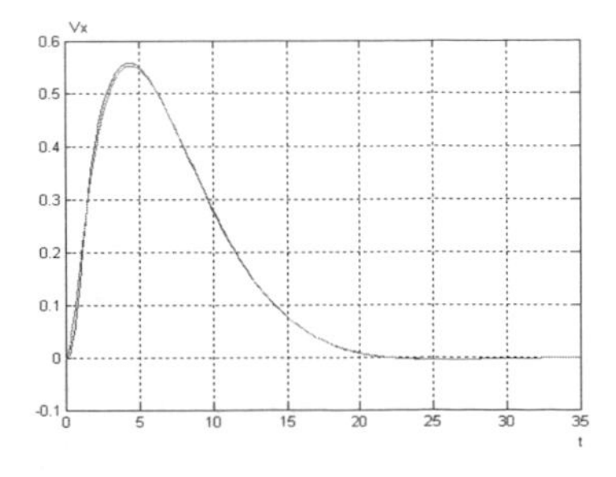

5 Field Testing of Control Algorithms

The automatic control algorithms were derived after successful study of ROV mathematical model. Derived control algorithms were fine tuned and optimized by simulation using MATLAB and Lab VIEW™ software by modifying the control parameters to get more steady performances. The final control algorithms were implemented in the main control software which will be executed from ROV compact field point processor.

5.1 Tank Test

Series of tests were carried out in a tank facility of dimensions 16 m x 9 m with 7 m depth to validate the functionality of control algorithms. After initial tests, the execution time and other critical control loop parameters were optimized for all three angular motions (Heading, Pitch and Roll).

The auto heading mode was enabled after the successful fine tuning tests and ROV is found to hold its set heading angle (i.e., 133.2 deg, 91.2 deg, …) within +/- 1.25 deg. After successful testing of holding mode, the ROV was physically disturbed from the set target angle. The control algorithm reacted to bring back the ROV to its targeted heading angle. In this case, the performance is satisfactory with in ± 2 deg. (Illustrated in Graph 2).

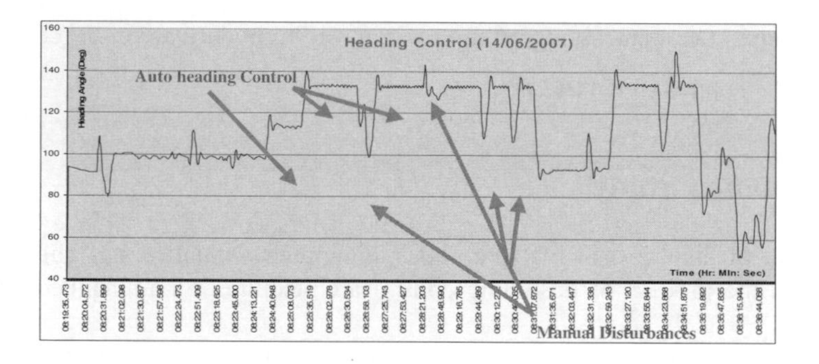

Graph 2: Results of Auto Heading Control in Test Tank

During the above auto heading control, the Pitch & Roll motion control were tested for their auto holding control. (Illustrated in Graph 3).

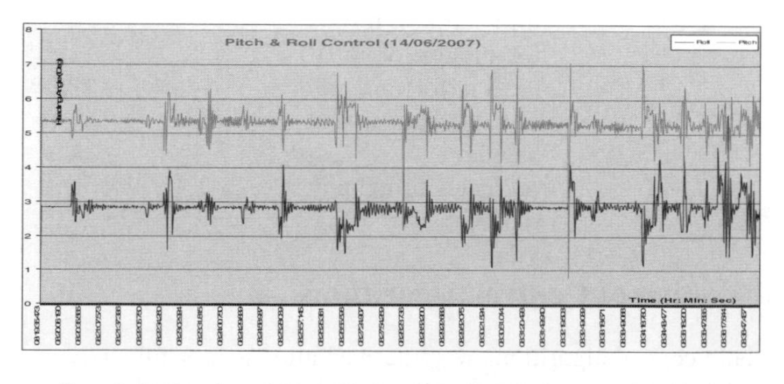

Graph 3: Results of Auto Pitch and Roll Control in Tank test

5.2 Sea Trials

Testing of heading loop algorithms was conducted successfully at pre planned survey lines using PDS 2000 navigation software during sea trial on Feb 2010, at Ennore port, Off Chennai. ROV was moving forward in the controlled heading within 2 degree heading correction. Testing of depth loop algorithms was also conducted within the available water depth of 7 m using the altitude data from Doppler Velocity Log (DVL) during the above sea trial and the ROV was held its depth within 20 cm of range.

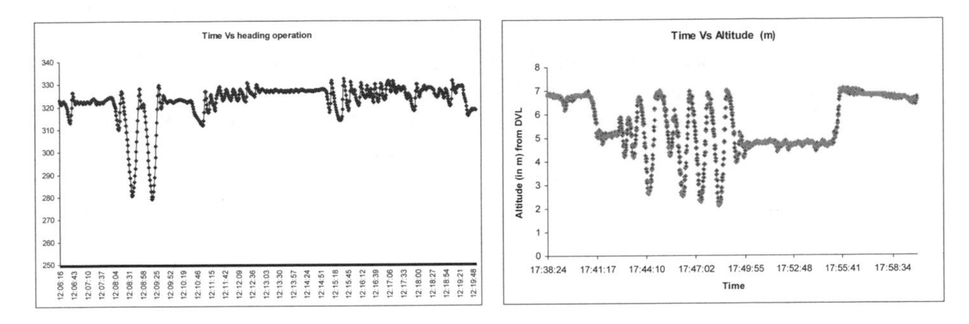

Graph 4: Results of Auto Heading, Auto Depth / Altitude (tuned) algorithm

6 Conclusion

The control algorithms for semi and full automatic modes were derived after detailed study of the mathematical modeling followed by simulation to fine tune the performance using MATLAB and Lab VIEW™ software. Simulations were done based on the initial derived parameters and coefficients. Control software was implemented on the ROV using Lab VIEW™ software supported by National Instruments real time hardware modules.

To obtain a good tuned model with more accurate model and control parameters, a series of tank tests were executed. The developed algorithms of automatic control have been embedded into main data acquisition and monitoring control system and are fully adapted for operation. The results of recent shallow water sea trials in which the automatic angular motion control algorithms were successfully tested are also presented.

Acknowledgement

The authors would like to thank the Secretary, Ministry of Earth Sciences and the Director, National Institute of Ocean Technology for their support and guidance, and they would also like to thank all the members of Submersibles & Gas Hydrates group of National Institute of Ocean Technology for their active support and participation in developmental and testing phases.

References

[1] Fossen, T.I.: Guidance and Control of Ocean Vehicles. John Wiley & Sons, Chichester (1994)

[2] Whitcomb, L.L., Yoerger, D.R.: A New distributed Real-Time control system for the JASON underwater Robot. In: Proceedings of the 1993 IEEE IROS Conference, Tokyo, Japan (1993)

[3] Manecius Selvakumar, J., Atmanand, M.A., ILya, I.: EDBOE.: Mathematical Modeling of Dynamics of ROV and its Control Simulation at National Systems Conference conference-NSC-2006 organized jointly by National Institute of Oceanography (NIO, Goa) and Systems Society of India from, November 2- 4 (2006)

[4] Manecius Selvakumar, J., Ramesh, R., Atmanand, M.A., Ramadass, G.A., Ramesh, S., Jayakumar, V.K., Muthukumaran, D.: Navigation Control System of Work lClass ROV – ROSUB 6000 at International Symposium on Ocean Electronics-SYMPOL 2007 Organized by Department of Electronics of Cochin University of Science and Technology Cochin from, December 11 - 14 (2007)

[5] Manecius Selvakumar, J., Atmanand, M.A., Ramadass, G.A., et al.: Technology Tool For Deep Ocean Exploration – Remotely Operated Vehicle. In: Twentieth International Offshore (Ocean) and Polar Engineering Conference, Beijing, China, June 20-26 (2010)

Distributed Real Time Control Systems for Deep Water ROV (ROSUB 6000)

R. Ramesh, V.K. Jayakumar, J. Manecius Selvakumar, V. Doss Prakash, G.A. Ramadass, and M.A. Atmanand

Submersibles & Gas Hydrates, National Institute of Ocean Technology,
Chennai – 600100, Tamil Nadu, India
{rramesh,v_k_jayakumar,manecius,doss,ramadass,atma}@niot.res.in

Abstract. An unmanned Remotely Operable Submersible (ROSUB 6000) capable of working up to 6000 meters water depth is being jointly developed by National Institute of Ocean Technology, India and Experimental Design Bureau of Oceanological Engineering, Russia. In general ROVs are developed to perform complex offshore operations in ever increasing water depths and accordingly reached a high level of technical design. These vehicles are designed to perform multi tasks in real time operation. The distributed real time control system was designed to provide modular architecture with reliable platform for data acquisition, control and logging. This paper aims at presenting the ROV hardware and software design architecture, which has been developed and tested using National Instruments ™ Real Time hardware and LabVIEW® software for data acquisition and control.

Keywords: Remotely Operated vehicle (ROV); Tether management system (TMS); Deep-water vehicles.

1 Introduction

ROSUB 6000 for deep sea operation is an electric work class ROV equipped with under water luminaries and two manipulators with pay load capabilities of 150kg for mounting scientific and mission oriented tasks. The system comprises of the following major sub system namely

- ✓ Remotely operated Vehicle
- ✓ Tether Management System
- ✓ Ship system

The ROV-TMS assembly is launched from the ship crane and the ROV to be released from the TMS after reaching the required operating depth. ROV has seven electric thrusters for maneuvering in manual or automatic mode. It has a Photonic Inertial navigation system (PHINS) supported by Doppler Velocity Log (DVL) and acoustic transponder/responder for position and navigation of ROV. Multibeam SONAR and depth sensor is also available for assisting navigation along with several video cameras and lights for monitoring. The TMS is docked with ROV as shown in Figure.1.

P. Vadakkepat et al. (Eds.): FIRA 2010, CCIS 103, pp. 33–40, 2010.

Fig. 1. TMS docked with ROV

Design and development of the distributed real time control system for work class ROV is one of the complex tasks that requires many specialized input and output (I/O) functions which are to be reliably performed in the hostile ocean environment[1]. The need of controller in the overall performance of ROV has been reported [2, 3]. In general ROV control system reveals at least the following two computational sub systems[4] with substantially different design requirements:

- User interface: An instrument and control panel for examining the system status, issuing system commands and data logging.
- Real time control: Performs scheduled control tasks with minimal delay with absolute reliability.

Implementation of above two systems requires a computer for user interface and data logging. When designing a control system, a single computational platform is preferred to implement all control functions to perform complex tasks. For any deep water ROV, TMS should be used to avoid drag in the ROV and also to act as a depressor. TMS need to do operations such as docking/undocking of ROV, reel in/reel out of tether. Hence to meet all the requirements a distributed system is needed. This paper describes the design architecture of the distributed real time control system of ROV and its software in detail.

2 System Description

The ROSUB control system is divided into three stages

- Ship control system
- TMS system
- ROV system

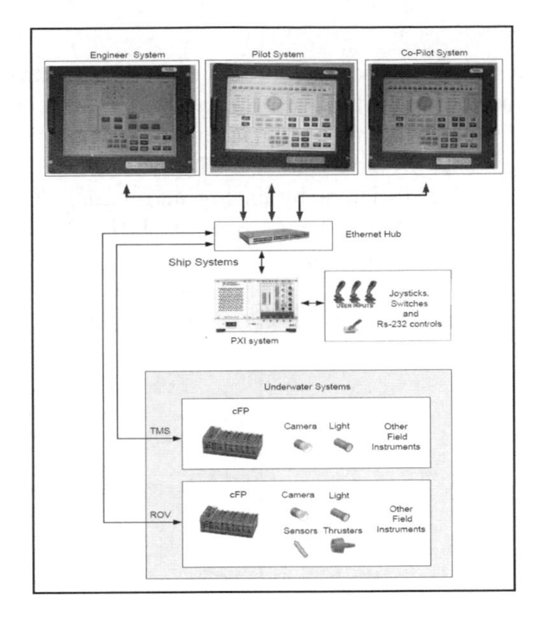

Fig. 2. Schematic of the control system

The above systems are distributed through Ethernet. Schematic diagram of the control system is shown in Figure 2. ROV and TMS are powered and communicated through Electro-Optic cable from the ship. The Fiber optic media converters have been used for bidirectional conversion of optical signal into electrical signal for data and video. The fiber optic media converters are used in console and remote units. The ROV is maneuvered either manually through navigational joystick or semi auto control by giving target from the ship control computer.

2.1 Ship Control System

The ship control system consists of PCI eXtensions for Instrumentation (PXI) system, three industrial computers, console fiber optic media converters and plasma displays. Three individual single mode fibers are used for ROV, TMS and SONAR. The ROV console fiber optic media converter is used to convert data signal into optical signal at 1550nm wavelength, which will be transmitted to ROV. It also converts received optical signal from ROV into data signal (100 MBPS Ethernet, 8 channel Serial port and 4 channel video) at 1310nm wavelength. The TMS console fiber optic media converter is used to convert data signal into optical signal at 1550nm wavelength, which will be sent to TMS. It also converts received optical signal from TMS into data signal (10 Mbps Ethernet and 2 channel video) at 1310nm wavelength. The received video signals of the ROV and TMS cameras are sent to plasma displays through video handling system. These optical signals are sent and received through 7000m Electro-optic umbilical cable used for communication with ROV and TMS. The video handling system consists of digital video recorders, matrixes and video overlay systems, which are interfaced to PXI system through RS-232 serial ports.

2.1.1 PXI System

The PXI system is a high-performance real time embedded controller, which has 2.2 GHz, P4-M processor with 256MB RAM. RT controllers deliver a flexible, rugged platform for deterministic, real-time test and control applications. The controller is interfaced with navigational joysticks, pan and tilt joysticks, foot pedal switches, and hard control of ROV and TMS lamps by analog and digital I/O modules. The PXI system communicates with other systems using TCP/IP. Detailed hardware block diagram of PXI system is shown in Figure 3.

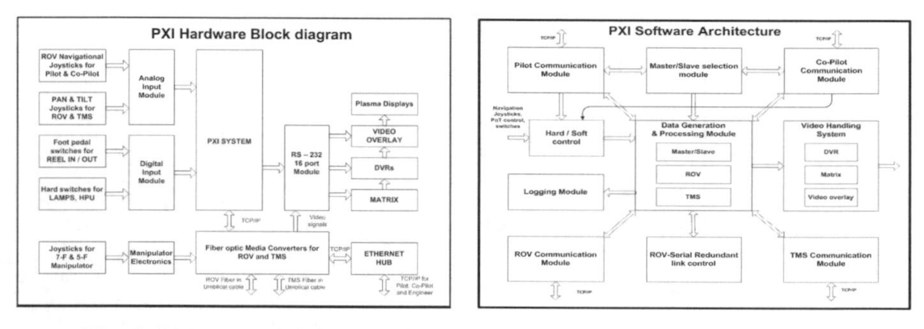

Fig. 3. PXI system hardware

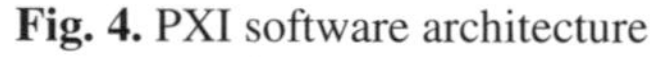

Fig. 4. PXI software architecture

The following hardware are used in the PXI System

1. ROV Fiber optic media converter
2. TMS Fiber optic Media Converter
3. PXI real time controller with
   a. 96 digital I/O module
   b. 64 channel Analog input
   c. 16 RS-232 port
4. Video handling system

Dedicated application software developed in LabVIEW® executing in NI-PXI-8186 system controls the ROSUB system. It accepts command from the user (hard controls/soft controls), generates commands and communicates with compact Field Point (cFP) controllers at ROV and TMS and also communicates with the pilot and co-pilot computers. It also performs error logging, data logging pertaining to the system. It acts as a bridge for communication between ship and sub-sea systems using TCP/IP Ethernet communication. It controls the video handling system for selecting the video channels, single or quad video and pan and tilt of deck cameras. Parameters acquired from ROV and TMS are sent to Video overlay system for merging with video signal. The detailed software modules are shown in the PXI software architecture in Figure 4.

2.1.2 Industrial Computers

Three industrial computers are used for Graphical User Interface (GUI) for Pilot, Co-Pilot and engineer station. Computers have 3.0 GHz processor s with 512 RAM and 80 GB hard disk, which is rack mounted. Three Industrial grade 20” LCD monitors has 1600 x 1200 resolution and integrated resistive touch screen to provide more user friendly front panel of GUI. The system has 19” rack mountable chassis CPU and 20” touch screen monitor. The industrial PC network in ship control console acts as windows based development platform for LabVIEW® application development.

User interface application software was developed using LabVIEW® and executing in Pilot and Co-Pilot Industrial Computers. The graphical user interface application software is identical for Pilot and Co-Pilot. These two systems will act as master and slave. Both systems are responsible for executing the tasks such as data initialization, data generation, data processing, data storage, communication and activity monitoring. The engineer station is used for diagnosis and configuration of ROV and TMS. The front panel GUI screen of Navigation is shown in Figure 5. Leak detection, pressure and temperature in pressure cases, subsea power supply status and thruster’s speed are monitored in GUI.

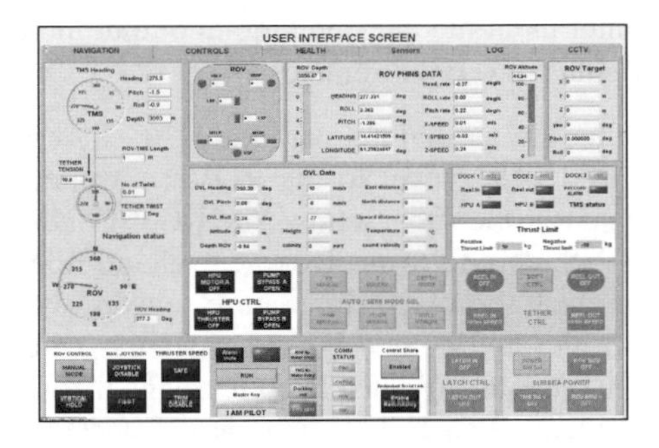

Fig. 5. Navigation graphical user interface screen

2.2 ROV System

Compact Field Point processor

The Compact field point processors are embedded real-time controller for data acquisition and control of ROV and TMS. It has an operating frequency of 400MHz. The I/O modules were selected according to the requirement and interfaced to the controller that is capable of managing eight modules. The modules are securely mounted on a metal backplane that provides a solid mounting surface for the Compact Field Point bank and forms the communication bus between the controller module and the I/O modules. Compact Field Point banks run LabVIEW® Real-Time, providing the functionality, connectivity and flexibility of envisaged operation on a small rugged, industrial platform.

ROV data acquisition system and remote fiber optic media converter are enclosed in watertight chamber along with the necessary electronics as shown in Figure 6. The controller acquires the signal from transducers, sensors and in turn sends it to ship control system. The acquired data is used for navigational control loops and local data logging.

Fig. 6. Hardware assembly view in the Frame

The following hardware is used at ROV system:

- RT controller with 8 slot back plane, Analog input modules, Analog Output modules, Counter input module and Digital output modules.
- Fiber Optic Media converters (remote)

DC/DC converter modules are used along with the DAC system in underwater enclosures to provide stabilized 24V power to the ROV controller and to serve the low voltage power requirement of the subsystems. The ROV Controller with I/O modules are interfaced with PHINS, depth sensor, Doppler velocity log, underwater cameras, Lamps, thrusters, water entry sensors, Temperature sensors and power status sensors. Ethernet signal of controller and composite video signals of cameras are interfaced with Fiber optic media converter to convert optical signal for communication to the ship system. Lamps, Cameras, Pan & Tilts units are protected with rated fuse to protect the complete system. Figure 7 gives hardware details in ROV system. Four shielded twisted pair cable in tether are interfaced between ROV and TMS for RS-485 serial redundant link.

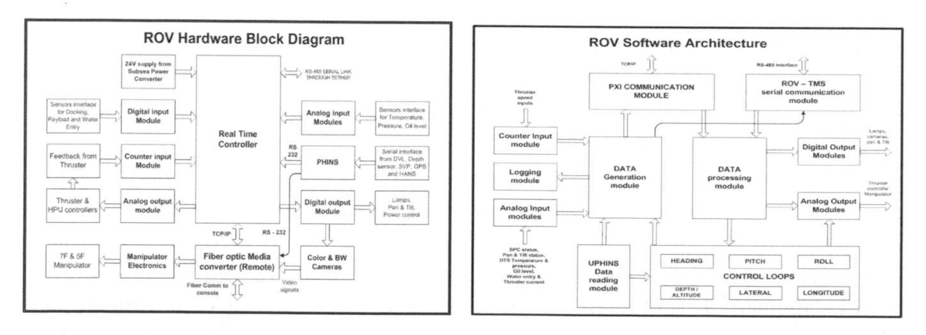

Fig. 7. ROV hardware block diagram

Fig. 8. ROV software architecture

ROV application software was developed in NI LabVIEW® RT. Software was downloaded into ROV controllers via Ethernet. The embedded codes were executed on LabVIEW® real-time operating system. This program is responsible for executing the tasks such as navigational control loops, data initialization, data acquisition and control, data generation, data storage, communication and activity monitoring. Data logging is enabled in embedded controller when ship communication fails. Program reads angular and linear position data from the PHINS and executes the six control loop functionalities such as Heading, Pitch, Roll, Forward, Lateral and Depth. Figure 8 shows the details of ROV software modules. RS-485 serial redundant link communication module will be enabled when Ethernet communication fails in ROV.

2.3 TMS System

TMS data acquisition system and Remote fiber optic media converter are similar to that of ROV system. The TMS Controller with I/O modules were interfaced with compass, depth sensor, tether load sensor, docking sensor, water entry sensor, cameras, lamps, hydraulic power units, temperature and pressure sensor.. Dedicated application software for TMS was developed with the NI LabVIEW® RT. Software was downloaded into TMS controllers via Ethernet.

3 Salient Features of Design

- Distributed systems are networked through Ethernet.
- The System is flexible such that additional module can be added and modules can be reused for further development.
- RT controllers in ROV and TMS are Compact, Rugged & Reliable.
- A Redundant RS-485 serial link is activated automatically for any failure in ROV and TMS fibers for safe retrieval of the system.
- Data, events and errors that will give a detailed description about the current status in the user interface screens and as log into text file. These files are stored at real-time controllers and hence no loss of data.
- Many interlocks and alarms are implemented for system safety and reliability.
- The graphical programming software is easier to modify and troubleshoot.

4 Results

The realized system hardware and software in ROV, TMS and Ship systems are validated and tested for the envisaged functionality. Qualification tests were conducted in different phases such as integrated dry test, in-house test facility and sea trials. Deep water sea trials of ROV performed with developed hardware and software at depth of

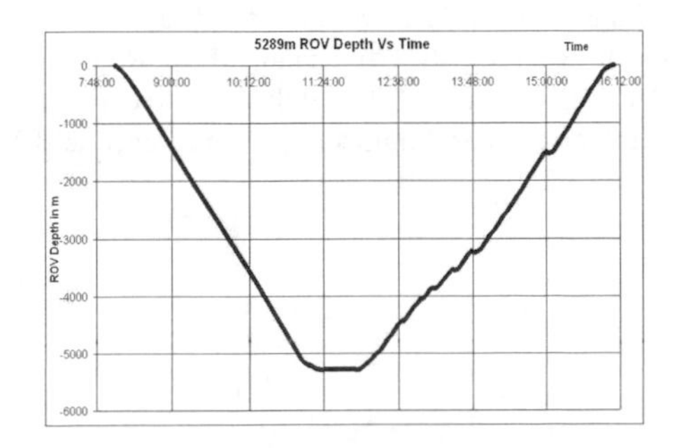

Fig. 9. Plot of acquired and logged depth data from ROV (5289 m)

5256m and 5289m using ORV SAGAR NIDHI at Central Indian Ocean Basin (CIOB). Plot of acquired and logged depth data from ROV during the deep sea trials are shown in Figure 9.

5 Conclusion

The distributed real time control system for 6000 meter depth rated ROV was designed and developed. Real-time controllers enabled us to build the application in a most effective manner. The system acquires all the signals from ROV and TMS and logs the required data in flash memory. It also provides online information on the status and feedback from the navigation unit. It controls the navigation of ROV in real time according to the user inputs from the ship control console. Since the system hardware and software is developed based on the open architecture it has the flexibility to provide option for future enhancement.

Acknowledgments. The authors gratefully acknowledge the support extended by Ministry of Earth Sciences, Government of India, in funding this research. The authors wish to thank Director of NIOT for his support in the project. The authors also wish to thank B.S. Binu, A. Nallayan and Submersibles and Gas Hydrates group for their contribution and support.

References

1. Chaffey, M., Pearce, A., Herlien, R.: Distributed data and computing system on an ROV designed for ocean science. In: IEEE Oceans 1993, Victoria, BC (October 1993)
2. Bowen, P.S., Chappel, S.G., Gonzalez, R.: Using common lisp in the eave autonomous underwater vehicle. IEEE Journal of Oceanic Engineering 15(5), 221–227 (1990)
3. Christy, R., Papoulis, F.A., Healey, A.J.: Adaptive sliding mode control of autonomous underwater vehicle in the dive plane. IEEE Journal of Oceanic Engineering 15(3), 152–160 (1990)
4. Whitcomb, L.L., Yoerger, D.R.: A New distributed Real-Time control system for the JASON underwater Robot. In: Proceedings of the 1993 IEEE IROS Conference, Tokyo, Japan (1993)
5. Atmanad, M.A., Shajahan, M.A.: Instrumentation for underwater crawler for minig in shallow water. In: Proceedings of the ISARA, Singapore (2000)
6. Manecius Selvakumar, J., Ramesh, R., Atmanand, M.A., Ramadass, G.A., et al.: Technology Tool For Deep Ocean Exploration – Remotely Operated Vehicle. In: Twentieth International Offshore (Ocean) and Polar Engineering Conference 2010, Beijing, China (June 2010)

Control Architecture for AUV-150: A Systems Approach

S.K. Das, D. Pal, S. Nandy, V. Kumar, S.N. Shome, and B. Mahanti

Robotics and Automation Group,
Central Mechanical Engineering Research Institute, Durgapur, CSIR,
M. G. Avenue, West Bengal, INDIA-713209
{s_das,dibyendu,snandy,vkumar,snshome}@cmeri.res.in,
biplab.mahanti2000@gmail.com

Abstract. The aim of this paper remains to introduce the various design and runtime aspects of a state-of-the-art control architecture adopted for the development of an autonomous underwater vehicle (AUV-150), capable of operating up to a depth of 150 meters to perform sea-bed mapping and collecting oceanographic data. The system control architecture has been presented as an ensemble of both hardware and software modules organized in a well-connected framework for effective operation. Various specifications, harnessing layout and design issues have been discussed in this paper.

Keywords: AUV, Autonomous, Underwater, Vehicle, Control, Architecture, Hardware, Software, Emergency.

1 Introduction

When it comes to the field of underwater explorations and experiments, Autonomous Underwater Vehicles (AUVs) have proved to be reliable and cost effective technological solution [1]. Usually, they are autonomous agents carrying specific payload deep underwater to carry out particular tasks like underwater bathymetry, imaging and other investigative and surveillance operations. Since, they generally work on their own, without any human guidance whatsoever, therefore it is essential for them to be supported with reliable and efficient control architecture, consisting of a robust hardware platform well coordinated by an intelligent software backbone.

The AUV-150 being modular in structure and functionality, it has been truly challenging to develop a hardware framework that remains well connected, distributed and reliable enough during underwater missions. A robust hardware backbone has been developed through careful selection of components and reliable integration, which in turn supports navigation and decision autonomy for the vehicle.

Navigational autonomy has been achieved on large scale through the effective coordinated operation of controller, navigational sensors, like INS and FLS, and actuators, altogether governed by the control software architecture. Considerable modularity in development of guidance control software has been achieved through data-acquisition over a highly reliable Ethernet backbone augmented with specific media-converters, which connects all the sensors, actuators and controllers throughout the system. This paper also serves to be a close inspection into the essential hybrid nature of the system

P. Vadakkepat et al. (Eds.): FIRA 2010, CCIS 103, pp. 41–48, 2010.

architecture, wherein the software organization consists of both deliberative as well as reactive modules, which exploit the integrity of the hardware framework, in order to give an edge to the performance of the system as a whole.

An effective coordination between the two functionalities is highlighted in order to establish the overall dynamic behavior of the system. Moreover, other characteristics like hardware abstraction, concurrent execution of threads and implementation of interfaces for modules are discussed. Necessary characterization of devices like actuators and sensors, together with their incorporation into software modules is also presented as part of the runtime coordination of the hardware-software architecture.

2 Hardware Framework

Although the hardware architecture of AUV-150 is somewhat similar to [2], the harnessing is typically suited to its overall modular hull design. It has consecutive cylindrical pressurized and wet modules connected together to give it a torpedo shape, consisting of a nose module and tail module housing the major electronics and devices, power module and 2 thrust modules. Each of the tail, nose and battery modules has both enclosed and wet chambers for housing electronic hardware as well as payload and navigational sensors. The AUV being thruster-actuated, each of the two thrust modules houses a pair of vertical and horizontal tunnel thrusters for controlling attitude and orientation of the vehicle. The tunnel thrusters are of Sea-Eye make delivering a thrust of 13 Kgf and 12.8 Kgf in the forward and reverse sense respectively. A brushless DC thruster of Tecnadyne make with 21.4Kgf and 14.5Kgf thrust in the forward and reverse sense and having 90-110 V, 1 KW power input, has been used for propulsion and is coaxially attached to the tail portion of the vehicle. In order to achieve roll stabilization mechanically, a pair of 4 static Fins has been attached to the hull of the AUV. A pack of 6 Battery Banks, each consisting of Lithium-Polymer Cells having a nominal voltage of 25 Volts and 70AH, have been used as energy source for the system. The banks reside inside the pressurized chamber of the power module. The battery banks have protection circuitry against short circuit as well as overheating.

2.1 Main Controller Backbone

An intelligent control computer of Skilligent make housed inside the tail module incorporates data acquisition from all the navigational sensors and generates control commands to the thrusters. A single board computer (DVS-350M) of Advantech make, which resides inside the pressurized nose module, logs data from pay-load sensors like camera, Side Scan SONAR and CTD profiler. It has a built in facility to interface 16 video channels. An uninterrupted and reliable Ethernet backbone connects the various modules as well as sensors and hardware drivers. Sensors having RS232 interface and thrusters communicating in RS485 mode are connected to controller modules through Ethernet via media converter NPORT-5450. A Photonic Inertial Navigation System (PHINS) made by IXSEA has been used for estimation of the internal state of AUV. It consists of 3 axis fiber optic gyro and magnetic compass along with a Kalman Filter based estimator. A DVL (WH-300) from RD Instruments, having 300 KHz as operating frequency and bottom range of 300 m, has been used as a secondary internal sensor in connection to the PHINS. Uninterrupted and stable communication with the surface control is established through a combination of radio frequency (RF) as well as acoustic modem. The RF modem (AWK-100) resides inside the tail module and is connected to the main Ethernet bus. The antenna is mounted on the hull and resides inside a Teflon based pressurized casing. Although having a low bandwidth and data rate, the acoustic modem transponder (TN 1510-B) of LINK-QUEST is in combination with USBL Acoustic Positioning and an auxiliary back-up battery attachment, which makes it quite cost-effective and a significant tracking system during emergency. An indigenous application-level protocol [3] has also been developed for the purpose of underwater data transmission with support for error detection as well as recovery.

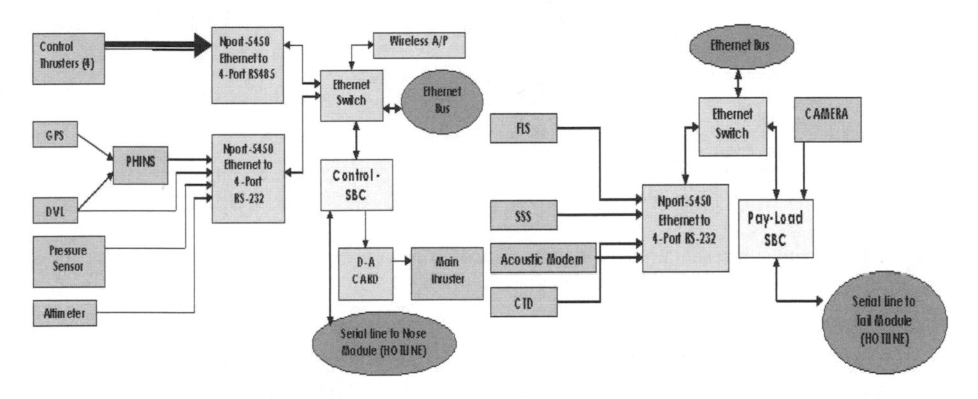

Fig. 1. Schematic for tail-side module (left); nose-side module (right)

3 Software Architecture

A hybrid architectural system [4] is featured with deliberative activities like planning, sequencing and communicating as well as reactive behavioral control, which is closely coupled with sensors and actuators. In this regard it should be mentioned that deliberative control is discrete in comparison to its continuous counterpart, i.e., reactive control module. It is only through coordination between the two components that the overall dynamic behavior of the AUV is established. In the subsequent sections hybrid software framework adopted for intelligent control of the AUV, shall be discussed. Hybrid model simulating the architectural behavior of the system shall also be presented.

3.1 Overview

A schematic representation of the architectural framework is shown in Fig. 2. The overall architecture consists of the deliberative and reactive modules, with two abstraction layers or schemas namely, the actuation model as well as the perceptual schema. The deliberative modules are responsible for planning, task sequencing as well as governing the dynamic behavior of the system as a whole. As shown in the figure, the broken lines from *State_Handler* to the reactive layer represent the various set points generated by the deliberative module for the continuous reactive control-loops. The *State Handler* is also responsible for deciding the execution status for each of the controller threads and other associated threads executing in the system-process context. The task sequencing and planning is achieved by updating a policy vector, which is a global data structure storing the execution status of all the threads running in the process context. The *Task_ Manager* needs to suspend, create and terminate the various threads of execution uses the policy vector thus updated.

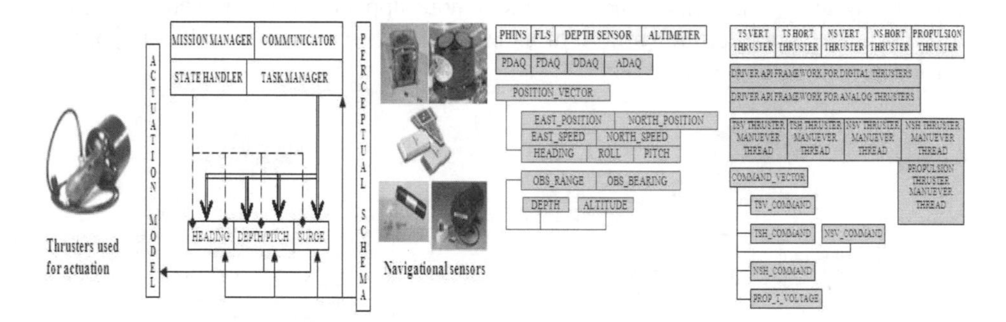

Fig. 2. Schematic representation of major control software architecture (left) with details of perceptual schema as well as actuation model (right)

Mission_manager is the entry point to the system context and is responsible for initiating all the other deliberate modules. *Communicator* sub-system is involved with hybrid communication with the surface control, i.e. supporting radio communication while the vehicle is on the surface as well as acoustic communication throughout the period for which the vehicle remains underwater. Contrastingly, the reactive modules are highly responsive towards environmental changes and work on a greater sampling rate in comparison to the deliberative modules. The reactive layer drives the system towards the desired operating set point, as updated by the goal driven deliberative layer. The continuous control-loops have been modeled and designed on the basis of system dynamics and using conventional control theory. Sensory feedback from the perceptual schema helps in evolving events, thereby triggering various state transitions inside the *State_Handler*. Table 1 represents the layers comprising both the perceptual schema as well as the actuation model. Both the schemas provide a layer of abstraction to the reactive and deliberative modules, so that the controllers do not have to communicate with the devices directly. Therefore, a change in the device driver or the schemas does not necessitate a modification in the coding for the controller modules. Thus it helps in maintaining and upgrading the architecture even if specifications for the devices (i.e. sensors and actuators) are changed.

Table 1. Representative layers of perceptual schema and actuation model

Layers	Perceptual Schema	Actuation Model
Device Layer (uppermost)	Internal sensors like INS, DVL, Depth Log, Altimeter	Analog thrusters for propulsion; Digital thrusters for orientation
Data interfacing layer (middlemost)	Threads contributing to data acquisition	Threads concerned with configuring actuator parameters and generating digital and analog signals for manipulating the thrusters
Data Structure layer ()	Vehicle position; orientation; depth and altimetry	Control signal required to be given to control as well as propulsion thrusters
Device Layer (uppermost)	Internal sensors like INS, DVL, Depth Log, Altimeter	Analog thrusters for propulsion; Digital thrusters for orientation
Data interfacing layer (middlemost)	Threads contributing to data acquisition	Threads concerned with configuring actuator parameters and generating digital and analog signals for manipulating the thrusters

3.2 Hybrid Modeling

Functionally, the proposed architecture closely resembles a hybrid system thereby, essentially consisting of two subsystems: (1) a discrete dynamic controller; (2) a set of continuous dynamic processes. The State_Handler aided by the Task_Manager essentially governs the overall time-varying behavior of the system by coordinating and controlling various threads of execution, which are involved with the reactive layer and operate in a more continuous domain of execution (having lower sampling time). It is therefore, important to develop a model that accurately describes the dynamic behavior of such a system. For this purpose, a hybrid model [5] characterized with differential equations as well as FSM based state-machine with the help of StateFlow/Simulink from Matlab has been developed and simulated. Fig. 3 represents a skeletal framework of the model in the simulating environment, wherein the *system_FSM* characterizes the *State_Handler* in a reduced form, whereas *Subsystem* is the model representative for the reactive domain. Since the proposed AUV has only 5 degrees of freedom (with roll balanced by mechanical design), the 6-DOF rigid body equation [6] has been adopted for modeling its transient behavior in the time-domain [7].

$$M\,(dV/dT)+C(V)\,V+D(V)\,V+g(\eta)=\tau \tag{1}$$

Where, M is the mass/inertia matrix of the AUV including hydrodynamic added mass/inertia terms, C is the coriolis and centripetal matrix, D is hydrodynamic damping matrix and g is gravity and buoyancy force vector. The hydrodynamic damping is estimated as follows:

$$\mathrm{D} = 0.5\,\rho\,\mathrm{A}_{\mathrm{F}}\mathrm{V}^{2}\mathrm{C}_{\mathrm{D}} \tag{2}$$

Where, ρ is seawater density, V is the design velocity of the AUV, AF is the frontal area and CD is the viscous pressure drag coefficient.

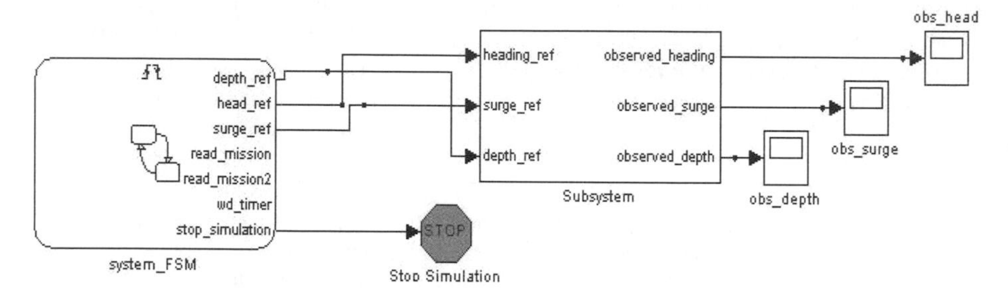

Fig. 3. Skeletal framework of the hybrid model

The deliberative component is modeled using a hybrid automaton M:

$$M = \{S, \delta, T, A, Op, I, F\} \tag{3}$$

Where,
S = {states governing the entire discrete unit}
T = {transitions for a particular state}
δ = state-transition function
Eo = {events generated by the Stateflow machine}
As = {actions being executed while entering /staying into a particular state}
At = {actions executed while a transition takes place}
Op = {output variables}
I = Initial state with default transition
F = Final state of termination

4 Emergency Components

4.1 Leak Detection Unit

An indigenously developed Leak-Detector is used in AUV-150 for detecting leakage of any sort. It consists of a long strip-like sensing part, strapped on the inside lower surface of each dry chamber of the hull. If any leakage occurs, it triggers the emergency handling card, which in turn sends a pulse to the intelligent control computer. The control computer immediately stops the entire mission program, generates a control signal for upward thrust over a short duration of time and sends an emergency signal through the active communication link (Acoustic link when underwater and RF link when on the water surface) to the surface console and automatically shuts down the entire system.

4.2 Computer Health Check-Up Unit

The control computer as well as the DVS-350M are also connected alternatively by a direct serial line namely the Hot-Line. If there is any congestion or bottleneck on the Ethernet bus, it helps the computers to stay connected. Both the computing units execute a health check up routine in parallel. If emergency happens like low battery power, break down of the internal Ethernet bus, malfunctioning of navigational sensors or some program execution fault, any or both of them executes the same emergency routine and triggers the emergency handling card for forced shut-down of the system by electrically disconnecting the main power supply line.

5 Discussion

In order to have flexibility and redundancy both at the hardware as well as software level, redundant access to hardware by software modules as well as device-level abstraction have been incorporated. Concurrent execution is supported with the help of user-level threads running in the same process context. UML tools from Rational Rose have been used for design of logical diagrams, classes and testing of modules during various phases of the development of control software. Control signals have been generated using estimated characterization polynomials obtained by testing the control as well as propulsion thrusters as shown in Fig. 4 and Fig. 5.

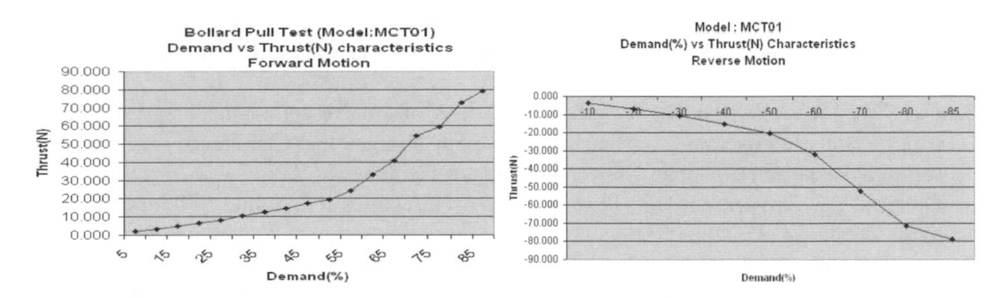

Fig. 4. Thrust characterization curves for control thruster: forward (left) and reverse (right)

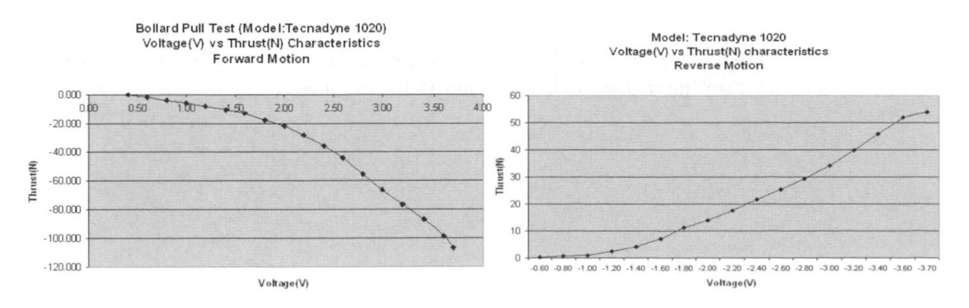

Fig. 5. Thrust characterization curves for propulsion thruster: forward (left) and reverse (right)

6 Conclusion

The present paper aims towards describing the overall control architecture required for autonomous navigation and guidance control of an autonomous underwater vehicle. A robust fault-tolerant hardware architecture coordinated with hybrid deliberative-reactive architecture provides reliable and effective functioning of the system. The aim of this paper has been to illustrate the hardware/software co-design and coordinated execution, in order to represent the system as a whole. Modeling the hybrid architecture qualitatively establishes the dynamic behavior of the system as shall be governed by the actual software framework. Towards the end, emergency handling as well as few design and hardware interfacing issues have been discussed.

Acknowledgments. The authors would like to express their gratitude to the Ministry of Earth Sciences (MoES), Govt. of India, for sponsoring this project, and to CSIR for extending all the necessary cooperation to CMERI, Durgapur. The authors would also like to acknowledge the support of the entire Robotics and Automation group for their cooperative and coordinated efforts towards making this project a success.

References

1. Yuh, J.: Design and Control of Autonomous Underwater Robots: A Survey. Autonomous Robots 8, 7–24 (2000)
2. Sangekar, M., Chitre, M., Koay, T.B.: Hardware Architecture for a Modular Autonomous Underwater Vehicle STARFISH. In: Proceedings of MTS/IEEE Oceans 2008, Quebec (2008)
3. Das, S.K., Saha, A., Pal, D., Shome, S.N.: Design of a naïve communication protocol for data transmission over an acoustic channel. In: 5th International Conference on Communication Systems and Computer Networks, Springer, Heidelberg (2008)
4. Oreback, A., Christensen, H.I.: Evaluation of Architectures for Mobile Robotics. Autonomous Robots 14, 33–49 (2003)
5. Das, S.K., Shome, S.N., Nandy, S., Pal, D.: Modeling a hybrid reactive-deliberative architecture towards realizing overall dynamic behavior of an AUV. Procedia Computer Science 1(1), 259–268 (2010)
6. Fossen, T.I.: Guidance and Control of Ocean Vehicles, pp. 48–54. John Wiley and Sons, UK (1994)
7. Shome, S.N., Nandy, S., Das, S.K., Pal, D.: AUV for shallow water applications-some design aspects. In: 18th International Offshore and Polar Engineering Conference, Vancouver (2008)

Autonomous Underwater Vehicle for 150m Depth–Development Phases and Hurdles Faced

S.N. Shome[1], S. Nandy[1], S.K. Das[1], D. Pal[1], B. Mahanty[2], V. Kumar[1], R. Ray[1], and D. Banerji[1]

[1] Scientist
[2] Project Assistant, Robotics and Automation Group,
Central Mechanical Engineering Research Institute (CMERI), Durgapur, CSIR, India
{snshome,snandy,s_das,dibyendu,vkumar, ranjitray,dbanerji}@cmeri.res.in, biplab.mahanti2000@gmail.com

Abstract. This paper describes the various development phases and the associated hurdles faced during the design, fabrication, sub-system level testing, assembly, integration and overall system testing of an Autonomous Underwater Vehicle (AUV). This AUV has been designed for a depth of 150 m with multi-thruster actuation for shallow water applications. The AUV is having onboard power, electronics and advanced control module, navigation and payload sensors and modular software architecture. During the development of AUV various hurdles like how to power on AUV from outside, loose connections, isolation and grounding, water leakage, battery tripping, etc. have been faced and resolved. The present paper describes the complete development aspects in brief and highlights the various hurdles with remedies throughout the development. The AUV has been tested successfully for various missions at Idukki Lake, Cochin, India up-to a depth of 5m.

Keywords: AUV; design; navigation; control; software architecture.

1 Introduction

Underwater vehicles help us to exploit the huge oceanic natural resources and monitoring of ocean parameters in a highly unstructured and hazardous environment. Autonomous Underwater Vehicles (AUV) and Remotely Operated Vehicles (ROVs) are the most commonly used underwater vehicles used for undersea exploration with advantages and disadvantages over each other. ROVs are tele-operated through a cable which also supplies power to the vehicle. The hydrodynamic drag on the cable and the power transmission losses through the cable increases with the increase in the operating depth of the ROVs. AUVs have a wider domain of operation as they are untethered and carry power supply on board. AUVs can be utilized for seabed mapping and oceanographic data collection, undersea exploration, inspection and many other purposes with suitable customization [1]. This paper mainly deals with overall developmental issues along-with hurdles faced with remedies. The complete development of AUV is mainly subdivided into mechanical design and analysis subsystem, power

P. Vadakkepat et al. (Eds.): FIRA 2010, CCIS 103, pp. 49–56, 2010.

subsystem, sensor subsystem, navigation, guidance & control subsystem, communication subsystem and emergency handling subsystem. Each subsystem is highly interlinked and complementary for the complete AUV system development. Though considerable literature is available for all the aspects mentioned above but almost no literature is available which deals with the various hurdles and their remedies throughout the development.

Mainly four factors are responsible for AUV design and development like speed, depth of operation, mission time and payload capacity. All these parameters are highly coupled and the configuration and size of the AUV is decided through an iterative procedure optimizing an objective function. The onboard power system of AUV comprised of high energy density batteries like lithium ion or lithium-polymer. Since the energy carried onboard an AUV is limited, any savings in the energy consumption can be translated into longer mission durations using efficient path planning as indicated in [2] & [3]. The motion in various directions of AUV are obtained through thrusters or/and controlled surfaces/hydroplanes/fins as mentioned in [4] & [5].

AUVs are expensive, and a dedicated team of scientists work together for years to make the system seaworthy. Systematic design approaches have been followed for the development of an efficient seaworthy AUV. During the development phases, various types of hurdles faced like system power on related issues, isolation and ground looping problems which are responsible for system malfunction, malfunctioning of digital pressure gauge which is responsible for damage of AUV pressure hull, remote switching of the Digital Video Server (DVS) etc. and all these are resolved judiciously by appropriate diagnosis and action plan.

The present paper mainly focuses the developmental issues of AUV along-with the hurdles faced during the development with remedies. The developed AUV is capable of working in shallow depth of 150 m with envisaged applications of seabed mapping and data collection. The initial experiments for various missions up-to a depth of 5m had been conducted at Idukki Lake, Cochin and results have been highlighted in [7].

2 Development Issues

The present AUV has been designed for operation up to a depth of 150 m with a maximum speed of 3 knot, mission time spanning 4-6 hours and payload capacity ranging from 20-30 kg. Key factors determining AUV characteristics are: maximum working speed, depth of operation, payload capacity and mission time, which are highly interrelated. Proper judicial selection of parameters is therefore very important. The various developmental issues considered during design of the present system are detailed below.

Configuration – The important parameters of the AUV with respect to shape and size are: minimum drag, minimal flow separation, improved vehicle stability adequate space for accommodating all necessary hardware. From the simple perspective of drag reduction, a form that promotes laminar flow within the boundary layer constitutes the appropriate choice. Taking this into consideration, a dome shaped fore and a hull form aft has been chosen.

Skin friction and form drag contributes to the overall vehicle drag. Friction drag varies with speed and exposed surface area. Form drag is a function of how well the hull shape minimizes flow separation. It is estimated based on the frontal area and is usually larger than the friction drag for typical AUV configurations. Longer and slender shapes are therefore better for frontal drag. Test results show that streamlined form with length to diameter ratio in between 7 and 10 performs well to minimize drag. The various hydrodynamic coefficients need to be calculated based on the current shape and size of the AUV and incorporated in the dynamics and control of AUV.

Modularity – The system should be as modular as possible for accommodating new hardware, replacement/addition of new modules, up-gradation of actuation or propelling systems, removal of equipment, easy handling, maintenance and accessibility. The entire software needs to be split into individual component modules with certain tasks assigned individually for improving the reliability and the overall performance of the software.

Degrees of freedom –Efficient AUV underwater operation without any constraint requires six degrees of freedom – three for positioning the vehicle at any particular location and three for orienting the vehicle for task completion. The present AUV has the specific task of seabed mapping and collection of environmental data. The system has therefore been designed with only five controllable degrees of freedom with constraint on rolling.

Stability – The AUV system needs to be modeled in detail for evaluation of center of gravity (CG), center of buoyancy (CB), mass moment of inertias etc. to predict on the stability of the vehicle. Sensors and other equipment need to be placed judiciously for proper system balancing. For static stability, the center of buoyancy should remain above the center of gravity for increasing the resisting arm against destabilization. Roll stabilization can be facilitated by providing an array of static fins mounted at the fore and aft of the vehicle.

Near neutral buoyancy - The AUV needs to be designed to operate at near-neutral buoyancy such that the hull remains almost horizontal when submerged, and helps in recovering the vehicle rapidly on the sea surface.

Hydrodynamic modeling – Hydrodynamic modeling plays an important role to minimize the propulsion energy requirement and the stability and maneuverability at various operating speeds and largely influences the overall control of AUV [8]. The pressure to which the AUV is subjected is directly proportional to depth. Therefore, some form of pressure hull needs to be provided for equipment that work in dry atmospheric environment. The depth rating of the vehicle determines the thickness and construction of the pressure hull.

Energy System - For optimized vehicle mass and size, the onboard energy system should possess high energy density, facilitate easy recharging and have sufficient life cycles in case of secondary batteries. Equipment safety should be kept in mind while selecting a particular type of battery or any other energy system that should provide power for the full mission duration and should have proper monitoring and mitigation system for continuing the operation in case of local failure for safe retrieval.

Communication - Hybrid communication, RF while AUV is on surface and through acoustic link while underwater, needs to be incorporated.

Navigation and control – Instantaneous positional information on the AUV following point-to-point control method needs to be evaluated for tracking any particular path or for movement from one location to another. For improved accuracy and better estimation of the location, sensor integration with Kalman filtering or any other method needs to be adopted with incorporation of more than one navigational sensor. The AUV should have the capability of intelligent detection and obstacle avoidance and of retracing the original vehicle path.

Payload Sensors - The payload sensors have to be placed judiciously for minimum disturbance and obstruction. Depending on the mission requirements, payload sensors have to be selected. Acoustic based sensors should be placed away from the thrusters.

Diagnostics feature – It is important to monitor condition of all internal equipment and the processes in operation through a stand-alone microcontroller based system for monitoring the computer network, sensors and thrusters.

3 System Description

Considering the above developmental issues, in the first phase, a full scale steel mock-up unit was developed as shown in Fig.1 to test several navigation & control algorithms and also to check any problems related to assembly and integration. In the second phase of development, the actual aluminium working prototype which is shown in Fig. 2 has been developed. The problems faced during integration and testing of steel mock-up unit have been incorporated into the design of actual prototype.

Fig. 1. Mock-up unit of AUV in 1:1 scale

Fig. 2. Actual working prototype of AUV

The developed AUV has its own power, propulsion system, and intelligent navigation and control modules. AUV is designed modularly to accommodate new payloads and enhancement of capabilities. The AUV has six modules from nose to tail. The propulsion system consists of five thrusters mounted at suitable locations to provide motion along surge, sway, heave, pitch & yaw directions. Roll is balanced through mechanical design as it needs to remain stable during the seabed mapping. The vehicle is programmed with a set of instructions that enable it to carry out an underwater mission without assistance from an operator on the surface. The various components like thrusters, battery packs, sensors, computational platforms and associated electronics have been judiciously placed inside alternate dry and weight chambers. The detailed design is described in [6]. All the sensors and other devices are mounted judiciously onboard AUV so as make the system a meta-centrically stable vehicle (through CG & CB adjustment). After several iterations the length and diameter of the AUV is fixed at 4.8m & 0.5m respectively. The weight of the AUV actual prototype made of special grade Al alloy (Al-6061), including all its onboard subsystems, is 490 kgf. The AUV has slight positive buoyancy (around 3 kgf) including its payload to facilitate its retrieval in case of a power failure. Various navigational sensors are placed in various modules for localization of AUV using appropriate navigation, guidance and control methodology [8].

The nose module houses the forward-looking sonar and camera with a light at the front to detect obstacles in front of the body as well as for obtaining environmental pictures. Side scan sonar is fitted at the bottom for mapping of the sea-bed. CTD is mounted in this section for recording the temperature, depth and conductivity at various operating locations. Acoustic modem with acoustic positioning system (USBL) is fitted at the top of this section for communicating with surface station and knowing the position of the system with respect to surface control station. The nose module also contains a pressure hull for housing computers like Single Board Computer (SBC) and Digital Video Server (DVS) to upload various data coming from different devices during underwater mission.

There are two identical wet thrust modules – each of which houses two (2) tunnel thrusters, one in the horizontal direction and other in the vertical direction. The propulsion thruster is mounted at the end of the tail module. The tail module also houses altimeter and depth sensor. A Xenon flasher mounted at the top of the tail side thrust module makes AUV visible at night. The power module is specifically designed to keep energy system and inertial navigation system in dry condition within the pressure hull and Doppler Velocity Log sensor outside the pressure hull i.e., in wet condition.

The computational module is the brain of the AUV. All the computational devices are housed in dry condition inside the pressure hull of the computation module. All the devices of the other modules are connected to this module through interconnection cables with proper harnessing. GPS and RF antenna are also fitted on the top of this module. The onboard software residing on the computation module helps AUV to take decision autonomously based on the navigational data obtained from different sensors and inbuilt control methodology.

4 Hurdles Faced with Remedies

During the development, the project team faced lot of hurdles. Some of these listed hurdles may look silly and some of them are very critical, but as per our experience we have observed that these kinds of problems take a substantial amount of project time. The details of the hurdles and their remedies are listed chronologically:

1. **Loose Connection:** Problem of loose connections are very common in electrical contacts, various connectors of computers, sensors, thrusters and interfacing cables and ports like serial & Ethernet port etc., which are very difficult to trace and hampers system reliability drastically. Proper connections (screw tightening, hard soldering etc.) with utmost care have been performed in each and every case, where there is a possibility of loose connections and a step by step approach have been followed to identify the faulty point. For example, we have used an internal LAN connection, which connects the multiple Single Board Computers (SBCs), sensors and other devices which are located at different modules of AUV. Once, due to loosening of RJ45 connector and SBC's Ethernet port, LAN connection was hampered and it was very difficult to identify the problem. Once, the problem has been identified the problem is solved through direct hard soldering of Ethernet cables with concerned device without using RJ-45 jack and LAN reliability increased.
2. **AUV Power on Switch:** After complete assembly and just before putting the system into water the system needs to be switched on externally by some means and vice versa. This was a challenging task to us. We have fitted a battery charging connector (24 pin, GISMA make underwater connector) upon the AUV body for charging 6 sets, 25V batteries. We have used two terminals of the battery charging connector through which with an appropriate circuit 24V is applied to activate a relay, which makes the system on. After the system is on the circuit is disconnected and the battery connector is covered with a dummy underwater mating connector before putting the AUV in water. After testing is over the system is made switched off through software driven internal switching.
3. **Water Leakage:** AUV system consists of no. of sealed water proof dry chambers and wet chambers, which are connected through underwater connectors and interconnection cables. As there are many no. of sealing points (aprox. 35 nos.) and all are sealed manually, there is a high possibility of water leakage through any of the joints due to manual fault, O-ring defect, uneven and dirty surfaces etc. If there is water leakage in any module, it should be identified immediately using leak detector and the total system power should be shut down immediately to avoid any damage of components. An indigenous water leak detection system as shown in Fig. 3 has been developed and placed at the bottom part of each module to solve this problem.
4. **Isolation and Grounding:** Isolation and ground looping problem caused malfunctioning of the thrusters and other components. This problem has been solved using isolated power supplies, cards and assigning Local Ground to each component individually and ultimately providing a common Single Ground Reference Point (SGRP) for all the components of the AUV system.

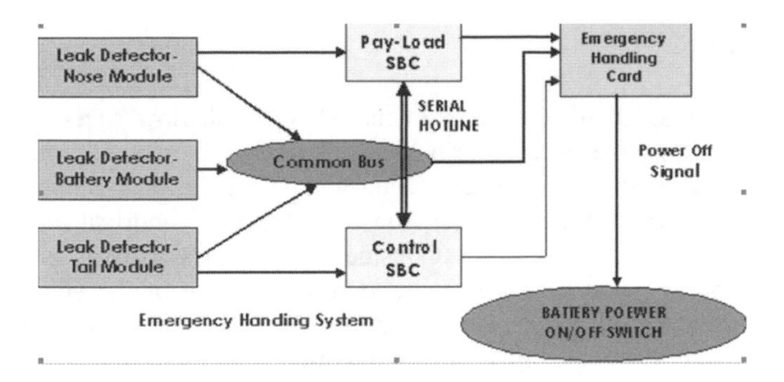

Fig. 3. Schematic representation for emergency handling sub-system

5. **Redundant pressure gauge:** AUV dry modules have been designed and then analyzed through Finite Element Modeling (FEM). A pressure testing chamber has been made to test the fabricated dry chambers up-to a pressure of 15 bar (equivalent to150 m depth) to simulate the equivalent sea condition and to measure the various deformations. Pressure chamber was fitted only with a calibrated digital pressure gauge. During testing of a chamber we have applied 15 bar pressure (digital pressure gauge was showing approx 15 bar) and keep the chamber for 30 mins. After opening the pressure chamber we have observed that the sealed dry chamber got completely deformed and after analyzing we are confirmed that the design was all right and damage occurred due to fault reading of digital pressure gauge. Actually, the digital pressure gauge was showing lower reading (at least 10 bar lower) than the actual pressure. After this experience, we have attached another analogue pressure gauge to the pressure chamber so that, we can tally both the gauges and take the decision accordingly.
6. **Failure of system software:** Due to very complicated nature of the AUV system (huge data and command) it was really very tough to come up with an efficient control and software architecture at the early stages of the development. In the early stages, system control software sometime hang-up which causes the non-stop motion of the thrusters. The problem had been analyzed and proper action had been taken and implemented in the software. Also a different relay operated power supply with software controlled switching for all thrusters are kept for emergency cut-off of power supply to the thrusters.
7. **Powering DVS-350M:** The payload computer DVS-350M requires a manual trigger in order to get powered on. Since, the DVS-350M stays inside the dry chamber of the nose module, therefore it had been difficult to switch it on from outside, or otherwise it had to be put inside keeping it in switched on mode. In order to overcome this difficulty, a special solenoid based switching board with a triggering voltage of 12 Volts, has been developed for powering the DVS-350M as and when required from outside.

5 Conclusion

The various phases of the development of the AUV for shallow water application up to a depth of 150m have been described in detail. The paper highlights the major hurdles faced during the development along-with the solutions. Being the first attempt, the shape of the vehicle has been considered to be cylindrical for better directional stability, is easier to construct. Considering the mission requirement, sensors and other equipment have been selected and rolling action of the vehicle has been stabilized. Modularity in mechanical structure, energy system and software codes were incorporated for reliability, maintainability, accessibility and subsequent up-gradation. The result of the initial trial at Idukki Lake, Cochin is very much promising. Extensive trials have been planned for making the system seaworthy.

Acknowledgement

The authors express their sincere gratitude and thanks to Ministry of Earth Sciences (MoES), Government of India, for sponsoring the project. Finally, the authors feel great pleasure in acknowledging the sincere efforts and devoted support of all the team members, who has made the AUV into a success up to this date.

References

1. Song, F., An, P.E., Folleco, A.: Modeling and Simulation of Autonomous Underwater Vehicles: Design and Implementation. IEEE J. of Oceanic Engg. 28(2), 283–296 (2003)
2. Antonelli, G., Chiaverini, S., Finotello, R., Morgavi, E.: Real-time path planning and obstacle avoidance for an autonomous underwater vehicle. In: Proceedings of the IEEE International Conference on Robotics and Automation, Detroit, Michigan, May 10-15, vol. 1, pp. 78–83 (1999)
3. Chyba, M., Leonard, N.E., Sontag, E.D.: Optimality for underwater vehicles. In: Proceedings of the 40th IEEE Conference on Decision and Control, Orlando, FL, December 4-7, vol. 5, pp. 4204–4209 (2001)
4. Licht, S., Polidoro, V., Flore, M., Hover, F.S., Triantafyllou, M.S.: Design and projected performance of a flapping foil AUV. IEEE J. Oceanic Engg. 29(3), 786–794 (2004)
5. Bogosyan, S., Arabyan, A.: High-order, sliding-mode-based precise control of direct-drive systems under heavy uncertainties. Proc. of the Institution of Mechanical Engineers, Part I: Journal of Systems and Control Engineering 221(5) (2007)
6. Shome, S.N., Nandy, S., Das, S.K., Pal, D.: AUV for shallow water applications-some design aspects. In: 18th International Offshore and Polar Engineering Conference, Vancouver (2008)
7. Das, S.K., Shome, S.N., Nandy, S., Pal, D.: Modeling a hybrid reactive-deliberative architecture towards realizing overall dynamic behavior of an AUV. Procedia Computer Science 1(1), 259–268 (2010)
8. Fossen, T.I.: Guidance and Control of Ocean Vehicles, pp. 48–54. John Wiley and Sons, UK (1994)

SOPC Based Human Biped Motion Tracking Control for Human-Sized Biped Robot

Su Yu-Te, Li Tzuu-Hseng S., Chen Wen-Chien, and Hu Jhen-Jia

aiRobots Laboratory, Department of Electrical Engineering
National Cheng Kung University, Tainan 70101, Taiwan, R.O.C.
{n2895109,thsli,n2695420,n2894112}@mail.ncku.edu.tw

Abstract. This paper presents the implementation of SOPC based human biped motion (HBM) tracking control for human-sized biped robot. aiRobot-HBR1 is a human-size biped robot with 110 cm height, 40 Kg weight, and has a total of 12 degree of freedom (D.O.F). The hardware of aiRobot-HBR1 is designed base on the human body model. According to the human body model, the dynamic model of integrated sensor control module (ISCM) is established. Using the sensor information, HBM can be recognized by the geometric relation. However the sensor information is easily affected by the noise, the discrete Kalman filter is applied in the HBM tracking control. Finally, the feasibility of Kalman filter based HBM tracking control is demonstrated by the experiments.

Keywords: Human-sized biped robot, SOPC, motion control, Kalman filter.

1 Introduction

The primary motivation behind the work is to create a platform for researching into the control of humanoid robot. In the robot research, humanoid robot is one of the popular research fields [1-3]. Comparing to other types of robot, humanoid robot is very flexible. Besides, humanoid robot is a highly integrated system which relates to many technical issues, and there are still many tough problems to be solved. Therefore, design a humanoid robot is full of challenge.

To researching into the robot control, we design a device in which human can control the biped robot in real time by human-body motion. The humanoid robots that perform tasks in the natural environments sometimes encounter environments which are extremely complex, unstructured, and hostile. The control system of the robot to deal with all the possible conditions for such a complicated environment is a difficult task. Moreover, the robots sometimes can be regarded as a substitute or an augmentor for human beings in some instance. For these reasons, the user to operate or help the robot when it is necessary is an acceptable idea. We design an integrated senor control module (ISCM) that can be used to control the biped robot. These devices are attached to human limbs so that the user is able to control the robot by the body motion. This paper focuses on the human body model firstly. Then an integrated sensor control module (ISCM) is developed to control the biped robot by the human biped motion directly. The HBM tracking control is depicted in Section 3. In section 4, the

P. Vadakkepat et al. (Eds.): FIRA 2010, CCIS 103, pp. 57–64, 2010.

experiment result is presented to demonstrate the effectiveness of the proposed control method. Conclusions are drawn in Section 5.

2 System Model of aiRobot-HBR1

In this section, we will first describe the human body model. According to this model, the system of human-sized biped robot is constructed. Then the dynamic model of ISCM is introduced to explain the mechanism of HBM tracking control. Finally, the kalman filter is applied to estimate the states of the ISCM and the complete procedure of HBM tracking control is proposed.

2.1 Human Body Model

For realizing human biped motion tracking, the human body model should be simplified, ease compute, and reliable. Actually, the structure of the human skeleton is so complicated that there may have several degrees of freedoms coupled in the same joint. The geometric human body model designed by Hanavan [4] in 1964 is wildly adopted. As shown in Fig. 1, Hanavan's model is a mathematical model that comprises 15 segments. He considered the segments as rigid bodies and assumed the segments are homogenous.

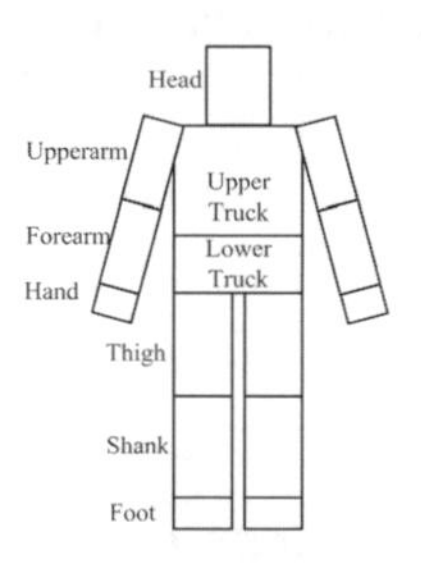

Fig. 1. Hanavan's human body model

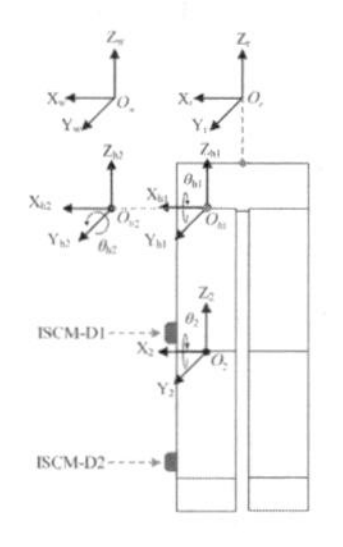

Fig. 2. Coordinate system definition of human leg

(a)

(b)

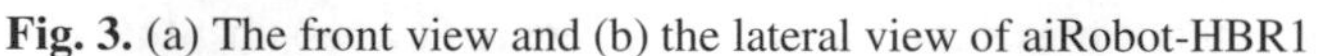

Fig. 3. (a) The front view and (b) the lateral view of aiRobot-HBR1

(a) (b)

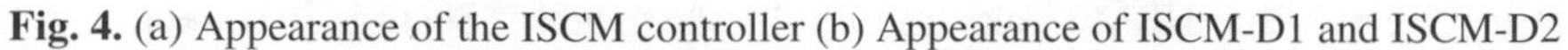

Fig. 4. (a) Appearance of the ISCM controller (b) Appearance of ISCM-D1 and ISCM-D2

As shown in Fig. 2, the human leg is divided into 3 segments for each leg and the joint arrangements of the robot were refer to define the coordinate systems of the human leg. The reference (base) coordinate system is attached in the truck of the human body. Moreover, the hip joint regard as a two-degree-of-freedom joint, so that there are two coordinate systems were established on the hip. About the knee joint, we consider it as a one degree of freedom joint. We did not established coordinate system in the foot, because only the information from the thigh and the shank was used to control the robot motion. Fig. 3 shows the appearance of aiRobot-HBR1.

2.2 Dynamic Model of the ISCM

The integrated sensor control module (ISCM) consists of a micro controller Unit: AT89C52, a three-axis accelerometer and a dual-axis gyroscope. Fig. 4. shows the appearance of ISCM. The dynamic model of the ISCM is on the basis of the transformation technologies in robotics [5] to fuse the information of accelerometer and gyroscope. For deriving the dynamic model of the ISCM, we define O_t coordinate system, which represents the coordinate system attached in segment i at time t, and $O_{t+\Delta t}$ coordinate system represents the same coordinate system attached in segment i at time $t+\Delta t$ .That is, the coordinate system O_t transform to the $O_{t+\Delta t}$ coordinate system at time interval $[t, t+\Delta t]$ owing to human motion. If the O_t coordinate system makes a rotation about an axis $r_1 = \begin{bmatrix} r_x & r_y & r_z \end{bmatrix}^T$ (normalized) in the O_t frame with $\Delta\phi$ angle and makes a translation by $d_t = \begin{bmatrix} d_x & d_y & d_z \end{bmatrix}^T$ later, the O_t coordinate and $O_{t+\Delta t}$ coordinate are coincident. Accordingly, the relationship between the O_t coordinate system and $O_{t+\Delta t}$ coordinate system is given by

$$\begin{bmatrix} x \\ y \\ z \end{bmatrix} = \begin{bmatrix} r_x^2 V_{\Delta\phi} + C_{\Delta\phi} & r_x r_y V_{\Delta\phi} - r_z S_{\Delta\phi} & r_x r_z V_{\Delta\phi} + r_y S_{\Delta\phi} \\ r_x r_y V_{\Delta\phi} + r_z S_{\Delta\phi} & r_y^2 V_{\Delta\phi} + C_{\Delta\phi} & r_y r_z V_{\Delta\phi} - r_x S_{\Delta\phi} \\ r_x r_z V_{\Delta\phi} - r_y S_{\Delta\phi} & r_y r_z V_{\Delta\phi} + r_x S_{\Delta\phi} & r_z^2 V_{\Delta\phi} + C_{\Delta\phi} \end{bmatrix} \begin{bmatrix} x_1 \\ y_1 \\ z_1 \end{bmatrix} + \begin{bmatrix} d_x \\ d_y \\ d_z \end{bmatrix} \quad (1)$$

where $V_{\Delta\phi} = 1 - \cos\Delta\phi$. $\begin{bmatrix} x & y & z \end{bmatrix}^T$ is a position vector with respect to the O_t coordinate system, and $\begin{bmatrix} x_1 & y_1 & z_1 \end{bmatrix}^T$ is the same vector with respect to the $O_{t+\Delta t}$ coordinate system. Now, we consider about the information from the sensors. The gravity is a vector field with fixed value. Accelerometer measures the component which is the projection of the acceleration value (sum of gravity acceleration and translation acceleration) onto the defined coordinate frames. Suppose the components of the measurement at time t and time $t+\Delta t$ are $\begin{bmatrix} g_x(t) & g_y(t) & g_z(t) \end{bmatrix}^T$ and $\begin{bmatrix} g_x(t+\Delta t) & g_y(t+\Delta t) & g_z(t+\Delta t) \end{bmatrix}^T$. As mentioned in above, the coordinates will make movements following the human leg motion. From Eq. (1), the relation between $\begin{bmatrix} g_x(t) & g_y(t) & g_z(t) \end{bmatrix}^{\mathrm{T}}$ and $\begin{bmatrix} g_x(t+\Delta t) & g_y(t+\Delta t) & g_z(t+\Delta t) \end{bmatrix}^{\mathrm{T}}$ can be expressed as follows:

$$\begin{bmatrix} g_x(t) \\ g_y(t) \\ g_z(t) \end{bmatrix} = \begin{bmatrix} r_x^2 V_{\Delta\phi} + C_{\Delta\phi} & r_x r_y V_{\Delta\phi} - r_z S_{\Delta\phi} & r_x r_z V_{\Delta\phi} + r_y S_{\Delta\phi} \\ r_x r_y V_{\Delta\phi} + r_z S_{\Delta\phi} & r_y^2 V_{\Delta\phi} + C_{\Delta\phi} & r_y r_z V_{\Delta\phi} - r_x S_{\Delta\phi} \\ r_x r_z V_{\Delta\phi} - r_y S_{\Delta\phi} & r_y r_z V_{\Delta\phi} + r_x S_{\Delta\phi} & r_z^2 V_{\Delta\phi} + C_{\Delta\phi} \end{bmatrix} \begin{bmatrix} g_x(t+\Delta t) \\ g_y(t+\Delta t) \\ g_z(t+\Delta t) \end{bmatrix} \tag{2}$$

The inverse of a rotation matrix is the transpose of the matrix. Eq. (2) becomes

$$\begin{bmatrix} g_x(t+\Delta t) \\ g_y(t+\Delta t) \\ g_z(t+\Delta t) \end{bmatrix} = \begin{bmatrix} r_x^2 V_{\Delta\phi} + C_{\Delta\phi} & r_x r_y V_{\Delta\phi} + r_z S_{\Delta\phi} & r_x r_z V_{\Delta\phi} - r_y S_{\Delta\phi} \\ r_x r_y V_{\Delta\phi} - r_z S_{\Delta\phi} & r_y^2 V_{\Delta\phi} + C_{\Delta\phi} & r_y r_z V_{\Delta\phi} + r_x S_{\Delta\phi} \\ r_x r_z V_{\Delta\phi} + r_y S_{\Delta\phi} & r_y r_z V_{\Delta\phi} - r_x S_{\Delta\phi} & r_z^2 V_{\Delta\phi} + C_{\Delta\phi} \end{bmatrix} \begin{bmatrix} g_x(t) \\ g_y(t) \\ g_z(t) \end{bmatrix}$$

Considering $\Delta t \to 0$ and $\Delta\phi \to 0$, above equation can be simplified as

$$\begin{bmatrix} g_x(t+\Delta t) \\ g_y(t+\Delta t) \\ g_z(t+\Delta t) \end{bmatrix} = \begin{bmatrix} 1 & r_z \cdot \Delta\phi & -r_y \cdot \Delta\phi \\ -r_z \cdot \Delta\phi & 1 & r_x \cdot \Delta\phi \\ r_y \cdot \Delta\phi & -r_x \cdot \Delta\phi & 1 \end{bmatrix} \begin{bmatrix} g_x(t) \\ g_y(t) \\ g_z(t) \end{bmatrix} \tag{3}$$

Moreover,

$$\begin{bmatrix} g_x(t) \\ g_y(t) \\ g_z(t) \end{bmatrix} = \begin{bmatrix} 1 & 0 & 0 \\ 0 & 1 & 0 \\ 0 & 0 & 1 \end{bmatrix} \begin{bmatrix} g_x(t) \\ g_y(t) \\ g_z(t) \end{bmatrix} \tag{4}$$

Subtract Eq. (4) from Eq. (3) and then divide by Δt, we get

$$\frac{r_x \cdot \Delta\phi}{\Delta t} = \frac{r_x}{\sqrt{r_x^{\ 2} + r_y^{\ 2} + r_z^{\ 2}}} \cdot \frac{\Delta\phi}{\Delta t}\bigg|_{\Delta t \to 0, \Delta\phi \to 0} \equiv \omega_x$$

$$\frac{r_y \cdot \Delta\phi}{\Delta t} = \frac{r_y}{\sqrt{r_x^{\ 2} + r_y^{\ 2} + r_z^{\ 2}}} \cdot \frac{\Delta\phi}{\Delta t}\bigg|_{\Delta t \to 0, \Delta\phi \to 0} \equiv \omega_y$$

$$\frac{r_z \cdot \Delta\phi}{\Delta t} = \frac{r_z}{\sqrt{r_x^{\ 2} + r_y^{\ 2} + r_z^{\ 2}}} \cdot \frac{\Delta\phi}{\Delta t}\bigg|_{\Delta t \to 0, \Delta\phi \to 0} \equiv \omega_z$$

Note that $\sqrt{r_x^{\ 2} + r_y^{\ 2} + r_z^{\ 2}} = 1$. ω_x, ω_y, and ω_z denote the components of the angular velocity measured by gyroscope. Thus, we get the dynamic model of the ISCM:

$$\begin{bmatrix} \dot{g}_x(t) \\ \dot{g}_y(t) \\ \dot{g}_z(t) \end{bmatrix} = \begin{bmatrix} 0 & \omega_z & -\omega_y \\ -\omega_z & 0 & \omega_z \\ \omega_y & -\omega_x & 0 \end{bmatrix} \begin{bmatrix} g_x(t) \\ g_y(t) \\ g_z(t) \end{bmatrix} + \begin{bmatrix} w_x(t) \\ w_y(t) \\ w_z(t) \end{bmatrix} \tag{5}$$

where $\begin{bmatrix} w_x(t) & w_y(t) & w_z(t) \end{bmatrix}^T$ is the noise existing in the system. The ISCM contains a two-axes gyroscope, thus we set $\omega_z = 0$. Discretizing Eq. (5), a discretized system equation is given by

$$\begin{bmatrix} g_x(k) \\ g_y(k) \\ g_z(k) \end{bmatrix} = \begin{bmatrix} a_{11} & a_{12} & a_{13} \\ a_{21} & a_{22} & a_{23} \\ a_{31} & a_{32} & a_{33} \end{bmatrix} \begin{bmatrix} g_x(k-1) \\ g_y(k-1) \\ g_z(k-1) \end{bmatrix} + \begin{bmatrix} w_x(k-1) \\ w_y(k-1) \\ w_z(k-1) \end{bmatrix} = A_d \cdot \begin{bmatrix} g_x(k-1) \\ g_y(k-1) \\ g_z(k-1) \end{bmatrix} + \begin{bmatrix} w_x(k-1) \\ w_y(k-1) \\ w_z(k-1) \end{bmatrix} \tag{6}$$

where A_d is the system matrix, $\begin{bmatrix} w_x(k-1) & w_y(k-1) & w_z(k-1) \end{bmatrix}^T$ represents the system noise, and $M_1 = e^{-T(-\omega_x^2-\omega_y^2)^{0.5}}$, $M_2 = e^{T(-\omega_x^2-\omega_y^2)^{0.5}}$, $a_{11} = 0.5(2\omega_x^2 + \omega_y^2 M_1) + \frac{1}{\omega_x^2+\omega_y^2}\omega_y^2 M_2$, $a_{12} = -0.5\omega_x\omega_y M_1 + \frac{1}{\omega_x^2+\omega_y^2}\cdot\frac{M_2}{e^2}$, $a_{13} = 0.5\omega_y M_1 - \frac{1}{(-\omega_x^2-\omega_y^2)^{0.5}} M_2$, $a_{21} = -0.5\omega_x\omega_y M_1 + \frac{1}{\omega_x^2+\omega_y^2}\omega_y^2 M_2$, $a_{22} = 0.5(2\omega_y^2 + \omega_x^2 M_1) + \frac{1}{\omega_x^2+\omega_y^2}\omega_x^2 M_2$, $a_{33} = 0.5M_1 + 0.5M_2$, $a_{23} = -0.5\omega_x M_1 - \frac{1}{(-\omega_x^2-\omega_y^2)^{0.5}} M_2$, $a_{31} = -0.5\omega_y M_1 - \frac{1}{(-\omega_x^2-\omega_y^2)^{0.5}}\omega_y^2 M_2$, $a_{32} = 0.5\omega_x M_1 - \frac{1}{(-\omega_x^2-\omega_y^2)^{0.5}} M_2$,.

The output equation in state-space representation is defined as follows:

$$\begin{bmatrix} a_{mx}(k) \\ a_{my}(k) \\ a_{mz}(k) \end{bmatrix} = \begin{bmatrix} 1 & 0 & 0 \\ 0 & 1 & 0 \\ 0 & 0 & 1 \end{bmatrix} \begin{bmatrix} g_x(k) \\ g_y(k) \\ g_z(k) \end{bmatrix} + \begin{bmatrix} v_x \\ v_y \\ v_z \end{bmatrix} \tag{7}$$

where v_x, v_y, and v_z are the measurement noise. Note that $\begin{bmatrix} a_{mx}(k) & a_{my}(k) & a_{mz}(k) \end{bmatrix}^{\mathrm{T}}$ is the measurement of the accelerometer. Kalman filter is used to estimate the state $\begin{bmatrix} g_x(k) & g_y(k) & g_z(k) \end{bmatrix}^T$ of the ISCM dynamic model, Eq. (6) and Eq. (7).

3 Kalman Filter Based HBM Tracking

Considering the dynamic model of the ISCM, system equation Eq. (6) and output equation Eq. (7), we can estimate the state $\left[g_x(k) \;\; g_y(k) \;\; g_z(k)\right]^{\mathrm{T}}$ by utilizing the discrete Kalman filter [6]. The estimated state is the gravity acceleration with respect to the sensor coordinate system, which is attached in the human leg. Define $\left[g_{x1}(k) \;\; g_{y1}(k) \;\; g_{z1}(k)\right]^{\mathrm{T}}$ as the estimated state of the ISCM-D1 and $\left[g_{x2}(k) \;\; g_{y2}(k) \;\; g_{z2}(k)\right]^{\mathrm{T}}$ as the estimated state of the ISCM-D2. The posture $\left[\theta_{\mathrm{h1}} \;\; \theta_{\mathrm{h2}} \;\; \theta_2\right]^{\mathrm{T}}$ can be obtained by the inverse kinematics, that is

$$\theta_{h1} = \tan^{-1}\left(\frac{g_{y1}}{\sqrt{g^2 - g_{y1}^{\;2}}}\right),\; \theta_{h2} = \tan^{-1}\left(\frac{g_{x1}}{-g_{z1}}\right),\; \theta_2 = \tan^{-1}\left(\frac{g_{y2}g_{z1} - g_{y1}g_{z2}}{g_{y1}g_{y2} + g_{z1}g_{z2}}\right)$$

where $g = \sqrt{g_x^{\;2} + g_y^{\;2} + g_z^{\;2}}$. The overall flow of the HBM tracking based on Kalman filter is depicted in Fig. 5. The parameter selection and tuning have much to do with the performance of the Kalman filter. Fig. 6 and Fig. 7 show the estimation result of the Kalman filter. Fig. 6 is measured in the condition that human expeditiously rotate the ISCM with a 90 degree. Fig. 7 is under a heavy vibration motion that human vibrates the ISCM in a horizontal plane. By the natural of the Kalman filter and well parameter tunings, the Kalman filter can reject the noise existing in the measurements efficiently. However, the dynamic model of the ISCM did not considered the compensation for the translation acceleration. We just consider the translation acceleration as the noise existing in the measurements. Although the estimate may become a little inaccuracy in the situation existing large translation acceleration, it is receivable. Based on these results, we can realize the posture of human thigh and shank with respect to the reference coordinate system by the ISCM. Fig. 8 depicts the control structure of the real-time HBM tracking control with ISCM for airobot-HBR1. The posture of the human thigh and shank is the control input of the real-time HBM tracking control. The posture of the human thigh and shank are mapped to the biped robot directly. In other words, if the rotation angle of the human knee joint is known, this angle is regard as the command of the knee joint of the robot, because the motion control system of the robot receives the absolute angle commands. Therefore, it is the "direct mapping" from the human body to airobot-HBR1.

4 Experiment Result

In this section, the results of the real-time HBM tracking control are presented. As shown in Fig. 9, ISCM-D1 and ISCM-D2 are attached in human thigh and shank. In the experiments, human swings his leg forward, backward, and right to demonstrate the validity and the performance of the real-time HBM tracking control. Fig. 10 shows the successive tracking results of the swinging forward. Fig. 11 shows the successive tracking results of the swinging backward, and Fig. 12 shows the successive tracking results of the swinging right. At the beginning of the tracking process, the foot of the robot has small angle delay from the human leg because of the friction force between the ground and the foot of the robot. The tracking algorithm takes less than 50 ms in the NIOS FPGA [7]. From the experiment results, the HBM tracking control works well in many practical experiments.

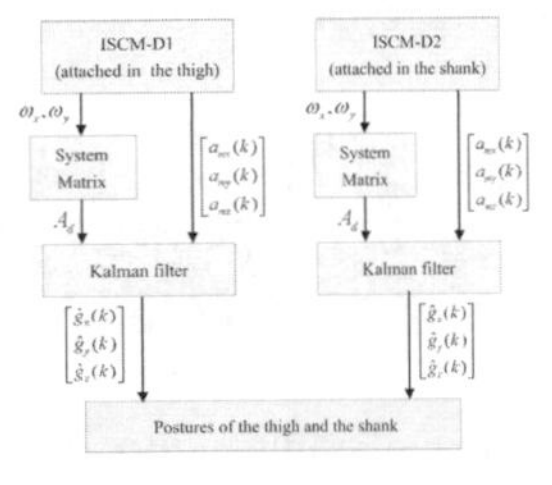

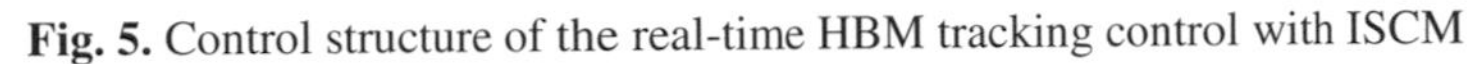

Fig. 5. Control structure of the real-time HBM tracking control with ISCM

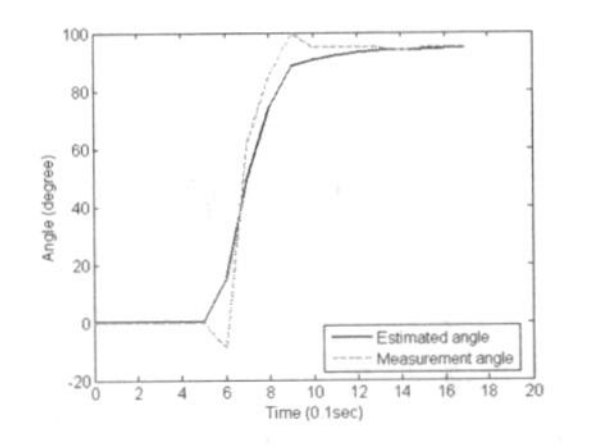

Fig. 6. Estimation result of the Kalman filter

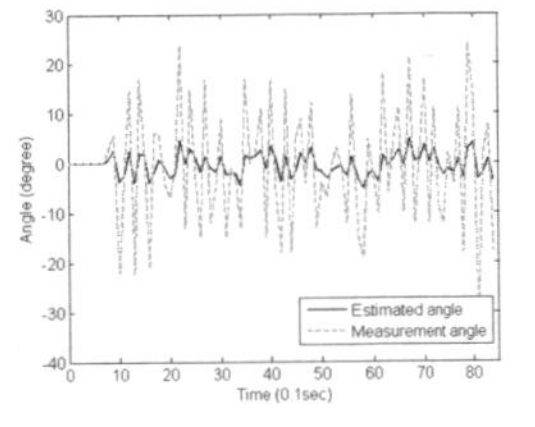

Fig. 7. Estimation result in a vibration motion

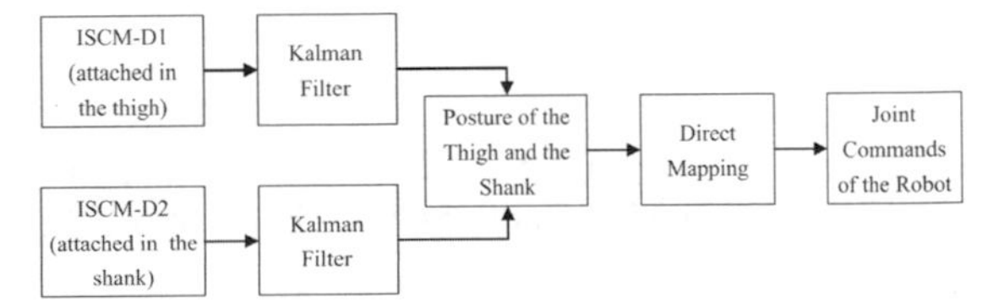

Fig. 8. Control structure of the real-time HBM tracking control

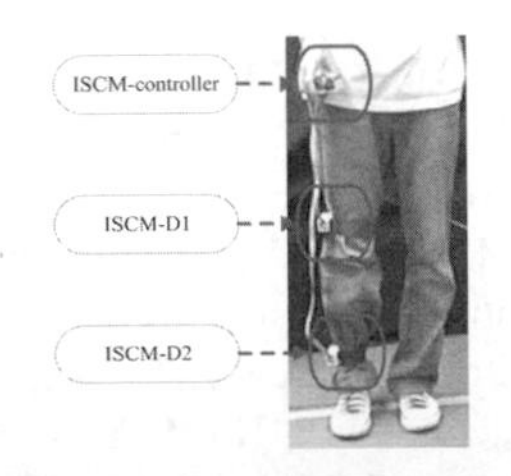

Fig. 9. Position of the ISCM in the experiment

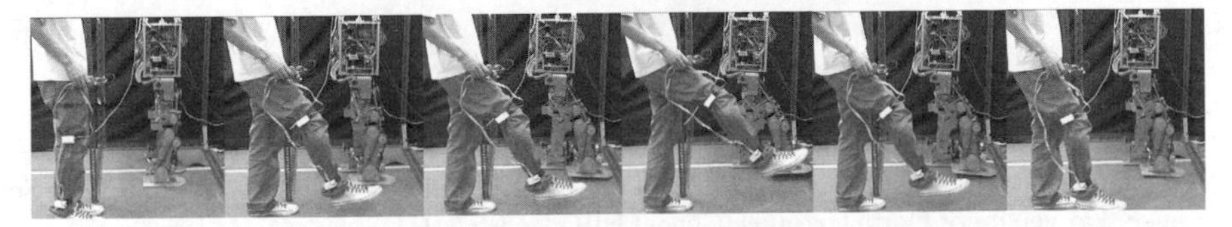

Fig. 10. Tracking results of the swinging forward

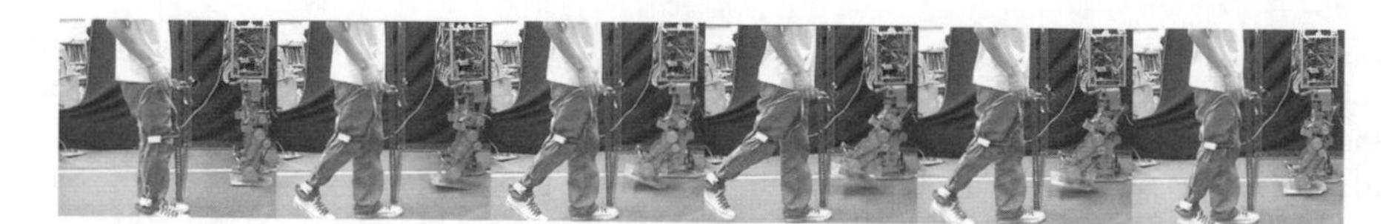

Fig. 11. Tracking results of the swinging backward

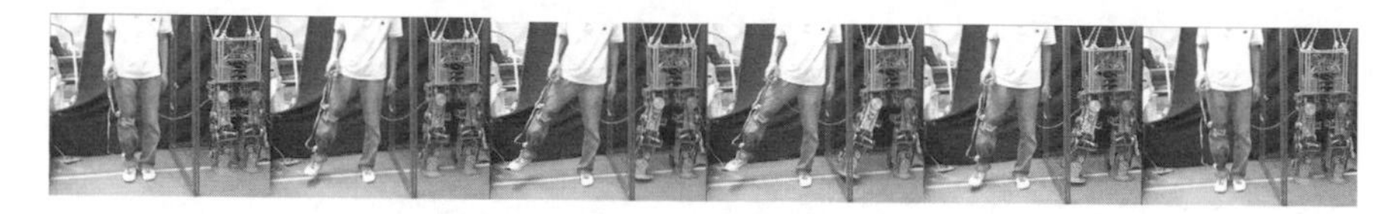

Fig. 12. Tracking results of the swinging right

5 Conclusion

This paper has presented the SOPC based HBM tracking control for the human-sized biped robot, aiRobot-HBR1. We have presented the dynamic model of the ISCM and the kinematics model of human leg. The Kalman filter is utilized to estimate the states of the ISCM model, and the kinematics model is to figure out the posture of human leg. Then the human posture maps to the joint commands of the robot directly to realize the real-time HBM tracking control. Finally, the experiment results show the feasibility of proposed HBM tracking control.

Acknowledgment

This work was supported by National Science Council of Taiwan, R.O.C, under Grants NSC97-2221-E006-0-160-MY3 and NSC97-2221-E006-0-172-MY3.

References

1. Su, Y.-T., Hu, C.-Y., Lu, M.F., Chang, C.-M., Lai, S.-W., Liu, S.-H., Li, T.-H.: Design and implementation of SOPC based image and control system for HuroCup. J. of Harbin Institute Tech (New Series) 15, 41–46 (2008)
2. Li, T.-H., Su, Y.-T., Liu, S.-H., Hsiao, M.-Y.: Design and Implementation of a Gait Pattern Generator Based on Fuzzy Control for Small-Sized Humanoid Robot by Using SOPC. In: 2008 CACS Int. Automatic Control Conference, Taiwan (November 2008)
3. Su, Y.-T., Li, T.-H., Hsu, C.L., Lu, M.F., Hu, C.Y., Liu, S.H.: Omni-Directional Vision-Based Control Strategy for Humanoid Soccer Robot. In: Proc. of 2007 IEEE IECON, pp. 2950–2955 (November 2007)
4. Hanavan, E.P.: A mathematical model of the human body. AMRL. Technical Report 64-102. Wright-Patterson Air Force Base, OH (1964)
5. Fu, K.S., Gonzales, R.C., Lee, C.S.G.: Robotics: control, sensing, vision, and intelligence. McGraw-Hill Book Company, New York (1987)
6. Welch, G., Bishop, G.: An introduction to the Kalman filter, Department of Computer Science, University of North Carolina, Chapel Hill, TR 95 – 041

Walking Pattern Generator Using an Evolutionary Central Pattern Generator

Chang-Soo Park, Jeong-Ki Yoo, Young-Dae Hong, Ki-Baek Lee,
Si-Jung Ryu, and Jong-Hawn Kim

Department of Electrical Engineering, KAIST, Daejeon, Korea
{cspark,jkyoo,ydhong,kblee,sjryu,johkim}@rit.kaist.ac.kr
http://rit.kaist.ac.kr

Abstract. For the generation of locomotion, such as walking, running or swimming, vertebrate and invertebrate animals use the Central Pattern Generator (CPG). In this paper, a walking pattern generator is proposed using an evolutionary optimized CPG. Sensory feedback pathways in CPG are proposed, which uses Force Sensing Resistor (FSR) signals. For the optimization of CPG parameters, quantum-inspired evolutionary algorithm is employed. Walking pattern generator is developed to generate trajectories of ankles and hip using CPG. The effectiveness of the proposed scheme is demonstrated by simulations and real experiments using a Webot dynamic simulator and a small sized humanoid robot, HSR-IX.

1 Introduction

Despite the complexity of high-DOF systems, these days many humanoid robots have been developed and their performance has been improved a lot[1]-[5]. However, their control algorithm still needs to be improved further to perform a practical task. In this regard, research on developing robust walking patterns of humanoid robots plays one of important roles in this field.

For generation of robust walking patterns of humanoid robots, there are two typical approaches, such as dynamic model based approach and biologically inspired approach. In the former, equations of motion are derived and utilized from mathematical model of the robot in the same way of conventional control researches of manipulator [6]-[8]. In the latter, Central Pattern Generator (CPG) is widely used. Animals are using CPG for generating locomotion, which consists of biological neural networks. It can produce coordinated rhythmic signals using simple input signals [9]. CPG was used to control an 8-link simulated planar biped model by generating the torque of each joint of its lower body [10].

The walking pattern algorithm based on the CPG has two problems. Firstly, the walking pattern algorithm based on the CPG requires much effort to make appropriate oscillation signals for biped locomotion. In order to overcome this, a position control method was presented [11]. However, it is not enough considering stable biped locomotion in 3D space. Secondly, the walking pattern generator based on the CPG is difficult to design by appropriate parameters for feedback pathways in neural oscillators. Therefore, genetic algorithm or reinforcement learning was applied to optimize the involved parameters in neural oscillators [12], [13]. However, these methods need a great number of iterations to optimize them.

P. Vadakkepat et al. (Eds.): FIRA 2010, CCIS 103, pp. 65–72, 2010.

This paper proposes a walking pattern generator using an evolutionary optimized CPG. Walking pattern generator is developed to generate trajectories of the ankles and hip in the Cartesian coordinate system using CPG. It is easy to set up parameters of CPG for generation of appropriate output signals for biped locomotion. The proposed scheme generates trajectory of the position of the center of hip along the lateral direction in addition to trajectories of the position of both ankles along the sagittal and vertical directions for stable biped locomotion in 3D space. The body posture for sensory feedback is obtained using signals of Force Sensing Resistor (FSR) sensors attached on the sole of foot. Also, the Quantum-inspired Evolutionary Algorithm (QEA) is employed to optimize parameters of CPG [14]. The effectiveness of the proposed scheme is demonstrated by computer simulations with the Webot model of a small sized humanoid robot in a dynamic environment and by real experiments with HSR-IX developed in the RIT Lab., KAIST.

This paper is organized as follows. In Section II, evolutionary CPG-based walking pattern generator is proposed. In Section III, simulation and experiment results are presented and finally concluding remarks follow in Section IV.

2 CPG-Based Walking Pattern Generator

This section presents the proposed CPG-based walking pattern generator architecture for stable biped locomotion. For biped locomotion, walking pattern generator generates position trajectories of both ankles and the hip using CPG. The body posture for sensory feedback is obtained using signals of FSR sensors attached on the sole of foot. Parameters of the CPG are optimized using the quantum-inspired evolutionary algorithm.

2.1 Neural Oscillator

In this paper, the neural oscillators as CPG are developed to generate rhythmic signals for humanoid robots. The neural oscillator is biologically inspired to generate a rhythmic signal, defined as follows [15]:

$$\tau \dot{u}_i = -u_i - \sum_{j=1}^{N} w_{ij} o_j - \beta v_i + u_0 + Feed_i, \tag{1}$$

$$\tau' \dot{v}_i = -v_i + o_i, \tag{2}$$

$$o_i = max(0, u_i) \tag{3}$$

where u_i is the inner state of the ith neuron, v_i is the self-inhibition state of the ith neuron, u_0 is the constant input signal, o_i is the output signal, w_{ij} is the connecting weight between ith and jth neuron, τ and τ' are time constants, β is the weight of the self-inhibition, and $Feed_i$ is the sensory feedback signal which is necessary for stable biped locomotion, of the ith neuron. u_0, τ, τ' and w_{ij} are constant parameters. τ and τ' decide the output wave shape and frequency, u_0 determines the output amplitude and w_{ij} determines the phase difference between ith and jth neurons.

2.2 Application to Walking Pattern Generator of CPG

In this paper, walking pattern generator generates trajectories of ankles and the center of hip, respectively, in the Cartesian coordinate system. This approach is easy to set up parameters of CPG, the initial states of neurons and connecting weight in neural oscillators, and feedback pathways, and to change step length or height. Walking pattern generator is provided to generate the trajectory of left and right ankles, respectively, along the sagittal and vertical directions as follows:

$$P_{L_X} = -A_X(o_1 - o_2), \tag{4}$$

$$P_{R_X} = A_X(o_1 - o_2), \tag{5}$$

$$P_{L_Z} = Z_c - A_Z(o_3 - o_4), \tag{6}$$

$$P_{R_Z} = Z_c + A_Z(o_3 - o_4) \tag{7}$$

where P_{L_X} and P_{R_X} are the distance between the center of hip and left and right ankles, respectively, along the sagittal direction, P_{L_Z} and P_{R_Z} are the distance between the center of hip and left and right ankles, respectively, along the vertical direction, A_X, A_Z are amplitude scaling factor and Z_c is offset factor.

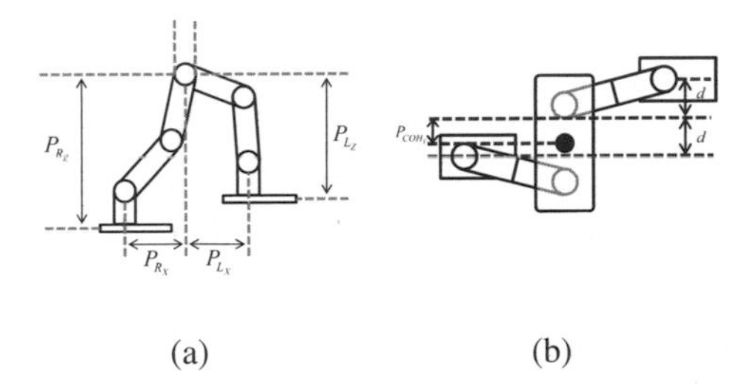

Fig. 1. Trajectories of ankles and the center of hip. (a) Trajectories of ankles along sagittal and vertical directions. (b) Trajectories of the center of hip along lateral direction.

In 3D space, positions along the lateral direction in addition to sagittal and vertical directions have to be considered for stable biped locomotion. Therefore, generation of position trajectory of the center of hip along the lateral direction is added. Walking pattern generator is provided to generate the trajectory of the center of hip along the lateral direction as follows:

$$P_{COH_Y} = A_Y(o_5 - o_6) \tag{8}$$

where P_{COH_Y} is the distance between the center of hip and the center position of both ankles along the lateral direction.

The projection of trajectories of swing leg's ankle on X-Z plane can be approximated as semi-ellipsoidal trajectories on the X-Z plane, such as $P_X = lcos\theta$ and $P_Z = lsin\theta$. Thus, the desired phase difference between ankle's vertical and horizontal oscillations should be $\pi/2$.

For stable biped locomotion along the lateral direction, the distance between the supported leg and the center of hip along the lateral direction is controlled to be minimum using neural oscillator. When humanoid robot is double supported, the position of the center of hip is given as the center position of both ankles along the lateral direction. When P_Z increases, the distance between the center of hip and the supported leg is designed to decrease such that P_{COH_Y} increases. On the other hand, when P_Z decreases, the distance between the center of hip and the supported leg is designed to increase such that P_{COH_Y} decreases. Consequently, phase difference between P_Z and P_{COH_Y} should be zero.

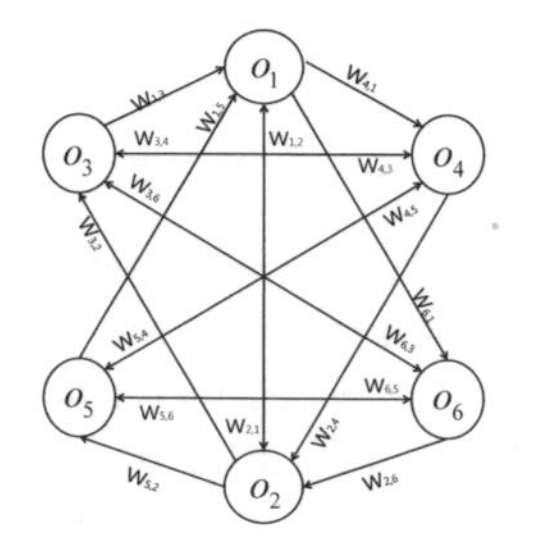

Fig. 2. The neural oscillator

Neural oscillators are designed for stable biped locomotion, as shown in Fig. 2, where $w_{1,2} = w_{2,1} = 2.0$, $w_{3,4} = w_{4,3} = w_{5,6} = w_{6,5} = 1.5$, $w_{1,3} = w_{1,4} = w_{2,4} = w_{2,6} = 0.25$, $w_{3,2} = w_{5,2} = w_{4,1} = w_{6,1} = 0.5$, $w_{3,6} = w6,3 = w_{4,5} = w_{5,4} = 0.5$ are used to satisfy phase difference conditions derived in this section.

2.3 Sensory Feedback Design

When humanoid robot walks using CPG, its balance is maintained using sensory feedback. For sensory feedback, the information on humanoid robot's body posture is needed such that FSR sensors are attached on the sole of foot to collect the information. The sensory feedback pathways are designed as follows:

$$Feed_1 = k_1((F_{L_f} - F_{L_b}) - (F_{R_f} - F_{R_b})), \tag{9}$$
$$Feed_2 = -Feed_1, \tag{10}$$
$$Feed_3 = k_2(F_L - F_R), \tag{11}$$
$$Feed_4 = -Feed_3, \tag{12}$$
$$Feed_5 = k_3((F_{L_l} - F_{L_r}) + (F_{R_l} - F_{R_r})), \tag{13}$$
$$Feed_6 = -Feed_5 \tag{14}$$

where F_L and F_R are left and right foot's vertical reaction forces, respectively, F_{L_f}, F_{R_f}, F_{L_b} and F_{R_b} are left and right foot's front and back vertical reaction forces, respectively, and F_{L_l}, F_{R_l}, F_{L_r} and F_{R_r} are left and right foot's left and right vertical reaction force, respectively.

2.4 Evolutionary Optimization for CPG Algorithm

The proposed CPG-based walking pattern generator has 8 parameters. In this paper, these parameters are evolutionary optimized by employing Quantum-inspired Evolutionary Algorithm (QEA) [14]. The goal to optimize the parameters is to make the humanoid robot approach the destination as soon as possible and to maintain its balance as stable as possible. The objective function is defined as follows:

$$f = \frac{k_x T_1}{d|_{t=T_1}} + k_y \sum_{kT=0}^{T_1} |P_{err}[kT]| + B_P \tag{15}$$

where k_x and k_y are constants and $T_1/d|_{t=T_1}$ is the speed of biped locomotion for T_1, which corresponds to the fitness function for fast biped locomotion. Also, $\sum_{kT=0}^{kT_1} |P_{err}[kT]|$ is the sum of position error along the lateral direction for T_1, which corresponds to the fitness function for decreasing position error along the lateral direction. When stability of biped locomotion increases, position error decreases. Therefore, decreasing position error means that stability of biped locomotion is improved. The last term is the penalty which is to be given if humanoid robot loses its balance and collapses, where B_P is assigned a priori as a constant value. .

3 Simulations and Experiment

The effectiveness of the proposed algorithm was demonstrated by computer simulations with the Webot model of a small sized humanoid robot, HSR-IX in a dynamic environment and real experiments with HSR-IX. In each simulation, if the humanoid robot walks n steps without falling down to the ground and n is lower than 15 steps, total number of steps is n steps, else total number of steps is 15 steps. $\tau/\tau' = 0.105/0.132$ and $\tau = 0.105A_\tau$ were used. The parameters on the CPG were optimized by QEA.

3.1 Effectiveness of Generating Trajectory of Position along the Lateral Direction

Fig. 3 shows the effectiveness of generating trajectories of the position of the center of hip along the lateral direction in addition to the position of left and right ankles, respectively, along the sagittal and vertical directions. $A_X = 2.0$, $A_\tau = 1.5$ and $A_Z = 1.0$ were used. When CPG-based walking pattern generator generated trajectories the position of the center of hip along the lateral direction in addition to the trajectories of the position of the left and right ankles, respectively, along the sagittal and vertical directions, total number of steps increased and position error along the lateral direction decreased. This result shows that it improves stability of biped locomotion if trajectory of the position of the center of hip along the lateral direction is generated using CPG in addition to the position of ankles along the sagittal and vertical directions.

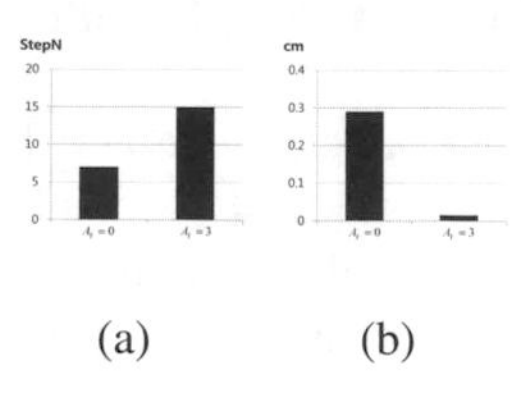

(a) (b)

Fig. 3. Effectivness of generating trajectory of position along the lateral direction. (a) Total number of steps. (b) Position error along the lateral direction.

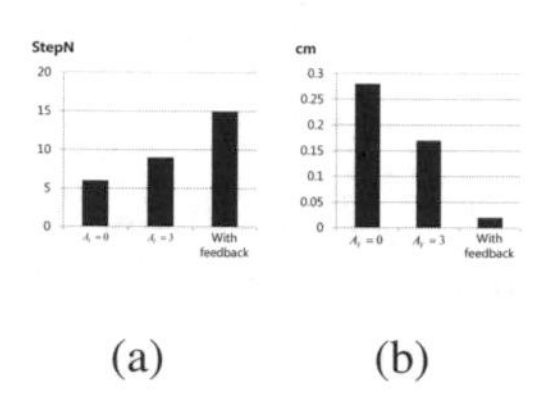

(a) (b)

Fig. 4. Effect of sensory feedback pathways. (a) Total number of steps. (b) Position error along the lateral direction.

3.2 Effectiveness of Sensory Feedback Pathways

Fig. 4 shows the effectiveness of including sensory feedback pathways. $A_X = 3.0$, $A_\tau = 1.5$ and $A_Z = 1.2$ were used. In this simulation, humanoid robot's walking speed was faster than that in Sec. 3.1, such that it was hard to maintain balance without sensory feedback pathways. Generating trajectory of the position with sensory feedback pathways, total number of steps increased and position error along the lateral direction decreased. This result shows that sensory feedback based on FSR sensor maintains humanoid robot's balance and prevents it from falling down to the ground.

3.3 Effectiveness of Evolutionary Optimization

In Fig. 5, the velocity and position error along were plotted, when k_x was fixed as 500, and k_y was changed. When k_y increased, the walking velocity was slower, but the position error along the lateral direction decreased. This result illustrates that when k_y decreases, the optimization of the parameters makes the humanoid robot walk faster, when k_y increases, the optimization of the parameters makes the biped locomotion more stable.

3.4 Experiment

Experiment was carried out with the actual humanoid robot, HSR-IX. $\tau/\tau' = 0.105/0.132$, $\tau = 0.105A_\tau$, $A_\tau = 4$, $A_x = 2.0$, $A_y = 3.0$ and $A_z = 0.5$ were used. For stable biped locomotion in real experiments, the parameters, $k_1 \sim k_3$, in sensory feedback pathways of the CPG, were evolutionary optimized using the computer simulation of

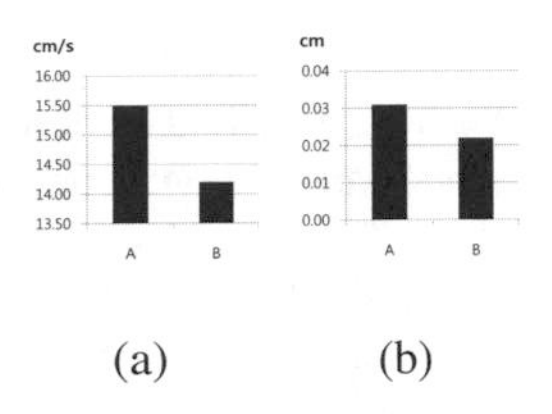

Fig. 5. Effectiveness of evolutionary optimization. A is $k_x = 500$, $k_y = 100$ and B is $k_x = 500$, $k_y = 200$. (a) The velocity. (b) Position error along the lateral direction.

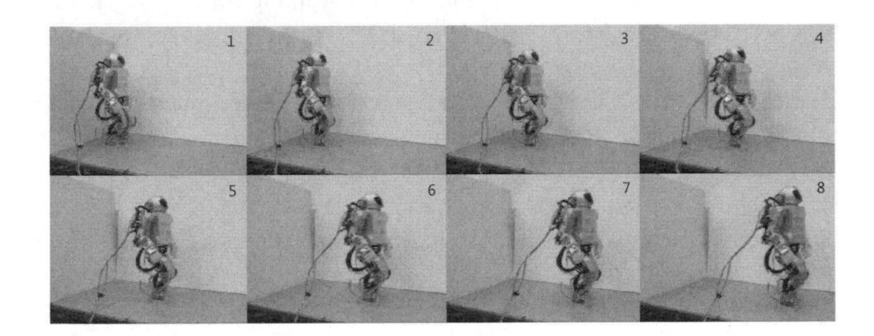

Fig. 6. Shapshots of experiment using HSR-IX

HSR-IX model by Webot. Then these parameters were tuned for proper walking by trial and error. Fig. 6 shows the humanoid robot achieved stable walking.

4 Conclusion

This paper proposed a walking pattern generator for stable biped locomotion based on evolutionary CPG. The neural oscillators in the CPG were developed to generate rhythmic signals. The proposed CPG-based walking pattern generator generated trajectory of the position of the center of hip along lateral direction in addition to trajectories of the position of ankles along the sagittal and vertical directions for stable biped locomotion in 3D space. The sensory feedback pathways in CPG were designed using FSR signals. Quantum-inspired Evolutionary Algorithm (QEA) was employed to optimize parameters of CPG for stable and fast biped locomotion. In order to demonstrate the performance of the proposed scheme, computer simulations were carried out with the Webot model of the small sized humanoid robot, HSR-IX in a dynamic environment and real experiments were carried out with HSR-IX developed in the RIT Lab., KAIST.

Acknowledgments

This research was supported by Basic Science Research Program through the National Research Foundation of Korea (NRF) funded by the Ministry of Education, Science and Technology (2010-0000831).

References

1. Hirai, K., Hirose, M., Haikawa, Y., Takenaka, T.: The development of honda humanoid robot. In: Proc. IEEE Int. Conf. on Robotics and Automations, Leuven, Belgium, pp. 1321–1326 (May 1998)
2. Ogura, Y., Aikawa, H., Shimomura, K., Kondo, H., Morishima, A., Lim, H.-O., Takanishi, A.: Development of a new humanoid robot WABIAN-2. In: Proc. IEEE Int. Conf. on Robotics and Automations,Orlando, Florida, pp. 76–81 (May 2006)
3. Akachi, K., Kaneko, K., Kanehira, N., Ota, S., Miyamori, G., Hirata, M., Kajita, S., Kanehiro, F.: Development of humanoid robot HRP-3P. In: Proc. IEEE-RAS Int. Conf. On Humanoid Robots, Tsukuba, Japan, pp. 50–55 (December 2005)
4. Park, I.-W., Kim, J.-Y., Lee, J., Oh, J.-H.: Online free walking trajectory generation for biped humanoid robot KHR-3(HUBO). In: Proc. IEEE Int. Conf. on Robotics and Automations, Orlando, Florida, pp. 1231–1236 (May 2006)
5. Kim, J.-H., Lee, K.-H., Kim, Y.-D., Lee, B.-J., Yoo, J.-K.: The origin of artificial species: Humanoid robot HanSaRam. In: Proc. 2nd International Conference on HNICEM 2005, Manila, Philippines (March 2005)
6. Kajita, S., Kanehiro, F., Kaneko, K., Fujiwara, K., Yokoi, K., Hirukawa, H.: A Realtime Pattern Generator for Biped Walking. In: Proc. IEEE Int. Conf. on Robotics and Automation, Washington, DC, pp. 31–37 (May 2002)
7. Lee, B.-J., Stonier, D., Kim, Y.-D., Yoo, J.-K., Kim, J.-H.: Modifiable Walking Pattern of a Humanoid Robot by Using Allowable ZMP Variation. IEEE Transaction on Robotics 24(4), 917–925 (2008)
8. Lee, B.-J., Stonier, D., Kim, Y.-D., Yoo, J.-K., Kim, J.-H.: Modifiable Walking Pattern of a Humanoid Robot by Using Allowable ZMP Variation. IEEE Transactions on Robotics 24(4), 917–923 (2008)
9. Grillner, S., et al.: Neural networks that co-ordinate locomotion and body orientation in lamprey. Trends in NeuroSciences 18(6), 270–279 (1995)
10. Taga, G.: Emergence of bipedal locomotion through entrainment among the neuro-musculo-skeletal system and the environment. Physica D: Nonlinear Phenomena 75(1.3), 190–208 (1994)
11. Endo, G., Morimoto, J., Matsubara, T., Nakanishi, J., Cheng, G.: Learning CPG Sensory Feedback with Policy Gradient for Biped Locomotion for a Full-body Humanoid. In: Proc. The 20th National Conference on Artificial Intelligence, vol. 3, pp. 1267–1273 (2005)
12. Hase, K., Yamazaki, N.: Computer simulation of the ontogeny of biped walking. Anthropological Science 106(4), 327–347 (1998)
13. Mori, T., Nakamura, Y., Sato, M., Ishii, S.: Reinforcement learning for a cpg-driven biped robot. In: Nineteenth National Conference on Artificial Intelligence, pp. 623–630 (2004)
14. Han, K.-H., Kim, J.-H.: Genetic quantum algorithm and its application to combinatorial optimization problem. In: Proc. 2000 Congress on Evolutionary Computation, vol. 2, pp. 1354–1360. IEEE Press, Piscataway (July 2000)
15. Matsuoka, K.: Sustained oscillations generated by mutually inhibiting neurons with adaptation. Biol. Cybern. 52(6), 367–376 (1985)

Analysis and Study of Human Joint Torque and Motion Energy during Walking on Various Grounds*

Kuo-Yang Tu and Wei-Cheng Lee

Institute of System Information and Control,
National Kaohsiung First University of Science and Technology
2, Juoyue Road, Nantsu District, Kaohsiung City, Taiwan, R.O.C.
tuky@ccms.nkfust.edu.tw

Abstract. It is interesting to calculate the joint torque and motion energy during human walking. In this paper, a two-link manipulator approach to human walking dynamics for estimating joint torque and motion energy is proposed. The accelerometers installed on human limbs measure the angle of time-sequent trajectories used to estimate joint torque and motion energy based on the approach dynamic equations. However, the joint torque and motion energy estimated by dynamic equations are very sensitive to their parameters such as link length and weight. It is hard to have accurate limb length and weight. The study can not measure absolute scale of the torque and energy, but only focuses on the comparison of walking on six different grounds, and the difference between left and right legs. Experiments of taking a load during walking are also included.

Keywords: Accelerometer; Torque; Human Walking.

1 Introduction

The analysis of walking gait is helpful to understand the joint torque and motion energy of a human walking. Such study often used to improve the walking of disable or old people [1, 2, 3]. In the recent years, the walking gait analysis makes use of many different methods. Using vision detects and records the markers pasted on human limbs [4]. Using accelerometers and gyroscope installed on the center of limb mass measures the human motion [1, 2, 5, 6]. Goniometer and ultrasound corrects the measurement of human walking [3, 4]. Using force sensor measures the soles during human walking. However, the facilities for human walking gait analysis are hard to apply to any place. They only applied in ideal environments such as laboratory. They are not easy to apply to the walking gait analysis in human daily life. Thus, in this paper using accelerometers to design portable and simple motion storage for human walking is proposed. For calculating the joint torque and motion energy, the human walking is approached to two-link manipulation. The dynamic equation of two-link manipulation is derived to calculate the torque spent by ankle and waist. However, the dynamics is very sensitive to parameter variation. Consequently, the accurate ankle

* This research was supported by National Science Council, Taiwan, Rep. of China under grant NSC 97-2221-E-327-019-.

P. Vadakkepat et al. (Eds.): FIRA 2010, CCIS 103, pp. 73–81, 2010.

and waist torque of human walking is hard to use the derived dynamics. Hence, the study of human walking only focuses on the comparison of ankle and waist torque wasted on various grounds. Experiments for the human walking on six different planes are thus included.

2 Two-Link Manipulation Approach to Human Walking Dynamics

The human walking dynamics is complicated and hard to derive. In this paper, two-link approach to the human walking dynamics is proposed. In this section, the walking of a human is first presented by some definition. Next, the reasonableness of the human walking dynamics approached to two-link manipulation is revealed. Then, the dynamics of two-link manipulation is derived.

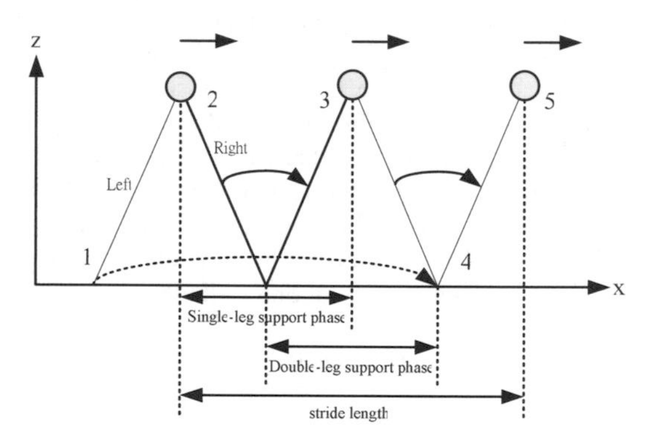

Fig. 1. The key terms of a human walking gait

Walking gaits maneuver human body. The walking gaits are composed by limb change between supporting and swing. Thus if we can record the trajectories of limbs angles with respect to time, it is possible to analyze the motion energy of human walking from dynamic equations. However, we need some descriptions of walking gait before analysis. Some definitions of walking gaits are presented as follows:

Definition 1: gait cycle
The gait cycle is the period of one leg to touch the ground from present to next. As shown in Fig. 1, the gait cycle is the phases from 1 to 5. However, in the uniform walking the period of one leg to leave the ground can be the gait cycle too.

Definition 2: stride length
The move distance of a gait cycle is stride length. As shown in Fig. 1, the stride length is the distance between phases 2 and 5.

Definition 3: Support leg and swing leg

During some time period, human uses only one leg to support body. The leg to support human body is called support leg. During the same period, the other leg swung forward for the next supporting position is called swing leg. As shown in Fig. 1, between phases 2 and 3, right leg is support leg, and left leg is swing leg.

During the gait cycle, right and left legs exchange the phases between support and swing. Either support or swing phases are achieved by the appropriate trajectories of limbs. Fig. 2 shows the limb trajectories of a three-link biped robot for four gait cycles [7].

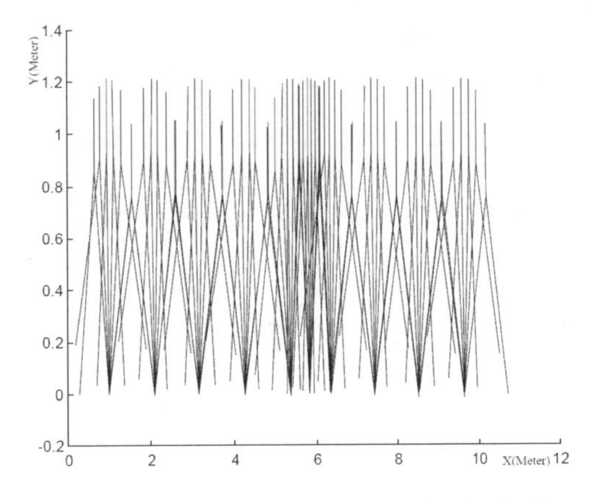

Fig. 2. The gait cycles of a three-link biped robot

During a gait cycle, the support leg supports human body to move forward. In slow walking, the knee between up and low limbs in the support leg almost does not change bending angle. Thus, the support leg can be regarded as a link during the gait cycle. Besides, the up body can be the other link. Therefore, in this paper a two-link model approach to the human walking is proposed. It is reasonable to neglect the dynamics of swing leg because its weight is very smaller than the up body. The dynamic equation of the two-link model formulates to calculate the motion energy and joint torque of human walking.

Based on the two-link model, a gait cycle can be separated into three phases: right single support phase, double support phase, and left single support phase. As shown in Fig. 3, the right single support phase is A, B and C^-, the double support phase is C, and the left single support phase is C^+, E, and D. The dynamic equation of this simple structure can calculates the torque used by waist and ankle during the human walking.

Assume that during the walking, human body is vertical to the ground. Fig. 3 also shows that the human body is vertical to the ground during walking. Therefore, the joint angles of the human limbs can be easily estimated as shown in Fig. 3. During right support phase, the angles of ankle and waist are equal to θ_{r1}. Similarly, those of ankle and waist are equal to θ_{l1} during left support phase. In addition, during double support phase, the angles between right and left legs are the sum of θ_{r1} and θ_{l1}. As a result, the waist angles can be estimated by measuring ankle angles. If we have the waist and ankle angles with respect to time, the dynamic equation of two-link model can calculates the joint torque and the motion energy wasted by human walking. Thus the dynamic equation of two-link model is derived as follows.

Fig. 4 shows the mechanism of a two-link manipulation. Let θ_1, and θ_2 be the joint angles, m_1 and m_2 be the mass of links 1 and 2, respectively, and l_1 and l_2 be length of links 1 and 2, respectively. Then, the following Lemma derives the torque used by ankle and waist (i. e. τ_1 and τ_2, respectively).

Lemma 1: For the two-link manipulation as shown in Fig. 4, the dynamic equation is

$$\begin{bmatrix}\tau_1\\ \tau_2\end{bmatrix}=\begin{bmatrix}\frac{1}{3}m_1l_1^2+\frac{1}{3}m_2l_2^2+m_2l_1^2+m_2l_1l_2C_2 & \frac{1}{3}m_2l_2^2+\frac{1}{2}m_2l_1l_2C_2\\ \frac{1}{3}m_2l_2^2+\frac{1}{2}m_2l_1l_2C_2 & \frac{1}{3}m_2l_2^2\end{bmatrix}\begin{bmatrix}\ddot{\theta}_1\\ \ddot{\theta}_2\end{bmatrix}$$
$$+\begin{bmatrix}-\frac{1}{2}m_2S_2l_1l_2\dot{\theta}_2^2-m_2S_2l_1l_2\dot{\theta}_1\dot{\theta}_2\\ \frac{1}{2}m_2S_2l_1l_2\dot{\theta}_1^2\end{bmatrix}+\begin{bmatrix}\frac{1}{2}m_1gl_1C_1+\frac{1}{2}m_2gl_2C_{12}+m_2gl_1C_2\\ \frac{1}{2}m_2gl_2C_{12}\end{bmatrix} \quad (1)$$

where S_i is $\sin\theta_i$ (for i = 1, 2), C_i is $\cos\theta_i$ (for i = 1, 2), and C_{12} is $\cos(\theta_1 + \theta_2)$.
Proof: Neglect.

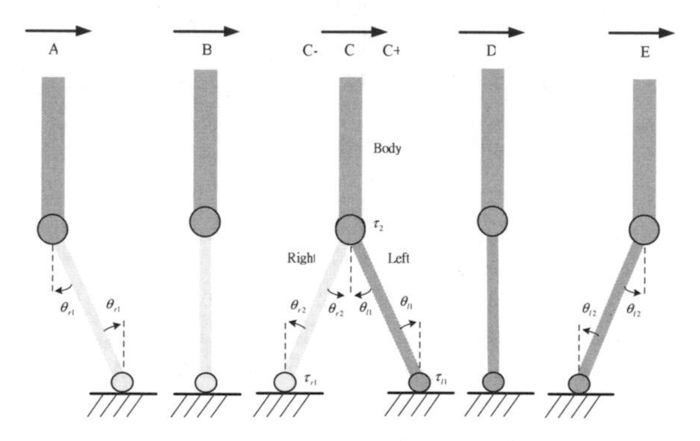

Fig. 3. The two-link model approach to a gait cycle of the human walking

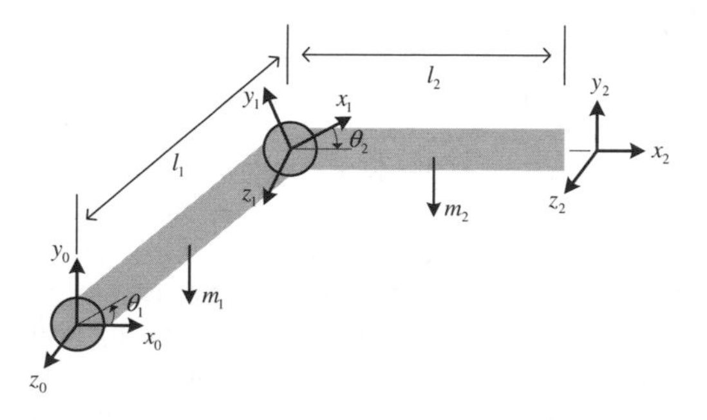

Fig. 4. The mechanism of a two-link model

If we have the angle trajectories of joints 1 and 2, Eq. (1) can calculate the torque trajectories of ankle and waist (τ_1 and τ_2) with corresponding mechanism parameters such as m_i and l_i (for i = 1, 2). After having the torque trajectories spent by human with respect to time, we can estimate the motion energy as follows:

$$E_i = \int \tau_i(t)\omega_i(t)dt = \sum_j^n \tau(j)\omega(j)\Delta T \quad \text{for } i = 1, 2 \cdot \tag{2}$$

where τ_1 and τ_2 are the torque of ankle and waist, respectively, ω_1 and ω_2 are the rotation angles of ankle and waist, respectively.

In this paper, the two-link model approach leads to develop motion sensors for $\theta_1(t)$ and $\theta_2(t)$. After measuring $\theta_1(t)$ and $\theta_2(t)$, Eq. (1) can be used to estimate the torque of ankle and waist $\tau_1(t)$ and $\tau_2(t)$. Eq. (2) can estimate the motion energy of ankle and waist.

3 Joint Angle Measure and Hardware Design

In this paper, we need to measure $\theta_1(t)$ (ankle) and $\theta_2(t)$ (waist) for Eqs. (1) and (2). Thus, the method of measuring ankle and waist angles is designed and the hardware for the measurement is developed in this section.

In this study, the limb angle measure makes use of accelerometer influenced by gravity. The accelerometer fixed on human limb will have the different scale influenced by gravity when the limb has different angle with respect to the ground plane. Assume that the accelerometer used to measure joint angle is only influenced by gravity. Thus the scale measured by the accelerometer corresponds to the angles between the limb and the ground plane. However, if the human maneuvers, the limb acceleration is the other force on the accelerometer. This method only measures the human walking on slow speed, so that the maneuver acceleration compared with gravity can be neglected.

The limb angle measure makes use of two-axis accelerometers. Let the acceleration scales measured on the directions X, and Z be g_x and g_z, respectively. Then the limb angle on the X and Z directions are

$$\theta_x = \sin^{-1}\left(g_x / g\right)$$
$$\theta_z = \sin^{-1}\left(g_z / g\right). \tag{3}$$

where g is gravity.

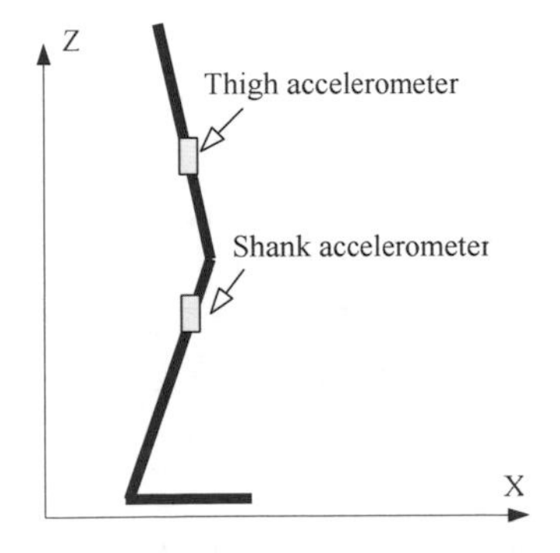

Fig. 5. The schema of accelerometers installation

There are four accelerometers installed on the shank and thigh of right and left legs, respectively. Fig. 5 shows the installation schema. The shank and thigh accelerometers measure the angles of waist and ankle, respectively. Because the dynamics neglects the swing leg, the ankle and waist angles need to be measured only if the leg is on support phase. Thus the measurement needs the rule to identify if the leg is in support phase or not. The rule defines a identification function as follows:

$$M(t)=\begin{cases}1, \text{if } e(t)<c\\ 0, \text{if } e(t)>c\end{cases} \quad (4)$$

where $e(t)=|\text{Thigh Angle - Shank Angle}|$, and c is a threshold value. If M(t) = 1 (0), then this leg is in support phase (swing phase). The threshold is defined to adjust for the people who have district posture.

Portable motion device for easy experiments during human walking is designed. This device not only captures the signals of four accelerometers, but also sends the accessed signals to PDA (Personal Digital Assistant). The signals of PDA will be transferred to a PC (Personal Computer) for filtering out noise, and calculating motion energy and torque. This function will be presented latter. The hardware is separated into sensor module and capture. The sensor module is designed by FPGA (Field Programmable Gate Array), and the capture module receives the signals sent from FPGA via USB.

4 Main Results

Fig. 6 is the angle trajectories of thigh and shank of left and right legs, in which blue color lines are the actual trajectories, and red color lines are the signals after filtering out noise.

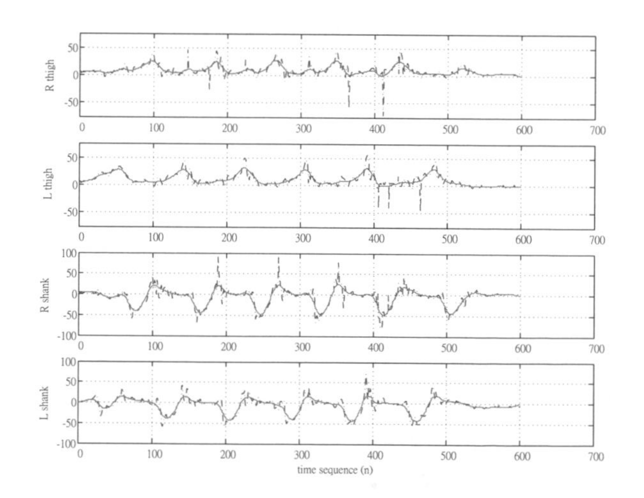

Fig. 6. The angle trajectories of thigh and shank of left and right legs (Blue trajectories are actual and red trajectories are the signals filtered out noise)

Table 1. The parameters of six testers

Tester	Sex	Tallness	Weight	Thigh	Shank
A	Boy	177	52	85	85
B	Girl	163	53	80	76
C	Boy	175	62	85	84
D	Boy	180	84	84	90
E	Boy	166	59	79	80
F	Boy	179	73	90	82

In this study, using accelerometers to measure the angles of thigh and shank is highly influenced by walking speed. The quantity of walking speed whose noise can be filtered out is important information in this study. Thus, an experiment is designed to analyze and then understand such walking speed. This experiment designs five walking speeds: 3~3.5, 2.5~3, 2~2.5, 1.5~2 and 1~1.5 (km/hour). The tester walks on the flat ground in five tries with these five speeds, respectively. Every try walks five meters. The examining results show that 1.5~2 km/hour is the proper speed that its noise can be filtered out easily. Therefore, the study makes use of 2 km/hour walking speed.

Table 2. The torque of ankle and waist as tester A walks in various ground

	τ_1 (ankle)		τ_2 (waist)	
A	Right leg (standard variance)	Left leg (standard variance)	Right leg (standard variance)	Left leg (standard variance)
Flat ground	139.44 (2.966)	120.31 (2.279)	24.788 (0.527)	22.388 (0.414)
Tiles trail	106.59 (2.422)	134.39 (3.054)	26.12 (0.593)	42.433 (0.964)
Asphalt	181.07 (4.3111)	577.73 (13.755)	90.475 (2.1542)	379.84 (9.043)
Ramp up	259.38 (6.8258)	189.84 (4.995)	79.659 (2.0963)	88.027 (2.316)
Ramp down	240.54 (6.6815)	206.42 (5.733)	71.92 (1.9978)	89.294 (2.48)
rugged road (Left high right low)	173.08 (4.5548)	237.82 (6.2584)	51.538 (1.3563)	68.759 (1.8095)

In the experiments, six testers examine motion energy during walking. The parameters of the six testers are show in Table 1. As shown in Table 1, there are one girl and five boys. Their ages are between 18 and 28 in the average 22.8±1.4.

In the experiments, the testers walk on six grounds: flat ground, tiles trail, asphalt, ramp and rugged road. After reading and recording the signals from accelerometers, we can get the signals like Fig. 7. Then we can use Eq. (1) to calculate τ_1 and τ_2, i. e. the torque of ankle and waist, and use Eq. (2) to calculate the walking motion energy.

The testers walk in the speed 1~2km/hour for 5 meters. Each ground examines five tries. Then, the torque and motion energy that the testers spend can be measured. Table 2 is the results of tester A. The results of all testers are analyzed to study the torque and motion energy that human spends in various grounds. The analysis makes use of flat ground walking as comparing reference. As walking on tiles trail, about ankle, three testers decrease motion energy, two testers increase, and one is almost equal. About waist, three increase, but three decrease. As walking on asphalt, five testers increase, about ankle, four testers increase, one decreases, and one is almost equal. About waist, five increase, but one decreases. About ankle on up ramp, five testers increase, but one decrease. About waist on up ramp, four increase, but two decreases. About the ankle on down ramp, five increase, but one decrease. About the waist on down ramp, five increase, but one decreases. In summary, the testers increase torque and motion energy in various grounds compared with the flat ground.

5 Conclusions

In this paper, experiments demonstrate that it is possible to estimate the joint angle trajectories by only using accelerometers. Thus, this idea will be used to estimate the joint torque and motion energy during human walking. Based on the experiments of six testers, we conclude the result that compared with the flat ground the other five grounds make testers increasing torque and motion energy. The experiment data also show a very special result that some testers use extremely different torque on left and right legs. For example, as shown in Table 2, tester A spends left leg very big torque as walking on asphalt. It's very interesting that the walking motion energy is personal data. A precise instrument to measure human walking motion energy that can detect human walking gait and give them some suggestion for walking is the further development.

Acknowledgments. This research was supported by National Science Council, Taiwan, Rep. of China under grant NSC 97-2221-E-327-019-.

References

1. Dejnabadi, H., Jolles, B.M., Casanova, E., Fua, P., Aminian, k.: Estimation and Visualization of Saggittal Kinematics of Lower Limbs Orientation Using Body-Fixed Sensors. IEEE Transactions on Biomedical Engineering 53(7), 1385–1393 (2006)
2. Lau, H., Tong, K.: The reliability of using accelerometer and gyroscope for gait event identification on persons with dropped foot. Gait & Posture 27(2), 248–257 (2007)
3. Veltink, P.H., Member I E E E, Franken, H.M.: Detection of Knee Unlock During Stance by Accelerometer. IEEE Transaction on Rehabilitation Engineering 4(4), 395–402 (1996)
4. Dejnabadi, H., Jolles, B.M., Aminian, k.: A New Approach to Accurate Measurement of Uniaxial Joint Angles Based on a Combination of Accelerometers and Gyroscopes. IEEE Transactions on Biomedical Engineering 52(8), 395–402 (2005)

5. Alvarez, D., González, R.C., López, A., Alvarez, J.C.: Comparison of Step Length Estimators from Wearable Accelerometer Devices. In: Proceedings of the 28th IEEE EMBS Annual International Conference New York City, USA, pp. 5964–5967 (2006)
6. Fong, D.T.W., Wong, J.C.Y., Lam, A.H.F., Lam, R.H.W., Li, W.J.: A Wireless Motion Sensing System Using ADXL MEMS Accelerometers for Sports Science Applications. In: Proceedings of the 5th World Congress on Intelligent Control and Automation, vol. 6, pp. 5635–5640 (2004)
7. Tu, K.-Y., Lee, T.-T.: Design of a Multi-Layer Fuzzy Logic Controller Using Pole Assignment for Biped Walking at Varying Speed. Journal of the Chinese Institute of Engineers 27(1), 55–68 (2004)

Applied Complex Motion Planning and Motion Control for Humanoid Robots in Vertical Motion Sceneries

Marco Wickrath

Dipl.-Inf. Marco Wickrath, Technical University of Dortmund,
Otto-Hahn-Str. 14, D-44221 Dortmund, Germany

Abstract. Robots moving up walls or similarly rampant terrain have been analyzed and developed for a longer time. Today systems can be found in real-world scenarios like window cleaning or inspection applications. However, with regard to the complexity of the motion control, research was usually dedicated to climbing mechanisms with vacuum, wheel or multi-leg systems. Biped humanoid robots have to overcome severe control problems and have therefore not been considered for these kind of applications. Nevertheless in this paper we describe a much more challenging scenario: A humanoid robot designed to climb up a freely configurable climbing wall totally autonomously.

Keywords: Humanoid Robots, Climbing Robots, Motion Planning, Motion Control, Autonomous System, Vertical Motion Sceneries.

1 Introduction

Robots climbing up a wall are not a new subject. There are many specialized robots endowed with the ability to climb up walls or rampant terrain by using vacuum, wheel or multi-leg systems. The project outlined here, however, contributes a new approach in this field by enlarging the capabilities of a humanoid robot to enable it to autonomously climb up a freely configurable climbing wall.

Previous developments in science improved skills like grabbing, robot vision, walking or climbing stairs. This led to the prototype of a new kind of motion for humanoid robots, one that brings their capabilities closer to those of human beings, a motion as a result of a combined usage of legs and arms, a motion into a vertical direction - the climbing.

The overall goal is a humanoid robot which is able to search a climbing wall, identify and analyze the grip configuration, walk to its starting position and climb up the wall. This project was split into two projects, one vision part and one motion planning and motion control part. Here we focus on the second part, the sequence of tasks is as follows: The positions of a current grip configuration need to be send to the robot, the robot calculates its way up using proper grips and its starting position, is placed there and finally autonomously climbs up the wall.

P. Vadakkepat et al. (Eds.): FIRA 2010, CCIS 103, pp. 82–89, 2010.

2 Preconditions and Preliminary Work

Since there is no archetype available, every part of this project had to be developed, starting with the decision for an appropriate humanoid robot, covering the design and implementation of the software to create its climbing ability and building a proper climbing wall.

2.1 Robot-Hardware

The hardware chosen was a Bioloid Comprehensive Kit manufactured by Robotis[1]. The humanoid robot which can be built from this kit has 18 degrees of freedom and was made able to climb up walls by adding some hardware add-ons considering the motion restrictions.

Fig. 1. Bioloid robot - construction schematic and final robot[2]

A. Climbing hand

The Bioloid robot does not have any grip mechanism but comes with an angle bracket as hands. Hence, the only chance to grab objects is by pressing its arms together. To be able to climb the robot needs a hook hand so it can hook into a grip, hold its position and pull itself upwards.

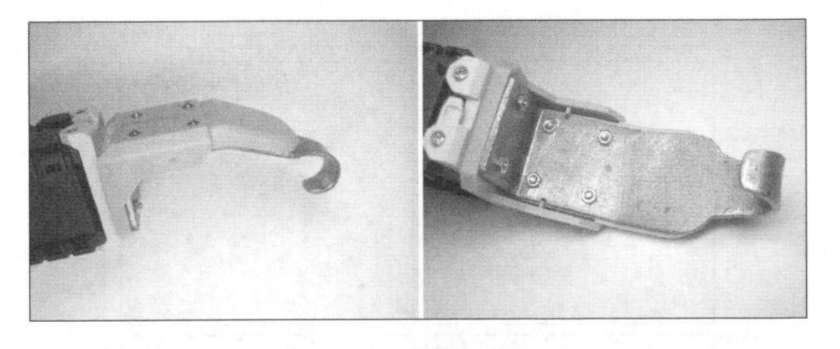

Fig. 2. Climbing hand

B. Foot extension

The feet of the Bioloid robot are short and cause trouble when the robots lifts one leg to place its foot onto a next grip during the climbing process. In this situation the knee

[1] `http://www.robotis.com` (26.04.2010)

[2] `http://www.robotis.com/zbxe/software_en/5418` (26.04.2010)

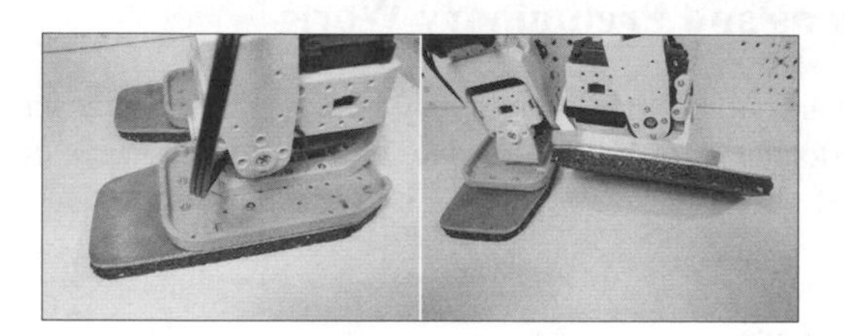

Fig. 3. Foot extension

collides with the wall and gets stuck. To overcome this problem the robot was equipped with a feet extension with an anti-slip sole to get more foothold, too.

C. Straight-leg-climbing technique

A constructional problem - the construction frames attached to the hip joint collide if both legs are turned outwards more than 35 degrees - led to the development of a special climbing technique..

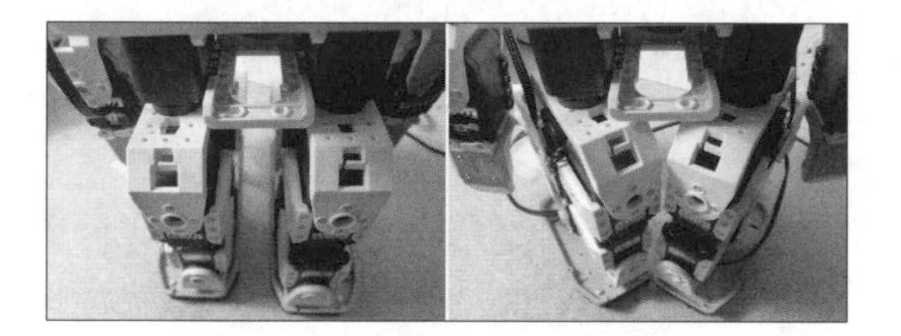

Fig. 4. Constructional problem

This problem results in the fact that the robot is not able to climb up the wall using a frog technique which allows full freedom of movement for every leg. The solution is a special technique named the straight-leg-climbing technique. This technique guaranties that one leg is always kept straight during the climbing process so that the second leg has full freedom of movement. The following image illustrates this climbing technique.

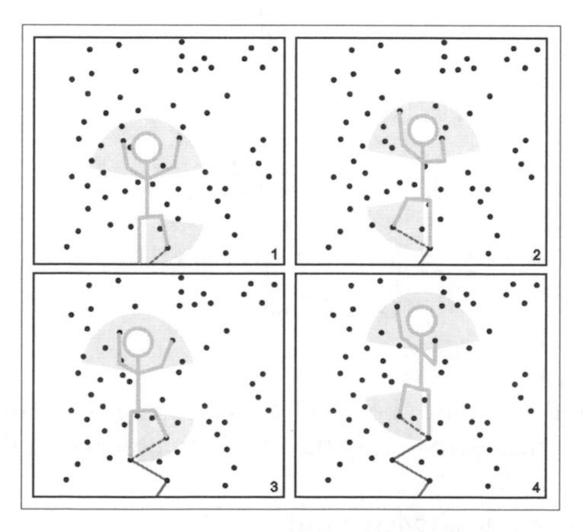

Fig. 5. Straight-leg-climbing technique

2.2 Climbing Wall

Finally, a proper climbing wall had to be built. The wall consists of a multi-hole wooden wall with special ring grips in which the new climbing hands can hook into and the feet can be placed onto.

Fig. 6. Climbing grips

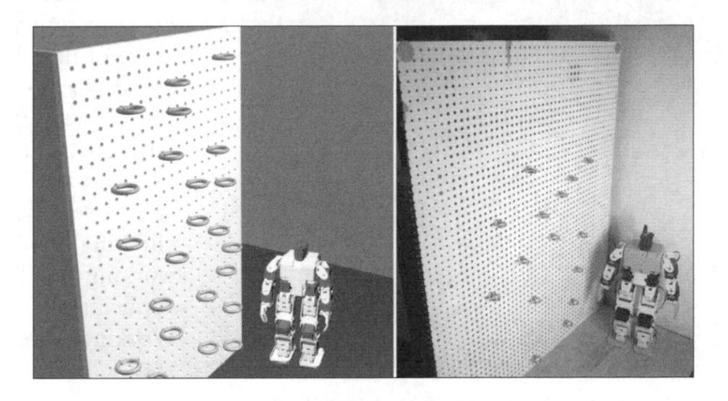

Fig. 7. Comparison of first idea and final climbing wall

3 Motion Planning

A special motion planning algorithm was developed. to identify ways up the wall while using a current grip configuration. After analyzing alternatives like a graph algorithm or a clustering into useful and not useful grips, finally an algorithm was developed that takes into account the specific abilities of the robot.

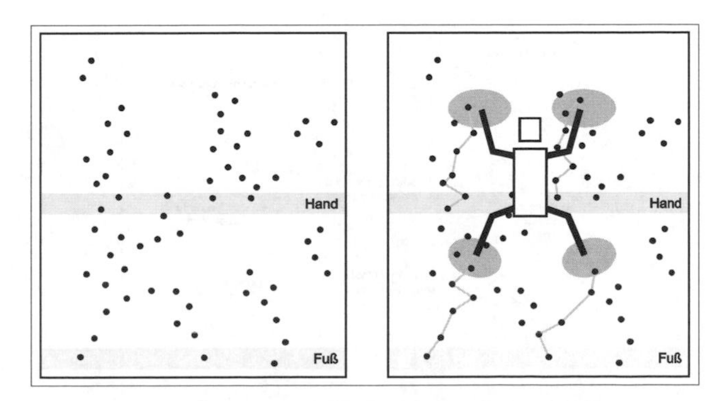

Fig. 8. Simple illustration of Robo-Shape-Algorithm

Labeled “Robo-Shape-Algorithm”, the basic idea is to consider only grips that can be reached by the robots hands or feet. These areas within reach are the “working areas”. Using a central reference point positioned in the exact centre of the robot the working areas for the hands and for the feet and their distance to the reference point constantly stay the same. Using this Robo-Shape consisting of the torso including the central reference point and the four working areas the shape can be moved along a virtual wall representing the current grip configuration to find valid climbing steps. A valid step is found if for a given reference point position at least one grip is located in one foot working area and also one grip in each of the two hand areas.

The basic for this algorithm is the exact definition of the Bioloid robot’s working areas. Fig. 9 visualizes the process of identifying these areas, Fig. 10 shows the final Robo-Shape.

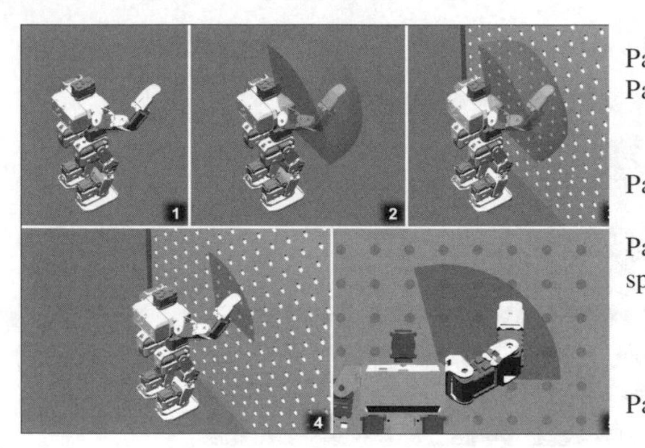

Part 1: Robot with raised arm
Part 2: Forward orientated working space of right hand, displayed as red sphere section
Part 3: Robot and working space positioned at climbing wall
Part 4: Intersection of working space and working face (i.e. climbing wall) results in working area for climbing application
Part 5: Working Area from back view

Fig. 9. Construction of working areas in theory

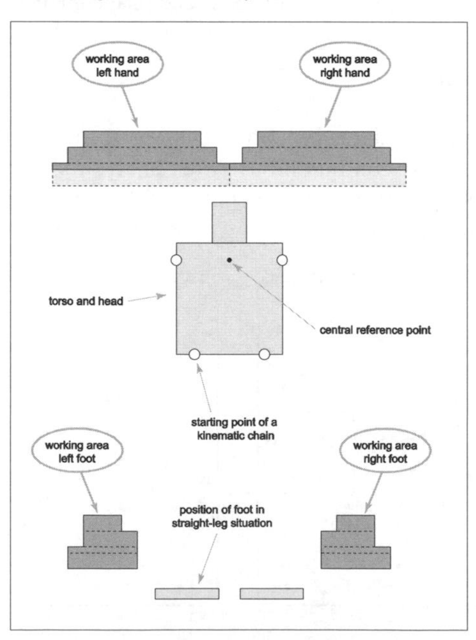

Fig. 10. Robo-Shape

After the working areas have been calculated the Robo-Shape-Algorithm works as follows:

1. Finding valid starting positions
The Robo-Shape is moved along the bottom line of a virtual wall representing a current grip configuration. Valid hand-foot-grip combinations taking physical aspects and the straight-leg-climbing technique into account are saved as possible starting positions.

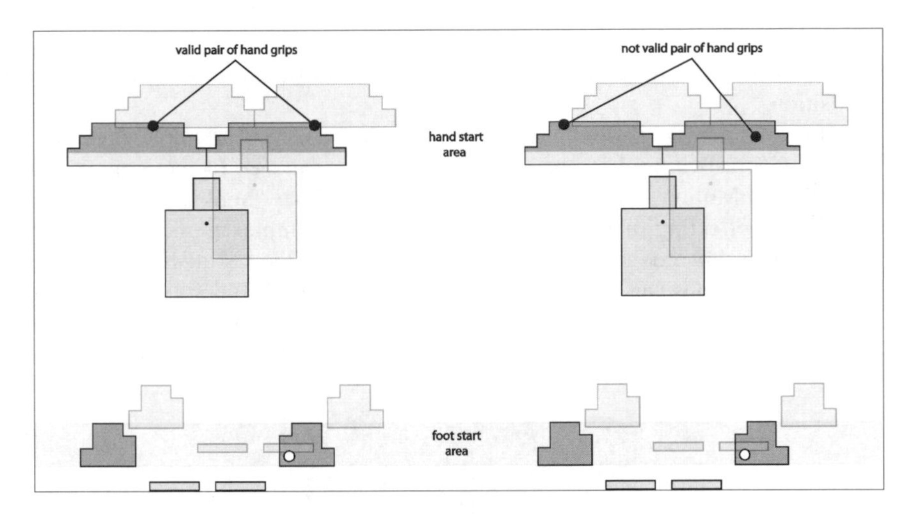

Fig. 11. Comparison of valid pair of hand grips and not valid pair. Grips being covered by both actual Robo-Shape position and simulated future straight-leg position of Robo-Shape (grey) are valid climbing grips.

2. Finding follow-up climbing steps
For each starting position the Robo-Shape is moved up to a straight leg position as if the real robot would climb up the wall reaching its final position by standing on the foot grip with its leg stretched straight. For this next position the algorithm searches for valid grips in the new areas of the wall covered by the current working areas of the robot. This step is repeated for every starting position until no more valid climbing steps can be found. Every starting position is saved with its follow-up steps as a path. The path that leads the robot to the highest position is chosen as the final climbing path.

4 Motion Control

The second part of this project is the execution of the climbing process. For this purpose an inverse kinematics for the Bioloid robot was developed. First the degrees of freedom are managed in an intelligent way to overcome the problem of kinematic redundancy and to get a unique result. Second and finally, projections and rotations that bring the robot's joints and the desired target point into one plane enable the calculation of necessary angles by simple trigonometric functions. The inverse

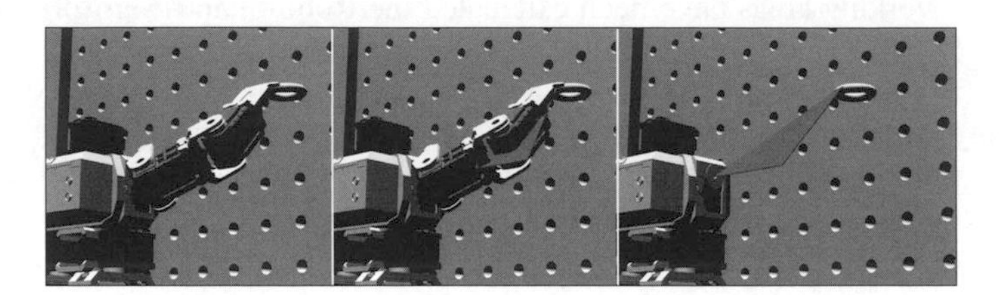

Fig. 12. Simulation of motion plane for right arm joints

kinematics enables the robot to move a kinematic chain (i.e. arm or leg) to a desired x/y/z-position.

To achieve climbing motions the one-step motion done by a simple execution of the inverse kinematics needs to be transferred into a sequence of motions, so called motion macros. Motion macros are simple sequences like "hook into grip", "place foot onto grip" effecting one kinematic chain or more complex motions to keep balance or to move upwards affecting every single joint of the robot. In this prototype these complex motions can be executed by simply moving the central reference point to its future position recalculating each foot and hand position, transferring the new joint positions and executing the whole step in one synchronized motion.

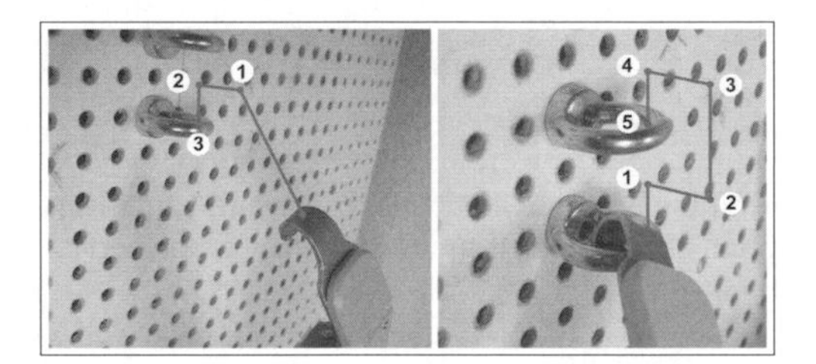

Fig. 13. Motion macro "hook into grip" for start and follow-up climbing step

5 Software PathSimulator

Motion planning and motion control have been implemented in the PathSimulator. This program displays a simulation of the real climbing wall which can be adjusted to the wall by special settings. The grip configuration of the real wall can be transferred into the program or special configurations containing interesting paths can be developed in the program and be rebuilt in reality. The program applies the algorithms developed in this project to the virtual grip configuration, shows all valid paths and can simulate the future climbing process by automatically moving a drawing of the Robo-Shape.

Finally the program can communicate with the robot. If the grip configuration of the real wall is equal to the simulated grip configuration one path can be chosen in the program. After that, the program asks the user to place the robot at the calculated starting position of the path and after a confirmation the program transfers the grip positions of the chosen path to the robot and starts the autonomous climbing process.

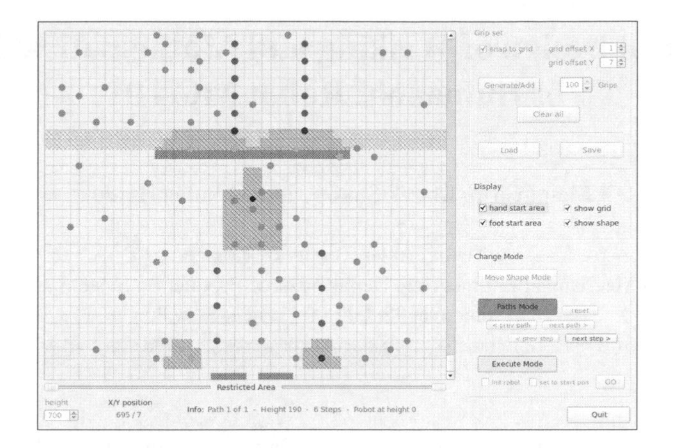

Fig. 14. Program PathSimulator showing simple climbing path

6 Summary

In this paper we have described the prototype of an autonomously operating humanoid robot endowed with the ability to climb up a wall. Choosing an adequate robot kit, changing the hardware and taking its characteristics into account to enable climbing abilities and finally building a climbing wall were the first steps. After that a motion planning algorithm and motion macros using a special inverse kinematics were developed. All these developments were implemented in a program called PathSimulator simulating the real climbing wall and transferring grip positions to the robot. This work adds a new skill to the capabilities of humanoid robots by enabling a combined and controlled usage of arms and legs to achieve a movement into a vertical direction, the climbing. Considering the fact that this goal has not been achieved before the results of this work are bringing humanoid robots one step closer to the capabilities of all human beings and enable them to overcome vertical obstacles and therefore to reach areas humanoid robots could not reach before.

References

1. Nassal, U.M.: Bewegungskoordination und reaktive Steuerung autonomer mobiler Mehrmanipulatorsysteme. VDI Verlag GmbH: Düsseldorf (1996)
2. Angeles, J.: Fundamentals of robotic mechanical systems. Springer, New York (2007)
3. Jazar, R.N.: Theory of applied robotics. Springer, New York (2007)
4. Siciliano, B.[u.a.].: Handbook of robotics. Springer, Berlin (2008)
5. Wickrath, M.: Angewandte komplexe Bewegungsplanung und -steuerung für humanoide Roboter in vertikalen Fortbewegungsszenarien, Diploma Thesis, Dortmund (2009)

Stuctural Design of Walking and Mathematical Model of Humanoid Robot MC - 01

Matej Čirip, Marek Sukop, and Mikuláš Hajduk

Technical University of Kosice
Faculty of Mechanical Engineering, Department of Production System and Robotic,
B. Nemcovej 32, 042 00 Kosice, Slovakia
matej.cirip@tuke.sk, marek.sukop@tuke.sk, mikulas.hajduk@tuke.sk

Abstract. This paper describes basic principles of the kinematic structure design for two-leg walking robots and also offers the model of walk for the concrete construction of the two-leg walking robot. The objective of this project is the design and description of the mathematical model of walk and experimental verification of the statically stabile movement of robot MC-01.

Keywords: Walk, mathematical model, humanoid robot, stability.

Introduction

In recent years the designers have began to focus mainly on the walking systems in design of the service robots. Major advantages of such systems lay in the fact that they are not limited by the height of the wheel diameters while overcoming the obstacles and they possess improved terrain adaptability for the highly demanding conditions. Versatility and terrain adaptability is limited up to certain level due to the complexity of the walking principle and the control of robot movement alone. This paper deals with the concrete solution of the principles problem and design of the mathematical model of walking of the designed two-leg walking robot.

1 Kinematic and Construction Design of Legs of Robot MC - 01

When designing the kinematic structure of legs for bipedal walking robot constructed according to human as model, it is necessary to take into the account the fact up to which level we wish to imitate the human walk and to simplify accordingly the actual kinematics of the biological model. We reduced the proposed kinematical structure of the model of lower extremities of robot shown in Fig.1 into the mechanism with 12 degrees of freedom of movement.

The hip joint possesses 3° of movement freedom in all planes of robot body with the axis of rotation intersecting in one common point. This joint operates as the ball hip (spherical) joint. Knee joint is realised with 1° of movement freedom. The ankle joint has got 2° of movement freedom and may rotate with the foot so in sagital as frontal plane of the system, while axis of rotation intersect in common point.

P. Vadakkepat et al. (Eds.): FIRA 2010, CCIS 103, pp. 90–97, 2010.

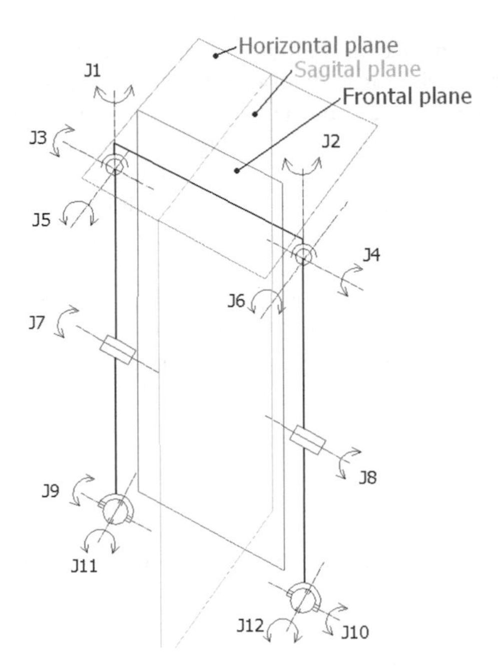

Letters Ji (Joint) designate the individual joints in the kinematic structure, where:

J1, J3, J5 form the spherical hip joint of the right leg

J2, J4, J6 form the spherical hip joint of the left leg

J7 is the knee joint of the right leg

J8 is the knee joint of the left leg

J9, J11 form two-axis joint of the ankle of the right leg

J10, J12 form two-axis joint of the ankle of the left leg

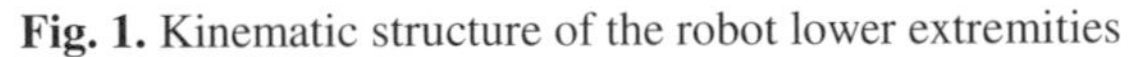

Fig. 1. Kinematic structure of the robot lower extremities

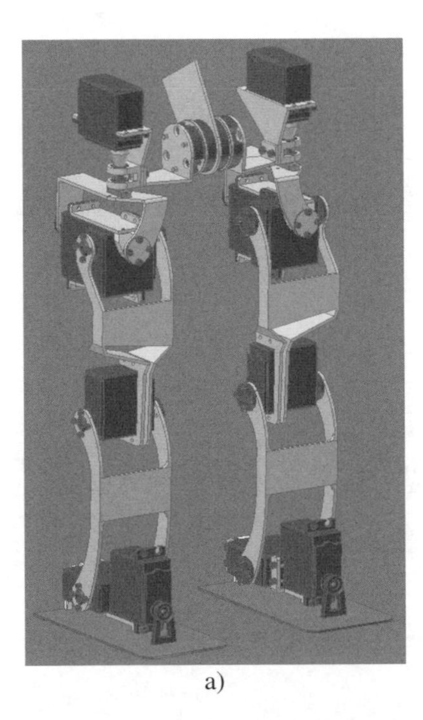

a)

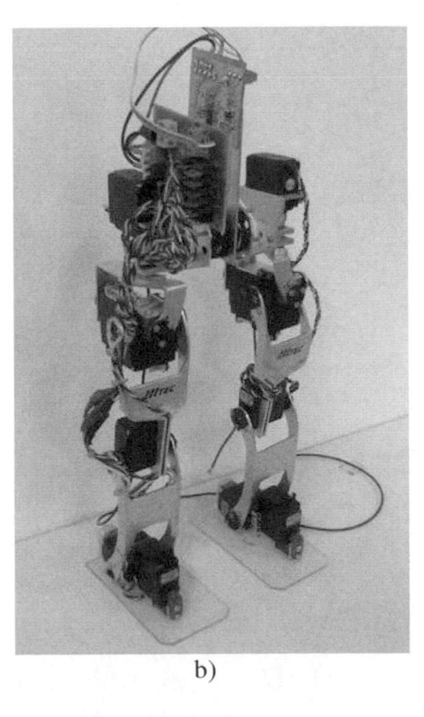

b)

Fig. 2. a) 3D model of robot, b) functional model of robot

The proposed kinematical system is connected in the point of the hipbone by the flexible rubber elements, see Fig.2a. These will partially absorb the inertial energy of the moving mass between robot legs and attenuate the unwanted oscillation transferred from one leg to other during the transferring phases of legs. This attenuating element is made of four rubber rotating bodies, which are pressed among five duraluminium plates. System of the flexible clutch of hip joints permits small attenuating movements only and changes the distance and mutual position of the right and left hip joint negligibly only.

Functional model of the lower extremities of two-leg walking robot is given in Fig.2b. It was constructed at the Department of Production Systems and Robotics TU in Košice. Functional robot is furnished with controlling and powering electronics to control the movement of twelve servo-motors.

2 Design of Structure of Step for MC - 01

The structure of walk of two-leg walking robot depends up to high degree on the complexity of applied kinematic chain of the given mechanism, i.e. mainly on the number of the degrees of freedom of robot movement.

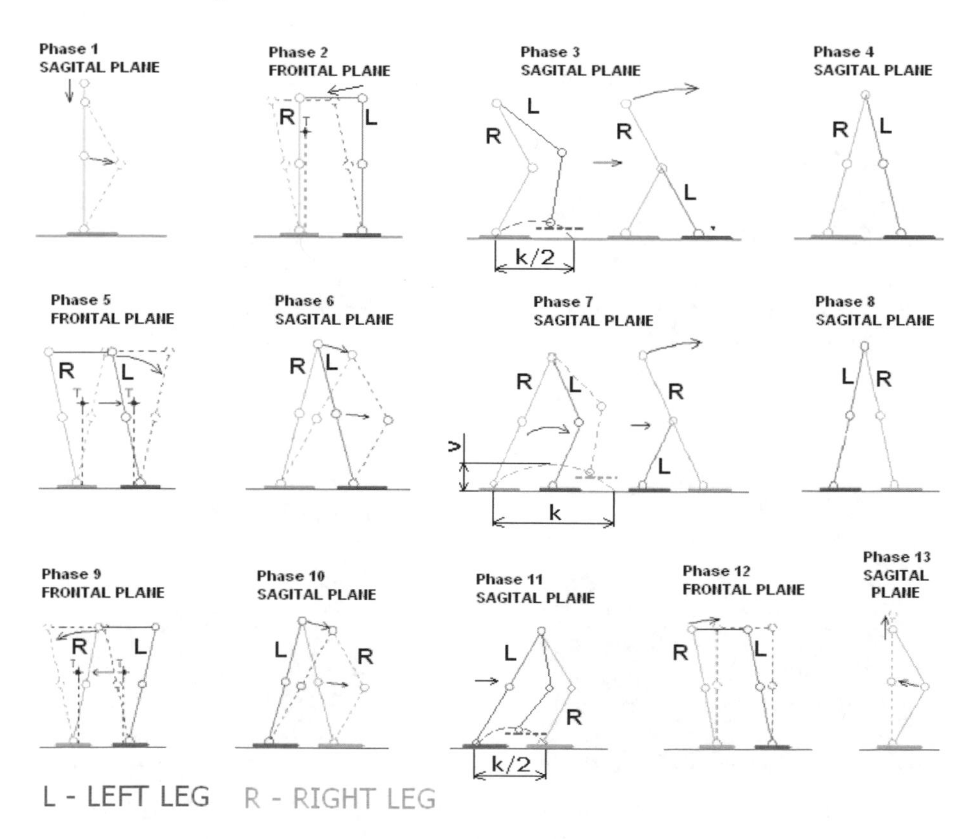

Fig. 3. Phases of step of modified robot walking

In Fig.3 there is shown design of the structure of the walking for robot MC-01 with kinematics of the lower extremities described in Fig.1. This is a static and stabile way of walking, where in each instant of the system movement the centre of gravity of robot is above the supporting area of foot. Such modified walk imposes significantly lower demands upon the control and mathematical description. According to this design the walk has got fixed parameters of step, as robot step length **k** = 240mm and maximum height of step **v** = 30mm. These parameters do not change in course of the walking. The cycle of step consists of 13 phases. In course of the individual phases of robot step the legs are either in supporting or transferring phases and that is when they change their position in space.

According to Fig.3 the robot movement starts in upright position so that both legs jointly go in knee bent in phase No.1, then in phase No.2 the transfer of point of gravity (CoG) over one of feet follows, in this case above right foot, making the movement in the frontal plane. Then the stepping out by left leg in phase No.3 follows, when the foot moves according to curve (ellipse), and then in phase No.4 both legs will stretch. The movement continues by transferring the point of gravity over left foot in phase No.5 and through the movement of hip joint along circle with the simultaneous bending of the left knee in phase No.6. Phase of the step No.7 describes the stepping out by right leg and its transfer along curve (ellipse) in front of the left leg, the follows the stretching out of both legs in phase No.8. In phase No.9 robot transfers the point of gravity over the right foot and in phase No.10 executes the movement of the hips joint and bending the knee of the right leg. Then follows the phase No.11, including the transfer of left leg by the length of half step and shifting it close to the right leg. The robot movement in phase No.12 means the shifting the point of gravity into the centre of the supporting area of both legs and in the last phase No.13 the stretching out of both robot legs into the upright position.

3 Mathematical Calculation of the Robot´s Leg Movement Trajectory

To calculate the individual robot movements within the space applied have been the calculations using the vector method of the inversion kinematics. Known are the parameters of the end element trajectories of the kinematic chain and applying the goniometrical functions and cosine theorem calculated can be the angular displacement of the individual robot joints. Application of the vector method calculation of the angular co-ordinates significantly simplified the overall calculations of the robot movements and the drives control as well.

The step length used in the calculation is **k** = 240 mm. During the transfer phases of legs the ankle with the foot move parallel with the support along ellipse with the shorter axis 30mm long, and therefore the step height is **v** = 30 mm. The length of the thighbone is **a** = 110 mm, length of calf bone is **c** = 110 mm and distance of the hip joint is **p** = 115 mm. The designation of the calculation angles needed for the drives movement and basic dimensions of the kinematic structure are given in Fig.4.

Main objective of the calculations is to determine the angles of the individual joints rotation designated in Fig.4 as φ_i. Relations for the individual angles calculations

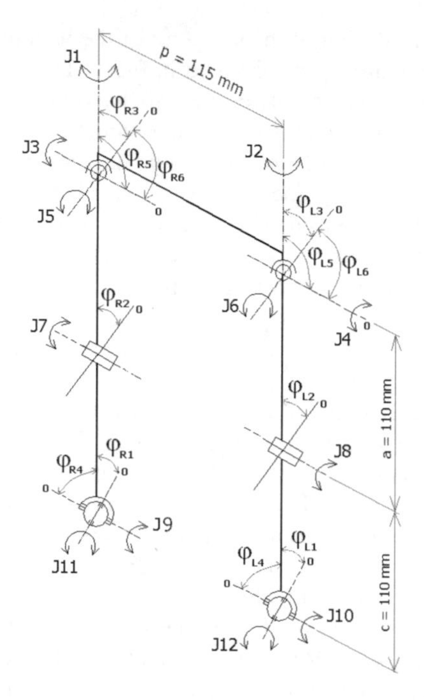

Fig. 4. Designation and orientation of the calculation angles of the individual joints, basic dimensions of robot

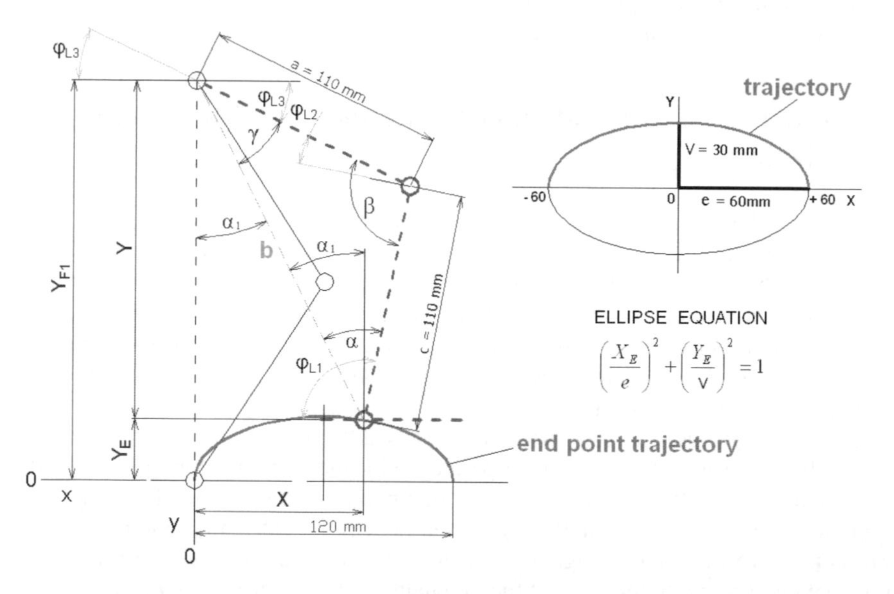

Fig. 5. Graphical demonstration of the movements in phase No.3 and description of the trajectory parameters

are based on the positioning of the local co-ordinate systems in Fig.4. The size and alternation of angles depend on the selected trajectory, along which the individual parts of robot move. These generalised angles are following their value calculation directly implemented into the controlling electronics of robot, which rotates the individual drives into the correct positions accordingly.

Demonstration of the trajectory calculation applying the vector method is represented for the step phase No.3. This is according to Fig.6 the stepping out with robot left leg by the half step length **k/2** = 120 mm. As the trajectory selected was the ellipse with the parameters given in Fig.5. The controlling parameter for the calculation of the angle co-ordinates is the value $X_E = (-60 \div +60)$.

Values **v** and **e** are the constants given by the ellipse dimensions. Value **a** represents the length of the thighbone and value **c** is the length of the shank bone. Value $\mathbf{Y_{F1}}$ is the last value of the co-ordinate Y in phase No.1.

To describe the trajectory applied is the equation of ellipse

$$\left(\frac{X_E}{e}\right)^2 + \left(\frac{Y_E}{v}\right)^2 = 1 \tag{3.1}$$

The relation for the leg lift during movement follows out from equation (3.1)

$$Y_E = v * \sqrt{1 - \left(\frac{X_E}{e}\right)^2} \tag{3.2}$$

while X_E reaches the values $X_E = (-60 \div +60)$

Based on Fig.5 for the co-ordinates of the ankle joint position hold equations

$$Y = Y_{F1} - Y_E \ ; \quad X = 60 + X_E \tag{3.3}$$

For the created triangle with sides **X, Y, b** according to Fig.5 holds the relation

$$b = \sqrt{Y^2 + X^2} \tag{3.4}$$

For the triangle with the sides **a, c, b** holds according to Fig.5 cosine theorem and following out relations for the individual angles α, β, γ

$$\alpha = \arccos\left(\frac{b^2 + c^2 - a^2}{2 * b * c}\right) \beta = \arccos\left(\frac{a^2 + c^2 - b^2}{2 * a * c}\right) \chi = \arccos\left(\frac{a^2 + b^2 - c^2}{2 * a * b}\right) \tag{3.5}$$

For the auxiliary angle α_1 in the triangle with the sides **X, Y, b** holds

$$\alpha_1 = \arccos\frac{Y}{b} \tag{3.6}$$

For the angles of the drives rotation according to Fig.4 and Fig.5 hold the following relations

$$\varphi_{L1} = 90 + \alpha - \alpha_1 \ ; \ \varphi_{L2} = \beta - 90 \ ; \ \varphi_{L3} = 90 - \chi - \alpha_1 \tag{3.7}$$

4 Movement of the Centre of Gravity of Robot for the Designed Step Structure

Model of robot walking designed in Fig.6 maintains in each time instant the static stability, which means that the perpendicular projection of the centre of gravity into the horizontal plane still exists above the supporting area formed by one or two feet as shown in Fig.9.

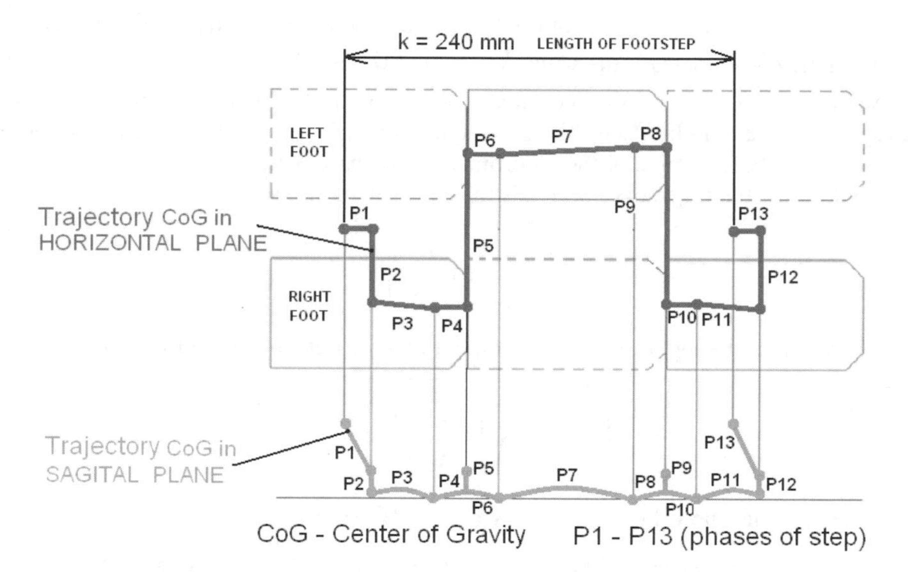

Fig. 6. Trajectories of the centre of gravity of robot movement according to designed step structure

Curves of the centre of gravity movement shown in Fig.6 are the approximate only; however they provide the review about the movement of the robot centre of gravity during the individual sections of step.

The designed model of robot walking is structuralised into 13 phases of step, which provide clearer picture of the partial movements within one step. When studying in details the individual phases of the step in Fig.3 it can be disclosed, that some phases of the step may be combined, i.e. executed simultaneously within the samre time interval. It is possible to combine phases No.1 and No.2, then phases No.4, No.5 and No.6 into further unit, similarly phases No.8, No.9 and No.10, and the last unit may be formed of phases No.12 and No.13. Movement of the centre of gravity of the walking robot in horizontal and sagital plane after the phases combination is shown in Fig.7.

Due to the combination of some step phases, the trajectory of the movement of the robot centre of gravity in sagital plane will partially change. We will deal with the combination of the step phases and movements optimisation following the initial experimental verifications of the designed way of robot walking.

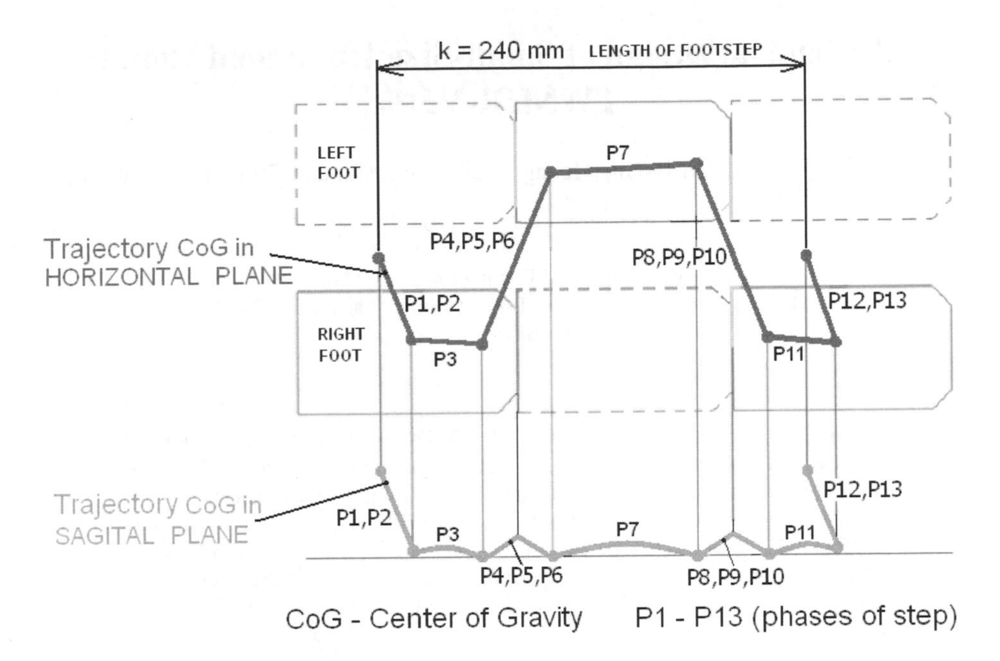

Fig. 7. Trajectories of the movement of robot's centre of gravity after modification including the combination of the step phases

5 Conclusion

The aim of this paper was to describe the principle of the locomotive apparatus based on the human biology, design of the appropriate model of walking, its mathematical description, which will be further exploited in programming of the robot electronics. Walk of two-leg humanoid robot is solved as the statically stabile in each instant of movement. Designed system of robot walling provides the sufficient stability in its movement by the centre of gravity transfer above the individual supporting areas of feet.

References

[1] Nemec, M.: Návrh bipedálneho kráčajúceho podvozku servisného robota – dizertačná práca (2006)
[2] ŽIvčák, J.: A KOLEKTÍV: Základy bioniky a biomechaniky. Grafotlač Prešov (2004)
[3] Smrček, J., Kárnik, L.: Robotika. Servisné roboty. Navrhovanie, konštrukcia, riešenie. Edícia vedeckej a odbornej literatúry SjF TU v Košiciach. Vydavateľstvo Vaško Prešov. Košice (2008)
[4] Skařupa, J., Mostýn, V.: Teorie průmyslových robotú, Vienala Košice (2000)

Mechanical Design of Small-Size Humanoid Robot: TWNHR-VI

Hsiang-Min Chan, Kai-Hsiang Huang, Yueh-Yang Hu, and Ching-Chang Wong

Department of Electrical Engineering, Tamkang University
151, Ying-Chuan Rd. Tamsui, Taipei County, Taiwan 25137, R.O.C.
wong@ee.tku.edu.tw

Abstract. A mechanical structure with 23 DOFs (degrees of freedom) is proposed so that an implemented small-size humanoid robot named TWNHR-VI is able to accomplish some human-like walking motions. The height and weight of the implemented robot is 55 cm and 3.7 kg, respectively. There are 2 DOFs on the head, 1 DOF on the trunk, 4 DOFs on each arm, and 6 DOFs on each leg. Three basic walking experiments of TWNHR-VI are presented to illustrate that the proposed mechanical structure lets the robot walk forward, turn, and slip effectively.

Keywords: Humanoid robot, Mechanical design, Autonomous mobile robot

1 Introduction

For the widely potential use, humanoid robot has been studied for decades by a lot of research groups. Researchers at Waseda University started the humanoid robot research since 1966, and they recently developed a biped humanoid robot WABIAN-2R [1]. The WABIAN-2R is developed to simulate human locomotion. It has led to the realization of emotional walking that is expressed by the parameterization of body motion, walking experiments based on an online pattern generation using obtained visual and auditory information [2-3]. Honda Corporation developed humanoid robots named P2, P3, and Asimo. The newest Asimo has 34 DOFs, 130 cm height, and 54 kg weight. The control method of Asimo is using the center-of-mass and ZMP to plan the pre-recorded joint trajectories and play them back with sensor-based compliant control [4-6]. Sony Corporation also developed several compact size humanoid robots including SDR-3X, SDR-4X, and QRIO. QRIO has 38 DOFs, 58 cm height, and 7 kg weight [7]. The Japanese National Institute of Advanced Industrial Science and Technology and Kawada Industries, Inc., have jointly developed HRP-2 and HRP-2P from 1998. HRP-2 that can walk, lie down, get up, open a door and through a door [8-11]. HRP-2 has 30 DOFs, 154 cm height, and 58 kg weight. Recently, Beijing Institute of Technology has developed a humanoid robot called BHR-03, and this robot is characterized by its light weight construction [12]. BHR-02 has 32 DOFs, 160 cm height, and 63 kg weight. Korea Advanced Institute of Science and Technology developed KHR-3 humanoid (HUBO) [13]. HUBO has 41 DOFs, 125 cm height, and 55 kg weight.

P. Vadakkepat et al. (Eds.): FIRA 2010, CCIS 103, pp. 98–105, 2010.

Although the robot has been investigated for many years, there are still many issues to be studied, especially in the humanoid robot area [14-16]. Hardware and software architectures, walking gait generation, and artificial intelligence are still the main research field of humanoid robot. In this paper, Intelligent Control Lab. of Tamkang University develops a small size humanoid robot called TWNHR. The Intelligent Control Lab. of Tamkang University develops a series humanoid robot for research and development of humanoid robot performing application tasks [17]. The newest humanoid robot called TWNHR-VI. The TWNHR-VI is the six generation humanoid robot of the Intelligent Control Lab. of Tamkang University. The objective of developing TWNHR-VI is to build a platform to investigate the walking gait generation and other artificial intelligence. The work is focusing on the static and dynamic walking on even and uneven ground. In this paper, a mechanical structure for TWNHR-VI with 23 DOFs is described. The rest of this paper is organized as follows: Section 2, a mechanical structure of TWNHR-VI is described. In Section 3, some experiment results are provided. Finally, some conclusions are made in Section 4.

2 Mechanical Design

The primary goal of the humanoid robot mechanical design is development of a robot that can imitate equivalent human motion. The main design concept of TWNHR-VI is light weight and compact size. The actuators of the TWNHR-VI are fabricated from ROBOTIS in order to promote the TWNHR-VI performance. The actuators of ROBOTIS have different output power model. Table 1 presents the parts of the TWNHR-VI actuators and the specifications of each actuator. The head and arms of TWNHR-VI use Ax-12 actuators to reduce the weight of the upper body. The waist, trunk and legs of TWNHR-VI use Rx-28 actuators to promote the output power of TWNHR-VI. The knees of TWNHR-VI use Rx-64 actuators to promote the high output power of TWNHR-VI. The head of TWNHR-VI is design to increase the view of the webcam. It has two degree of freedom to provide increase two dimension of the webcam. The waist and trunk of TWNHR-VI is design for the bend down motion. The arms of the TWNHR-VI are design to manipulation object. For example: to pick up and shoot the ping pang ball as basketball, or hold a stick with compact disks to lift up as weight lifting. The legs are design to imitate the human motion, For example: walking forward and backward, sideways shuffle and kick the ball.

Table 1. The parts of the TWNHR-VI actuators and the specifications of each actuator

Parts of TWNHR-VI	Actuator	Weight	Torque	Speed
Head	Ax-12	55g	16.5 kg·cm	0.196 sec/60°
Waist and trunk	Rx-28	72g	37.7 kg·cm	0.126 sec/60°
Arms	Ax-12	55g	16.5 kg·cm	0.196 sec/60°
Legs	Rx-28	72g	37.7 kg·cm	0.126 sec/60°
Knees	Rx-64	125g	64 kg·cm	0.162 sec/60°

The frameworks of TWNHR-VI are mainly fabricated from aluminum alloy 5052 and alloy 6061. The complex materials will realize the concepts of light weight, high stiffness and wide movable range. Each joint of TWNHR-VI use the ROBOTIS high torque actuator in order to promote the TWNHR-VI performance. In this paper, the developments of the head, arms, waist, trunk and legs are presented. Fig. 1 is the photograph of the TWNHR-VI. It has 23 DOFs and the height of the robot is 55 cm and the weight is 3.7 kg with batteries.

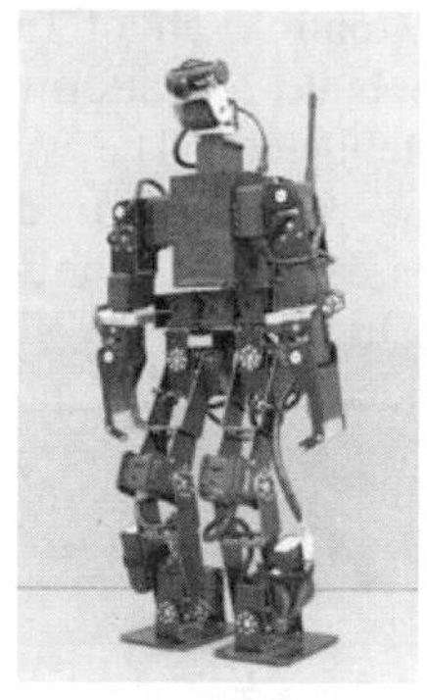

Fig. 1. Photo of TWNHR-VI

Fig. 2 shows the 3D Solidworks design of TWNHR-VI. It is the view from of the front, side, back and the 45degrees in Solidworks. Solidworks is a well-known powerful 3D software of the computer-aided design. It has been used in many fields and has been accept in many industry standards. The Solidworks software can check the collision of mechanism and simulation the design concept of TWNHR-VI. Through the check and the simulation procedure can speed up the design of TWNHR-VI. The solidworks can also check the stress test of the mechanism. Use the stress test to choose the material of the alloy and the design of TWNHR-VI. Finally a human-machine interface is designed to manipulate the actuators. The details of each part are described as follows:

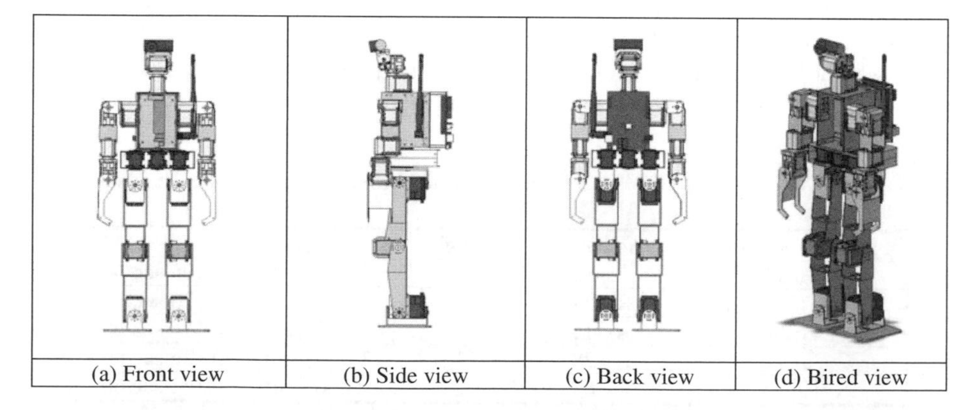

(a) Front view | (b) Side view | (c) Back view | (d) Bired view

Fig. 2. Solidworks design of TWNHR-VI

2.1 Head

The head of TWNHR-VI has 2 DOFs. Fig.3 shows the 3D mechanism design of the head. The head is designed based on the concept that the head of the robot can accomplish the roll and pitch motion. The first actuator is take charge for increase the width of the webcam. The second actuator is take charge for increase the depth of the webcam. These 2 DOFs can provide the head of TWNHR-VI as the head of human that can turn not only left and right but up and down.

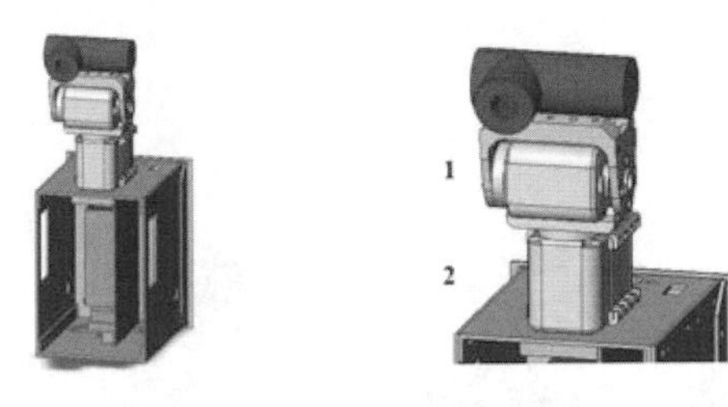

(a) Head 3D design (b) DOFs diagram

Fig. 3. Head mechanism

2.2 Waist and Trunk

The waist and trunk of TWNHR-VI have 1 DOF. Fig. 4 shows the 3D mechanism design of the waist and trunk. The actuator of the waist is in the middle of the three actuators. The trunk is designed based on the concept that robot can turn the body over when the robot is lie down or fall down. These DOF can provide the waist and trunk turn to increase the ambit of the arms and the head. When the robot is falling down from the side of the robot, This DOF can provide the TWNHR-VI turn the body over to the front side or back side to wake up. The TWNHR-VI reduced 1 DOF on waist and trunk. This is because the control circuit board is minimized so that the length of TWNHR-VI can be short. Finally, the DOF of the hip can replace the waist and trunk to do the same motion.

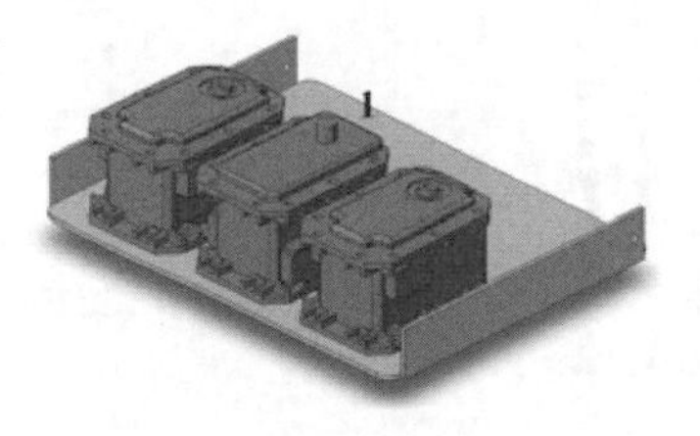

Waist 3D design

Fig. 4. Waist and trunk mechanism

2.3 Arms

Each arm of TWNHR-VI has 4 DOFs. Fig. 5 shows the 3D mechanism design of the arms. The arms are designed based on the concept of size of the general human arms and light weight. The arms of the robot can hold the objects such as a ping pong ball or hold a stick with compact disks. There are 2 DOFs in the shoulder, 1 DOF in the elbow, and 1 DOF in the wrist. These DOFs can provide the TWNHR-VI to hold on objects.

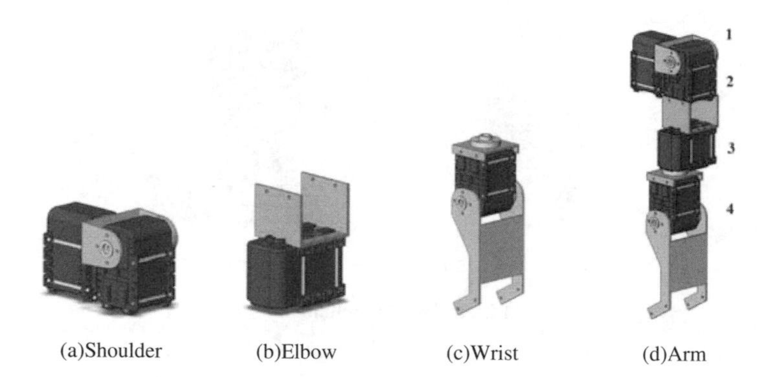

(a)Shoulder (b)Elbow (c)Wrist (d)Arm

Fig. 5. Left arm mechanism

2.4 Legs

Each leg of TWNHR-VI has 6 DOFs. Fig. 6 shows the 3D mechanism design of the legs. The legs are designed based on the concept that robot can accomplish the human walking motion. There are 2 DOFs in the hip, 1 DOF in the knee, and 2 DOFs in the ankle. These DOFs can provide the TWNHR-VI to accomplish the human walking motion. The TWNHR-VI reduced 1 DOF on knee. That is because the actuators technology is improved.

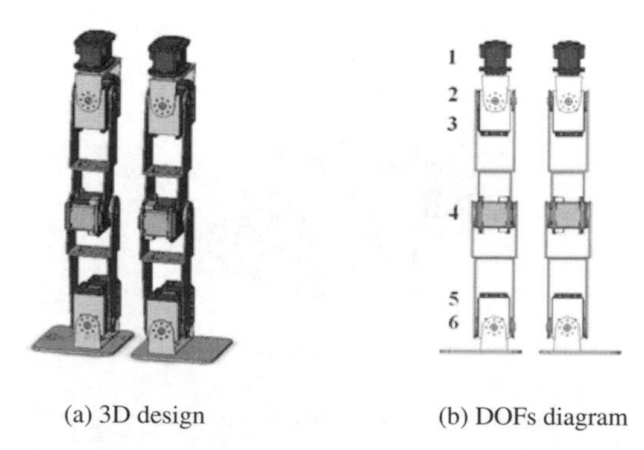

(a) 3D design (b) DOFs diagram

Fig. 6. Legs mechanism

3 Experiments

In order to verify the effectiveness of the implemented humanoid robot, three basic walking skills: straight walking, turning, and slipping are carried out on a horizontal even plane and described as follows:

3.1 Straight Walking

Fig. 7 shows the snapshots of straight walking for TWNHR-VI. The distance between every line in the picture is 2.5cm. Every cycle of the straight walking is able to walk forward 12.5 cm. And the time of each cycle is 1 sec. Compare with the result of TWNHR-IV [17], the straight walking performance of the TWNHR-VI is improved 5cm per second.

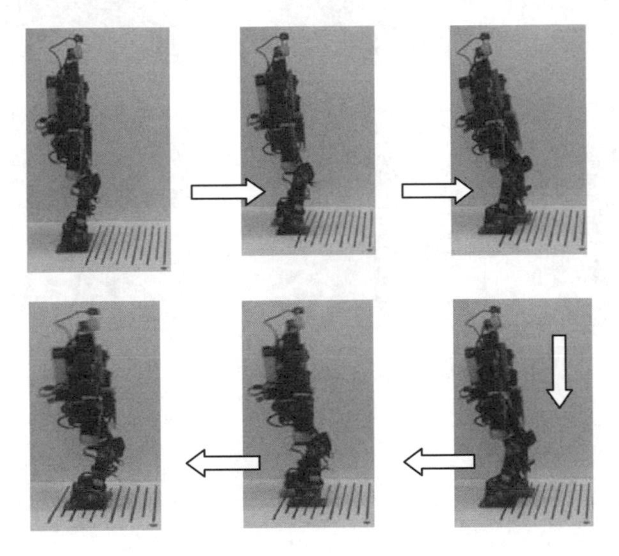

Fig.7. Straight walking

3.2 Turning

Fig. 8 shows the snapshots of right turning for TWNHR-VI. The angle between every line in the picture is 11.25 degrees. Each time of the robot turning is able to turn 22.5 degrees. And the time of each cycle is 1 sec. Compare with the result of TWNHR-IV [17], the turning performance of the TWNHR-VI is accurate than the TWNHR-IV.

Fig. 8. 22.5 degrees right turning

3.3 Sideways shuffle

Fig. 9 shows the snapshots of sideway shuffle to the right for TWNHR-VI. The distance between every line in the picture is 2.5cm. The robot is able to sideway shuffle 2 cm in each step. And the time of each step is less than 1 sec. Compare with the result of TWNHR-IV [17], the turning performance of the TWNHR-VI is accurate than the TWNHR-IV.

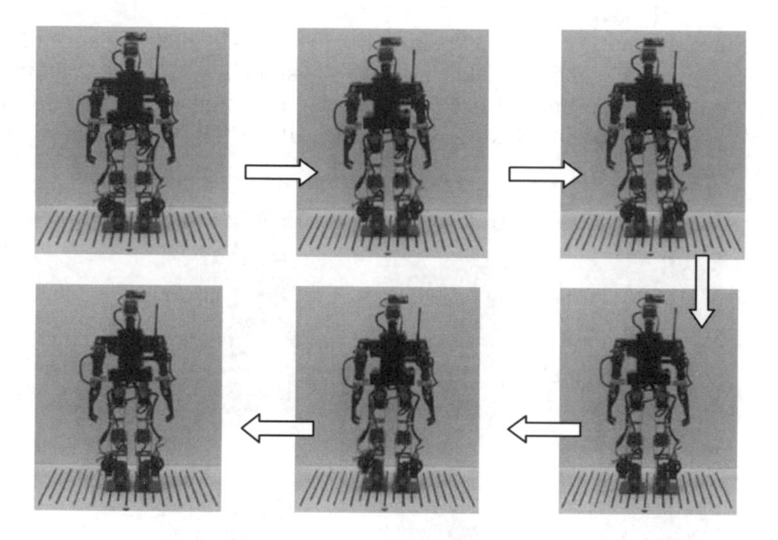

Fig. 9. Sideway shuffle to the right side

4 Conclusions

A mechanical structure with 23 DOFs is proposed to implement a humanoid robot in this paper. This robot has 2 DOFs on the head, 1 DOF on the waist and trunk, 4 DOFs on each arm, and 6 DOFs on each leg. From the experiment results, we can see that the proposed mechanical structure can let the implemented robot walk forward, turn, and slip effectively. One webcam is installed on TWNHR-VI so that it can be a vision-based soccer robot to find a ball and kick a ball autonomously. In the future, some artificial intelligence will be carried on TWNHR-VI to make it to be an intelligent robot.

Acknowledgement

This research was supported in part by the National Science Council (NSC) of the Republic of China under contract NSC 98-2218-E-032-005.

References

1. Omer, A.M.M., Ghorbani, R., Lim, H., Takanishi, A.: Semi-passive dynamic walking for biped walking robot using controllable joint stiffness based on dynamic simulation. In: IEEE/ASME Int. Conf. on Advanced Intelligent Mechatronics, July 2009, pp. 1600–1605 (2009)
2. Ogura, Y., Kataoka, T., Aikawa, H., Shimomura, K., Lim, H., Takanishi, A.: Evaluation of various walking patterns of biped humanoid robot. In: IEEE Int. Conf. on Robotics and Automation, April 2005, pp. 605–610 (2005)
3. Ogura, Y., Aikawa, H., Shimomura, K., Kondo, H., Morishima, A., Lim, H.O., Takanishi, A.: Development of a new humanoid robot WABIAN-2. In: IEEE Int. Conf. on Robotics and Automation, February 2006, pp. 835–840 (2006)
4. Hirai, K., Hirose, M., Haikawa, Y., Takenaka, T.: The development of Honda humanoid robot. In: IEEE Int. Conf. on Robotics and Automation, May 1998, vol. 2, pp. 1321–1326 (1998)
5. Sakagami, Y., Watanabe, R., Aoyama, C., Matsunaga, S., Higaki, N., Fujimura, K.: The intelligent ASIMO: system overview and integration. In: IEEE/RSJ Int. Conf. on Intelligent Robots and Systems, September 2002, vol. 3, pp. 2478–2483 (2002)
6. Sentis, L., Park, J., Khatib, O.: Compliant Control of Multicontact and Center-of-Mass Behaviors in Humanoid Robots. In: IEEE Int. Conf. on Robotics, vol. 26, pp. 1–19 (2010)
7. Kuroki, Y., Fujita, M., Ishida, T., Nagasaka, K., Yamaguchi, J.: A small biped entertainment robot exploring attractive applications. In: IEEE Int. Conf. on Robotics and Automation, September 2003, vol. 1, pp. 471–476 (2003)
8. Kaneko, K., Kanehiro, F., Kajita, S., Yokoyama, K., Akachi, K., Kawasaki, T., Ota, S., Isozumi, T.: Design of prototype humanoid robotics platform for HRP. In: IEEE/RSJ Int. Conf. on Intelligent Robots and Systems, September 2002, vol. 3, pp. 2431–2436 (2002)
9. Fujiwara, K., Kanehiro, F., Kajita, S., Yokoi, K., Saito, H., Kaneko, K., Harada, K., Hirukawa, H.: The first human-size humanoid that can fall over safely and stand-up again. In: IEEE/RSJ Int. Conf. on Intelligent Robots and Systems, vol. 2, pp. 1920–1926 (2003)
10. Kaneko, K., Kanehiro, F., Kajita, S., Hirukawa, H., Kawasaki, T., Hirata, M., Akachi, K., Isozumi, T.: Humanoid robot HRP-2. In: IEEE Int. Conf. on Robotics and Automation, April 2004, vol. 2, pp. 1083–1090 (2004)
11. Arisumi, H., Chardonnet, J.R., Yokoi, K.: Whole-body motion of a humanoid robot for passing through a door opening a door by impulsive force. In: IEEE/RSJ Int. Conf. on Intelligent Robots and Systems, October 2009, pp. 428–434 (2009)
12. Zhang, W., Huang, Q., Jia, D., Xin, H., Li, M., Li, K.: Mechanical design of a light weight and high stiffness humanoid arm of BHR-03. In: IEEE Int. Conf. on Robotics and Biomimetics (ROBIO), December 2009, pp. 1681–1686 (2009)
13. Park, I.W., Kim, J.Y., Oh, J.H.: Online biped walking pattern generation for humanoid robot KHR-3(KAIST Humanoid Robot - 3: HUBO). In: IEEE-RAS Int. Conf. on Humanoid Robots, December 2006, pp. 398–403 (2006)
14. Kitano, H., Asada, M.: RoboCup humanoid challenge: that's one small step for a robot, one giant leap for mankind. In: IEEE/RSJ Int. Conf. on Intelligent Robots and Systems, vol. 1, pp. 419–424 (October 1998)
15. Burkhard, H.D., Duhaut, D., Fujita, M., Lima, P., Murphy, R., Rojas, R.: The road to RoboCup 2050. IEEE Robotics & Automation Magazine 9, 31–38 (2002)
16. Behnke, S., Schreiber, M., Stuckler, J., Renner, R., Strasdat, H.: See, walk, and kick: humanoid robots start to play soccer. In: IEEE-RAS Int. Conf. on Humanoid Robots, December 2006, pp. 497–503 (2006)
17. Wong, C.C., Cheng, C.T., Huang, K.H., Yang, Y.T., Chan, H.M., Hu, Y.Y., Chen, H.C.: Mechanical design of small-size humanoid robot: TWNHR-IV. Journal of Harbin Institute of Technology, vol 15(sup. 2), 21–34 (2008)

Imitation Learning from Humanoids in a Heterogeneous Setting

Jeff Allen, John Anderson, and Jacky Baltes

Autonomous Agents Laboratory
Dept. of Computer Science, University of Manitoba
Winnipeg, Manitoba, Canada R3T2N2
{jallen,andersj,jacky}@cs.umanitoba.ca
http://aalab.cs.umanitoba.ca

Abstract. Humanoid robots are increasingly popular due to their flexible nature and how easy it is for humans to relate to them. Imitation learning has been used to transfer skills and experience between robots, however, transferring skills from a humanoid demonstrator to a non-humanoid such as a wheeled robot is no simple task. The imitator must be able to abstract the behaviour it observes and in addition translate this behaviour to actions that it is able to perform. Since humanoid robots' have much larger ranges of motion than most robots, this is can be quite difficult. This work describes an approach for a non-humanoid robot to learn a task by observing demonstrations of a humanoid robot using global vision. We use a combination of tracking sequences of primitives and predicting future primitive actions from existing combinations using forward models to construct more abstract behaviours that bridge the differences between the different robot types. To evaluate the success of learning from a humanoid demonstrator, we also evaluate how well a wheeled robot can learn from a physically identical wheeled robot, as well as a much smaller wheeled robot.

1 Introduction

Learning how to perform activities by observing and imitating others is a powerful mechanism seen in humans, primates, and other creatures [1,2,3]. There are many promising reasons to apply such techniques to learning in robots. Artificial intelligence has previously developed many successful approaches through biological inspiration, and work on mirror neurons, which map observed actions to the structures of the brain that produce those same actions, shows similar promise here [4,1,2].

An imitating robot must be able to understand its own actions in terms of their visual outcomes, as well as understanding the actions of others in terms of the imitator's own available actions. In the case of a humanoid robot learning by watching a humanoid demonstrator, there is a reasonably close mapping between the actions of learner and demonstrator because of the common physical structure. When a wheeled robot watches a humanoid demonstrator, on the other hand, there are actions that simply cannot be duplicated precisely: a humanoid might step over an obstacle, for example, while a wheeled robot would have no alternative but to drive around it. Similar differences in capability can be seen with other forms of heterogeneity (e.g. size differences) as well.

P. Vadakkepat et al. (Eds.): FIRA 2010, CCIS 103, pp. 106–113, 2010.

In this paper, we present recent work for non-humanoid robots to learn from humanoid robots using global vision. We use a combination of tracking sequences of primitives and predicting future actions from existing combinations using forward models to construct more abstract behaviours that bridge the differences between our wheeled imitator robot and our humanoid demonstrator.

The particular problem we use to evaluate our approach is a soccer free-kick: the imitation robot must learn from a series of demonstrators to position itself properly behind the ball and kick it into an empty goal. We demonstrate and evaluate our approach using a selection of robots that have previously been used in a number of different RoboCup leagues. Our robot imitator is a wheeled robot built from a Lego MindStorms kit. To determine if our system will allow a humanoid robot to demonstrate adequately for a wheeled robot, we use a Bioloid humanoid robot which uses a cellular phone for vision and processing. The non-humanoid demonstrators we use for comparison are a robot that is physically identical to our imitator, and a Citizen Eco-Be robot which is about $1/10$ the size of the imitator. This smaller size will provide a comparison between a demonstrator that is physically different in size, but similar in its primitive motions.

The remainder of this paper describes our imitation learning approach in detail, followed by details of its implementation on the robots we employ to validate this approach, and an evaluation as described above. Before this, we provide a brief overview of related work in imitation learning in robots and multi-robot systems.

2 Related Work

Prior work in imitation learning has often used a series of demonstrations from demonstrators that are similar in skill level and physiologies. The approach presented in this paper is designed from the bottom up to learn from multiple demonstrators that vary physically, in underlying control programs and skill levels. Our approach also differs from most prior work in that rather than defining primitives abstractly (e.g. avoidance, following, homing, aggregation, dispersion, navigation, foraging, and flocking [5,2]), we operate with the lowest-level primitives possible given the design of the robot.

The work of [6] is previous research involving imitation using multiple demonstrators of varying skill levels, falling back on reinforcement learning when no demonstrators were obviously better than others. Our work differs from theirs in how the demonstrators are compared. In our work the basis for this comparison is the actual behaviours learned from each demonstrator, and also unlike their work, our imitators can learn elements of good behaviours even from demonstrators that are not well-skilled overall. [7] also used imitation learning with multiple demonstrators and reinforcement learning, though their imitator extracts information about the demonstrators simultaneously and combines it without comparing their abilities. While this is a faster method of learning from multiple sources, it allows poor demonstrators, or those that cannot be imitated directly due to physiological differences, to have too strong an influence. Our work on the other hand, takes the varying skill levels of demonstrators into account.

Some recent work in humanoid robots imitating humans has used many demonstrations, but not necessarily different demonstrators, and very few have modelled each demonstrator separately. Those that do have different demonstrators such as [8] often have demonstrators of similar skills and physiologies (in this work all humans performing simple drawing tasks) that also share the same end-effector physiology as the imitator (in this case the imitator was a humanoid robot learning how to draw letters). In general, few prior approaches to imitation learning with heterogeneous demonstrators have focused on non-humanoids learning from humanoid robots.

3 An Approach to Imitation Learning from Physically Different Demonstrators

Unlike other imitation learning work, which use higher level behavioural primitives such as following, homing, or dispersion (e.g. [5]), our approach assumes only the lowest level fixed duration motor actions as primitives: 'forward', 'backward', 'turn left', and 'turn right'. The imitator must be able to convert the visual effects of demonstrators into terms the imitator can understand. To do this we employ discrete Hidden Markov Models (HMMs) to recognize the imitator's own primitives. The same HMMs are used to convert the visual representations of the demonstrations into sequences of the imitator's primitives. HMMs are very useful for recognizing sequences of events that have underlying stochastic elements [9]. We have compared a number of different variations on HMMs in previous work [10] to deal with recognizing and classify primitive actions from visual data. The imitator has one HMM for each primitive, trained by the robot processing data from our global vision server (Ergo [11]), which provides the locations, identities, and motions of objects on the field over time. The imitator can use these HMMs to convert the vision data into sequences of symbols, where each symbol represents one of the imitator's primitive motor commands.

An imitator must be able to approximate the intent of a sequence of actions, which requires the same facilities for abstraction and generalization necessary to deal with differences between demonstrator skill levels and demonstrator physiologies: the ability to recognize behaviours that account for multiple primitives. We use forward models to learn demonstrator behaviours, which have been used previously in imitation [1,12]. Our work differs because we use our forward models to model and predict entire behaviour repertoires of demonstrators, whereas they used a forward model for each individual behaviour. We maintain a forward model for each individual demonstrator and a single model for the overall behaviour of the imitator. The training and use of these forward models is detailed in section 3.1. The actual robots used in our implementation and evaluation are depicted in Fig. 1.

3.1 Prediction and Creation of Behaviours

Behaviours are composed of multiple primitives, and since the number of these combinations is essentially infinite, it is necessary to keep the number of existing behaviours reasonable. We define a *permanency* attribute for all behaviours. A behaviour with a permanency attribute of zero is removed, while exceeding an upper threshold allows the behaviour to be considered permanent (i.e. it is no longer evaluated for permanency).

The imitator maintains one main forward model, which contains behaviours it has learned from the demonstrations (as well as the primitives it started with). The imitator

Fig. 1. Two views of the heterogeneous robots used in this work (Right, with visual tracking markers for the global vision system). The Bioloid humanoid robot, a Lego MindStorms wheeled robot, and (at the humanoid's feet) a Citizen Eco-Be robot.

also has an additional separate forward model for each demonstrator it observes. While imitating a demonstrator, the imitator updates the forward model specific to that particular demonstrator. When the model makes frequent, accurate predictions about two primitives occurring in sequence (or a primitive and a behaviour, or two behaviours), it triggers the creation of a new behaviour to represent this sequence. We keep separate forward models for each demonstrator so the imitator can compare the overall quality of the demonstrators when deciding which demonstrator behaviours to move into the main imitator forward model.

The imitator's forward model uses a matrix to maintain a link from each behaviour/primitive to each other one. Each element in the matrix corresponds to the *confidence* that the forward model has that a behaviour (or primitive) will follow another. To make a prediction of the demonstrator's next action, the forward model uses the last recognized behaviour/primitive and looks up its row in the matrix. The behaviour/primitive with the highest confidence in that row is then predicted to occur next. If the prediction is correct, the confidence value linking the two behaviours/primitives is increased, otherwise it is decreased. When a behaviour is predicted successfully, its permanency is also increased by a small amount. The next section details the process of evaluating the quality of demonstrators and the process of removing behaviours.

3.2 Humanoid and Other Demonstrator Types

The quality of a demonstrator is represented by its *learning preference*, a value between 0 and 1. The LP is a learning rate used to increase the confidence values in its matrix faster for good demonstrators (higher LP) and slower for poor demonstrators. The imitator updates the LP of each demonstrator based on the accuracy of their forward model predictions (increase for correct, decrease otherwise). The imitator also updates the LP based on the outcome of a demonstrator's predicted behaviours. Scoring a goal, moving the ball closer to the goal, or moving closer to the ball increases the

LP. The opposite of these usefulness criteria will decrease the LP. Each demonstrator also has a *decay rate* which is simply $1 - LP$, so a good demonstrator has a low decay rate, and a bad demonstrator has a high decay rate. The decay rate is applied to the permanency attribute of all behaviours in the demonstrator's forward model at every prediction step. A good demonstrator's behaviours are correctly predicted (and their permanency increased) frequently enough that their increases in permanency overcomes their decreases from decay.

When a demonstrator's behaviour becomes permanent (permanency past an upper threshold), it is permanent *to the forward model for that demonstrator*, not permanent to the imitator. Behaviours that achieve permanency within their demonstrator specific forward models are copied into the imitator's main forward model. The behaviour's permanency attribute is then reset to the default value (the midpoint between min and max values for permanency). This equalizes the chance behaviours have of becoming permanent once they reach the imitator's main forward model. If a behaviour that one of the demonstrator's originally added to the imitator's model becomes permanent there, that demonstrator has its LP increased. If a behaviour is removed from the imitator's main forward model its demonstrator has its LP decreased.

3.3 Evaluation

To evaluate whether a humanoid robot can appropriately demonstrate for a wheeled robot imitator, we employed the robots previously shown in Fig. 1. This allows us to compare the viability of a humanoid demonstrator to demonstrators physically similar to the imitator of varying sizes (one the same as the imitator, one much smaller). The field size was adjusted for the Citizen robot because it could not complete demonstrations on a large-size field given its low available battery charge. The Humanoid and Lego MindStorms robot were demonstrated on a 102 x 81 cm field, while the Citizen was demonstrated on a 56 x 34.5 cm field.

In terms of the placement of the robot and ball on the demonstration fields, some consistency was required for proper comparison of results. Rather than placing the ball and robot anywhere, we limited ourselves to two field configurations shown in Fig. 2. In the first (a), the demonstrator is positioned for a direct approach to the ball. As a more challenging scenario, we also used a more degenerate configuration (b), where the demonstrator is positioned for a direct approach to the ball, but the ball is lined up to its own goal – risking putting the ball in one's own net while manoeuvering.

The individual demonstrators were recorded by a global vision system while they performed 25 goal kicks for each field configuration. The global vision system continually captures the x and y motion and orientation of the demonstrating robot and the ball. The demonstrations were filtered manually for simple vision problems such as when the vision server was unable to track the robot, or when the robot broke down (falls/loses power). The individual demonstrations were considered complete when the ball or robot left the field. Since all demonstrator types used IR control for hardware, it was also possible to record the command streams for reference, to be able to examine places where the demonstrator's intent differed from the imitator's interpretation.

One learning trial consists of each demonstrator forward model training on the full set of kick demonstrations for that particular demonstrator, presented in random order.

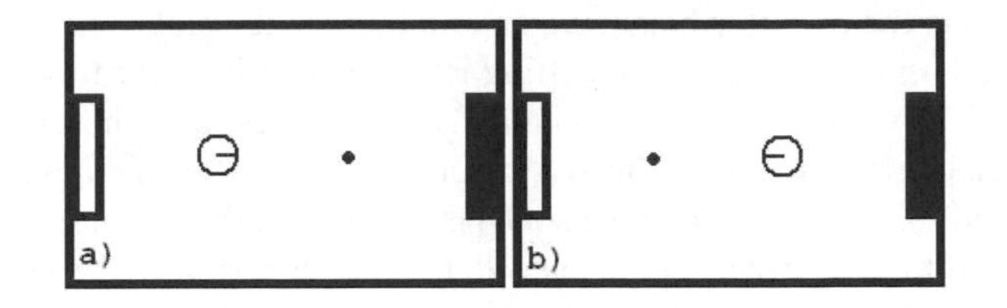

Fig. 2. Field configurations. The demonstrator is represented by a circle with a line that indicates the robot's orientation. The target goal is indicated by a black rectangle, the demonstrator's own goal is white.

Once the three demonstrator models are trained, the imitator's main forward model begins training. At this point all the demonstrators' models have been trained for their own data, and have provided the imitator with behaviours. The imitator now takes all the demonstrations for each of the two field configurations (a total of 150 attempted goal kicks) in random order. All of the demonstrator forward models predict and update their models at this time, one step ahead of the imitator. This essentially allows each forward model to offer advice to the imitator. As the imitator updates its forward models from each demonstration, we record the number of behaviours created, deleted, and made permanent (moved to the imitator's model) for each individual demonstrator. Since the order in which demonstrations are provided may influence the learning process, we evaluated the behaviour learning abilities on all permutations of demonstrator ordering (6 in total) and then averaged the results.

To evaluate the segmentation and classification of the demonstration sequences, we compared each segment's classification to the actual underlying commands that were sent by the demonstrator during that time. Each segment contains multiple vision frames, and each frame records the command that a demonstrator was sending at the time. Since the control programs are vastly different among the demonstrators, there is rarely a direct comparison between the commands recorded and the primitive symbols used to classify each segment. We hand-compiled a reference list which classified each command type by observing the visual outcome and choosing the imitator primitive that visually matched it the best. We then used this reference list to automatically evaluate the accuracy of the segments' classifications (imitator primitives) matching the underlying demonstrator commands. In this work, a segment that matches at least 75% of its frames' commands to its primitive symbol is considered an accurate match. We also recorded the number of segments that did not even receive a classification. These results are shown in Table 1.

In absolute terms, these matching percentages appear to be extremely low, especially for the humanoid, but it must be recalled that the command range and types for a humanoid are far different than a simple wheeled robot. The control program often uses command types relating to the movement of its specific body parts. When the Bioloid is scanning for the ball, this behaviour appears as a *stop* primitive in the visual stream (a stop primitive merely sends no commands, i.e. the robot is stationary). When the

Bioloid is kicking the ball, it is balancing on one foot, so the *kick* command appears as a backward motion due to the robot shifting its weight to its back foot. These results simply indicate how well the actual commands used match with the visual classifications. The actual visual output may match the primitive that a segment was classified as (e.g. the Bioloid's *kick* command may match perfectly with the imitator's *backward* behaviour). Even if the underlying commands do not match perfectly, the visual outcomes can still be learned by a physically different robot, and the humanoid's skills can still be learned by a non-humanoid imitator.

Table 1. Command Match Percentages

Demonstrator Type	Commands Matched
Citizen	27.97%
RC2004	33.15%
Bioloid	10.35%

To evaluate the ability of our wheeled imitator to learn from a humanoid demonstrator, we compare the number of permanent behaviours each forward model retains, as well as the final learning preference for each forward model. The results are shown in Table 2. The values are the averages across the 6 demonstrator ordering permutations. The values in each row show the average number of permanent behaviours created within each specific demonstrator forward model, and their average learning preference (LP).

Table 2. Number of permanent behaviours created and/or retained, and final learning preference (LP), by demonstrator, averaged over 6 trials of varying demonstrator ordering

Citizen		RC2004		Bioloid	
Permanent	LP	Permanent	LP	Permanent	LP
59.8	0.71	55.3	0.88	60.8	0.65

4 Conclusion

Our results show that humanoid demonstrators can be used to train physically different robots, in this case a wheeled robot. The wheeled imitator robot was able to learn from a physically similar wheeled robot, a much smaller wheeled robot, and the humanoid robot at a roughly equal level. This work shows that humanoid robots can pass on their skills through imitation learning not only to other humanoids, but also to other physical robot types as well. Since humanoid robots are much more flexible than most robot architectures it can be difficult, but very valuable for other robots to be able to learn from them.

References

1. Demiris, J., Hayes, G.: Imitation as a dual-route process featuring predictive and learning components: A biologically plausible computational model. In: Dautenhahn, K., Nehaniv, C. (eds.) Imitation in Animals and Artifacts, pp. 327–361. MIT Press, Cambridge (2002)
2. Matarić, M.J.: Sensory-motor primitives as a basis for imitation: linking perception to action and biology to robotics. In: Dautenhahn, K., Nehaniv, C. (eds.) Imitation in Animals and Artifacts, pp. 391–422. MIT Press, Cambridge (2002)
3. Billard, A., Matarić, M.J.: A biologically inspired robotic model for learning by imitation. In: Proceedings, Autonomous Agents 2000, Barcelona, Spain, pp. 373–380 (2000)
4. Weber, S., Matarić, M.J., Jenkins, O.: Experiments in imitation using perceptuo-motor primitives. In: Sierra, C., Gini, M., Rosenschein, J.S. (eds.) Proceedings of the fourth international conference on Autonomous agents, Barcelona, Catalonia, Spain, June 2000, pp. 136–137. ACM Press, New York (2000)
5. Matarić, M.J.: Getting humanoids to move and imitate. IEEE Intelligent Systems, pp. 18–24 (July 2000)
6. Yamaguchi, T., Tanaka, Y., Yachida, M.: Speed up reinforcement learning between two agents with adaptive mimetism. In: Proceedings of the 1997 IEEE/RSJ International Conference on Intelligent Robots and Systems, 1997. IROS 1997., vol. 2, pp. 594–600 (1997)
7. Price, B., Boutilier, C.: Accelerating reinforcement learning through implicit imitation. Journal of Artificial Intelligence Research 19, 569–629 (2003)
8. Calinon, S., Billard, A.: Learning of Gestures by Imitation in a Humanoid Robot. In: Dautenhahn, K., Nehaniv, C.L. (eds.) Imitation and Social Learning in Robots, Humans and Animals: Behavioural, Social and Communicative Dimensions, pp. 153–177. Cambridge University Press, Cambridge (2007)
9. Rabiner, L., Juang, B.: An introduction to hidden markov models. IEEE ASSP Magazine, 4–16 (1986)
10. Allen, J., Anderson, J.: A vision-based approach to imitation using heterogeneous demonstrators. In: Geib, C., Pynadath, D. (eds.) Proceedings of the AAAI Workshop on Plan and Intent Recognition, Vancouver, Canada, July 2007, pp. 9–16. AAAI Press, Menlo Park (2007)
11. Furgale, P., Anderson, J., Baltes, J.: Real-time vision-based pattern tracking without predefined colors. In: Proceedings of the Third International Conference on Computational Intelligence, Robotics, and Autonomous Systems (CIRAS), Singapore (December 2005)
12. Dearden, A., Demiris, Y.: Learning forward models for robots. In: Proceedings of IJCAI 2005, Edinburgh Scotland, pp. 1440–1445 (August 2005)

Here Comes the Robotic Brain !

Adalberto Llarena

Bio-Robotics Laboratory, Department of Electrical Engineering
Universidad Nacional Autónoma de México, UNAM
adallarena@aol.com

Abstract. Year after year computers increase their processing power. This has been an advantage for the implementation of new and powerful robotic algorithms. While every day more complex algorithms appear, most of the increased processing capabilities of the newest microprocessors get easily exhausted. By analyzing the Artificial Neuron Model, this paper predicts that with the current tendencies in the increase of the microprocessor's power in 2050 the processing power of computers could reach the processing power of the human brain. Besides that, while the robotic systems integration becomes harder and no standards exist in this matter, the use of Artificial Neural Networks becomes more popular. This work studies the possibility of having a Neural Processing Architecture in the future, i.e. a Robotic Brain, under the perspective of pure neural processing power, as a feasible alternative in the development and integration of more complex architectures and algorithms in robotics.

Keywords: Artificial Neural Networks, Artificial Intelligence, Robotic Brain.

1 Introduction

Day after day computers are increasing their processing capacity. Moravec [1] has done a prediction based on the Moore's law [2] extrapolation. As a result, by 2040 the processing speed of the computers (and autonomous robots by consequence) could reach the surprising speed of 10^{14} instructions per second (100 millions of MIPS), similar to the computational power of the human brain[1,2].

In recent years, with the integration of multiple cores in the computer's main boards, extensive work has been done in multiprocessing. Despite this, the single-processor architecture (or at most with a couple of processors) has been the most used. Based on this premise, it is highly possible that by the time the processing power of microprocessors reach the human brain, computers still use a single main processor. By that time, such microprocessors must be capable of emulating the simultaneous multiprocessing tasks the animal brain is capable of performing.

While different approaches [3,4] present architectures based on a modular system integration, the progressive addition of new modules, behaviours and knowledge is

[1] In this study, Moravec does not analyze an *Artificial Neuron Model*.

[2] The human brain computational power was estimated qualitatively based on supposing the existence of very efficient programs that could replace thousands of neuron connections.

P. Vadakkepat et al. (Eds.): FIRA 2010, CCIS 103, pp. 114–121, 2010.

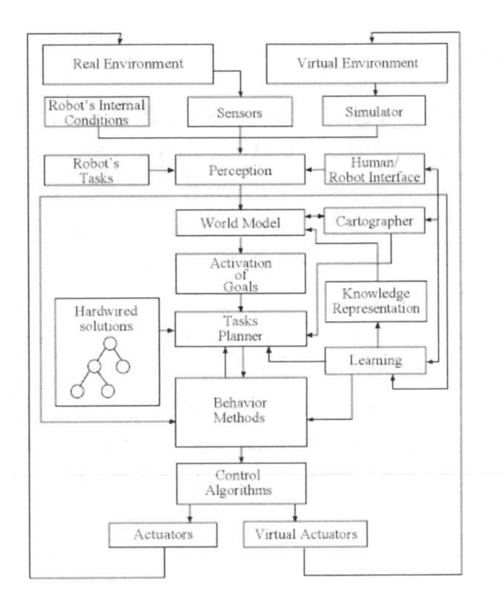

Fig. 1. ViRbot Architecture

not straightforward and the final system complexity, programming and testing times increase dramatically with the addition of new modules to existing architectures.

By far, the most important problem with such modular systems is the increase of time dedicated to transmission and control signals with the addition of new modules. A lot of processing and transmission time become necessary to get the robot's sensor information, transmit it to each specific module and also to realize the communication among them. This is the reason why it is difficult to have multiple microprocessors on a single-board computer [5].

For this reason, *Modular Programming* for practical real-time applications results too complex for the programmer and too "heavy" for running all the subsystems on a single processor, especially if a lot of sensors are used (such as a laser range finder or vision system giving many readings per second while moving in unconstrained environments). This leads to the impossibility of making an accurate model of the static environment, planning a well-defined trajectory, having *fixed-priority* based decision selection or performing just a simple movement in a straight trajectory between two points because the system awareness must be total at all times, as living beings are.

2 The Neural Approach

There is a natural fear of the *unknown*; this could be the most basic fear every human being has. Men feel comfortable with proven algorithms where all its states are perfectly "known". Basically, *Artificial Neural Networks* are the *Dark Side* of robotics because we cannot *exactly* interpret the stored knowledge inside a huge set of numbers. An output is only known until an input is given. We will never know the output for a specific input until feed it to the network. Moreover, ANNs are capable of giving responses to absent inputs during training, so they are capable of interpolating, extrapolating and generalizing. Maybe this could be the reason why Nature has chosen this kind of processing; it *always* gives a response.

2.1 Natural Approach

Nature has solved the motion, feeding, understanding, learning, and many other problems thousands of years ago. While scientific efforts are mainly focused on *methods*, we have never considered the *architecture* itself as a problem to solve (something that nature has solved, generation by generation, through Natural Selection) and DNA has the important role of carrying this architecture codification.

If we want robots to have similar behaviors and capabilities as humans, we could spend a lot of time and effort searching and testing for the best methods and algorithms to do so. Science has more than 20 years in this search and it is highly possible that the processor speed reach the human brain capacity before we find the correct method for implementing all those capabilities. Certainly, although a lot of work has been done, we are at the very beginning of the robotics age.

2.2 Emulating Living Beings

Perhaps the fastest way of achieving such talents is emulating some of the processes and systems that nature has, in particular, the architecture and the way to perfect it. Maybe the best way of building robots could be letting them to evolve in one way or another.

2.3 Brain's Processing Power Estimation

The brain is the main processing unit in higher living beings. Its operation is based in the overall performance of billions of individual cells called *neurons*. Each neuron generates a binary output (called *activation*) that is fed to other neurons through connections called *axons* (for the neuron's output) and *dendrites* (for the neuron inputs). See figure 2.

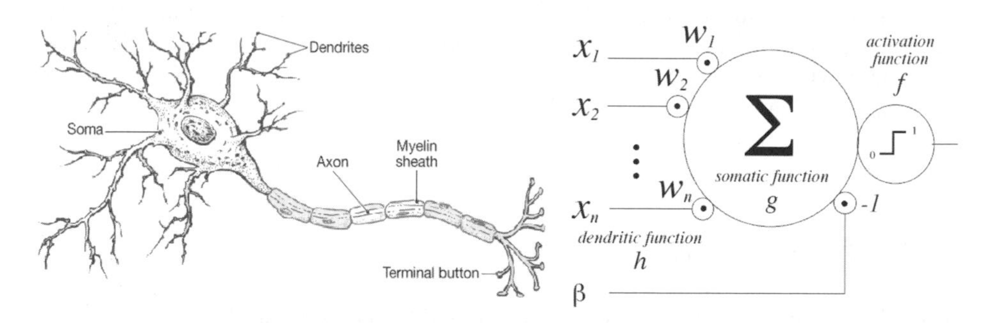

Fig. 2. Biological Neuron from [6] (left) and Artificial Model (right)

2.3.1 Brain Facts

Based on the estimated number of neurons of a human brain, it is possible to get an estimated amount of processing power as follows:

"The human brain cortex has approximately from 15 to 33 billion neurons, depending on gender and age"[7].

If we consider 3.3×10^{10} neurons of binary output, then we have 3.3×10^{10} binary neuron states; this would be 4.125 Gigabytes of storage if a binary state is taken for each neuron[3].

"In a digital computer, information is processed by a small set of registers that operate at speeds of billions of cycles per second. In a brain, information is processed by billions of neurons all operating simultaneously, but only at speeds around 100 cycles per second"[8].

100 cycles per second by 3.3×10^{10} neurons means 3.3×10^{12} neuron activations per second.

2.3.2 Artificial Neuron Model and Computing Power Estimation

The brain's basic processing unit is the *neuron*. Derived from a simplification of the biological factors involved in its operation, a neuron's output can be characterized as follows [9] (see figure 2):

$$y_r = f(g(h(\overline{x}))) \tag{1}$$

where

$$\textit{Dendritic function}: h_i = x_i \cdot w_i \tag{2}$$

$$\textit{Somatic function}: g = \sum_{i=1}^{n} h_i \tag{3}$$

$$\textit{Activation function}: f = \begin{cases} 1: \textit{if } g \geq \beta \\ 0: \textit{otherwise} \end{cases} \tag{4}$$

Based on (1), (2) and (3), the number N of floating point operations for calculating a single neuron output is given by:

$$N = 2n + 1 \tag{5}$$

where n is the number of inputs (connections) of the neuron with its neighbourhood.

"... Each one of them is interconnected with up to 10,000 synaptic connections. Each cubic millimeter of brain cortex has approximately 1 billion synapses" [10].

If we take $n = 10^4$ in Eq. (5) then we have $2 \times 10^4 + 1$ operations by single neuron activation. If we consider 100 activations per second per neuron then we need

$$2.0001 \times 10^6 \text{ operations per second} = 2 \text{ MFLOPS} \tag{6}$$

Finally, we can calculate the total brain processing power:

---

[3] An instantaneous brain's state could be now easily stored into a hard disk drive or even into a portable USB storage card (but not the whole network architecture of course).

$$3.3 \times 10^{10} \text{ neurons by } 2 \times 10^{6} \text{ FLOPS} = 6.6 \times 10^{16} \text{ FLOPS} = 66 \text{ peta FLOPS}^{4} \quad (7)$$

2.3.3 Moore's Law

"Moore's law [2] *describes a long-term trend in the history of computing hardware, in which the number of transistors that can be placed inexpensively on an integrated circuit has doubled approximately every two years. Although originally calculated as a doubling every year, Moore later refined the period to two years. It is often incorrectly quoted as a doubling of transistors every 18 months, as David House, an Intel Executive, gave that period to chip performance increase. The actual period was about 20 months"* [11].

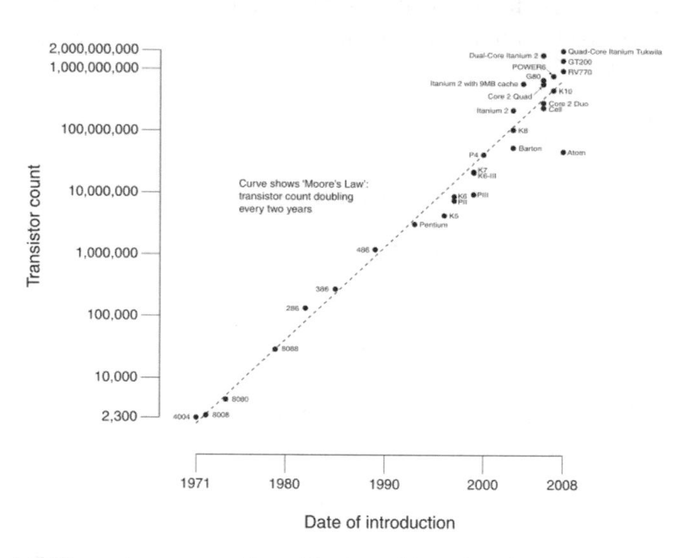

Fig. 3. Plot of CPU transistor counts (logarithmic scale) against dates of introduction [from 13]

2.3.4 Computer Power Projection

According to [13], an Intel Core 2 Duo SL9300 processor has a processing power of 12.8 gigaflops.

$$\text{Intel Core 2 Duo SL9300} = 1.28 \times 10^{10} \text{ FLOPS} \quad (8)$$

3) Computer's Power Prediction Formula: Taking (8) as a reference point, we can state the Moore's Law as follows (doubling the processor's power every two years and Intel Core 2 Duo speed as a known capacity released on 2008):

$$Capacity_{(FLOPS)} = 12.8 \times 10^{9} \cdot 2^{\left[\frac{year-2008}{2}\right]} \quad (9)$$

[4] In this estimation we supposed all the brain's neurons are of the same type (perceptrons [9]). In fact, in the animal brain there are several kinds of even more complex neurons.

If we solve (9) for the *year* variable, then we have

$$year = 2008 + \log_2\left(\frac{Capacity}{1.28\times10^{10}}\right) \tag{10}$$

Substituting (7) in (10)

$$year = 2008 + \frac{2}{\log(2)}\cdot\log\left(\frac{6.6\times10^{16}}{1.28\times10^{10}}\right)$$

$$year = 2053^{1} \tag{11}$$

Now suppose we take several animal brain volumes [14] and apply repeatedly Eq. (10), calculating the estimated processing reaching dates. A linear interpolation was done based on the human brain's processing power and the animal brain's volume[5,6].

Table 1. Animal Brain Capacity and Estimated Reaching Dates

Volume *cc*	Name	Processing Power *FLOPS*	Estimated Reaching Date *year*
1700	Dolphin	8.01×10^{16}	2053
1400	Human	6.60×10^{16}	2053
1000	Homo Erectus	4.71×10^{16}	2052
700	Homo Habilis	3.30×10^{16}	2051
470	Australopithecus Africanus	2.22×10^{16}	2050
415	Australopithecus Afarinsis	1.96×10^{16}	2049
390	Chimpanzee	1.84×10^{16}	2049
200	Dog	9.43×10^{15}	2047
20	Alligator	9.43×10^{14}	2040
10	Crow	4.71×10^{14}	2038
2	Rat	9.43×10^{13}	2034
1	Bat	4.71×10^{13}	2032
0.1	Goldfish	4.71×10^{12}	2025
0.001	Honeybee	4.71×10^{10}	2012

Table 1 takes in account only the total number of neurons and its connections to realize a linear interpolation. Recent studies [15] show that there is a linear relation between the body and brain sizes (Fig. 5). This could be related both to the increased amount of effectors (muscles and tendons) and sensorial inputs (skin and nerve connections). For this reason, some authors [16,17] propose that the brain's size in animals is not as important as the existence of some specialized and very complex structures like the cerebellum, thalamus, hypothalamus, etc.

---

[5] This result supposes that Moore's Law could continue valid until 2053; this cannot be warranted but it give us a good approximation if the current tendencies continue.

[6] More accurate estimations can be done taking in account the decrease in the estimated number of synapses among neurons in smaller animal's brains.

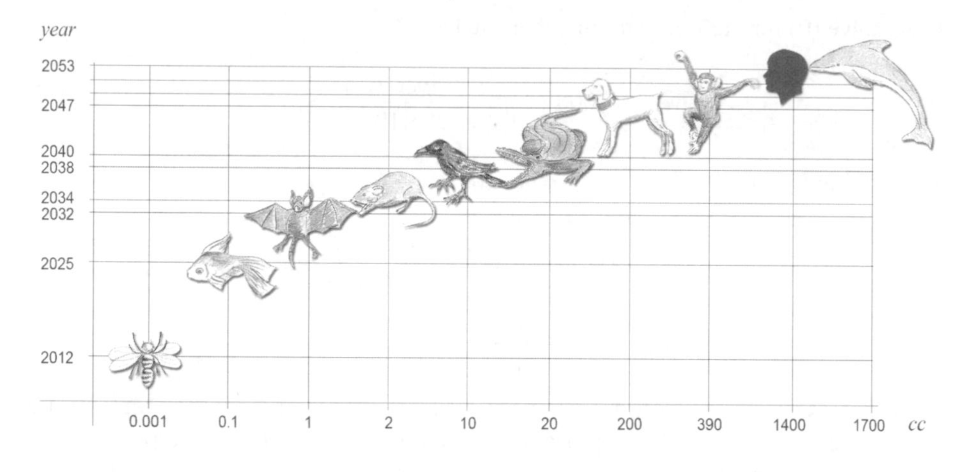

Fig. 4. Plot of estimated processing reaching dates for animal and human brain sizes

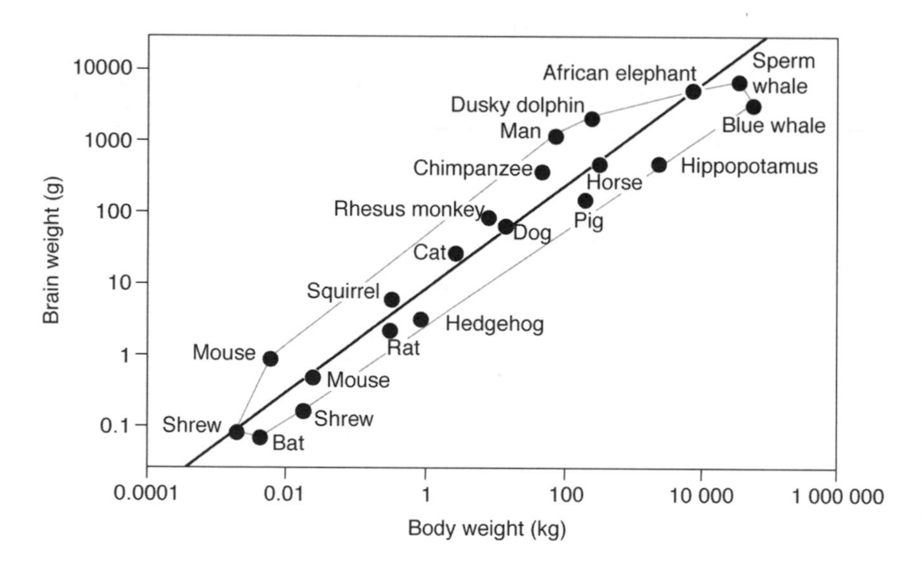

Fig. 5. Animal brain weight and body weight in vertebrates (taken from [15])

3 Conclusions and Discussion

This paper makes an analysis of the processing power of animal's brain in terms of Artificial Neural Networks. This analysis makes an approximation of such processing power in terms of processor operations per second (FLOPs), by using the Perceptron Model of an artificial neuron (supposing all the neurons of the same type) and assuming the Moore's Law valid by four more decades (something difficult to suppose with certainty).

In conclusion, the original estimated date of reaching the human brain's processing power by a computer [1] is delayed one entire decade. This confirms the idea of some robotics contests, predicting that by 2050 a team of robots could play against humans and defeat them. If computing techniques and algorithms do not solve the progressive computing problems at the same rhythm as Moore's Law, it is highly possible that in 2050 robots will be still solving very basic tasks of living beings. Perhaps the right path would be imitating nature; letting robots evolve and learn.

More studies are in progress to determine the best way to do it. By analyzing some brain structures, a basic artificial brain model will be presented in the future.

References

1. Moravec, H.: Mind Children. Harvard University Press, Cambridge (1988)
2. Moore, G.E.: Cramming more components onto integrated circuits. Electronics Magazine 38(8) (April 19,1965)
3. Savage, J., LLarena, A., Carrera, G., Cuellar, S., Esparza, D., Minami, Y., Peñuelas, U.: ViRbot: A System for the Operation of Mobile Robots. In: Visser, U., Ribeiro, F., Ohashi, T., Dellaert, F. (eds.) RoboCup 2007: Robot Soccer World Cup XI. LNCS (LNAI), vol. 5001, pp. 512–519. Springer, Heidelberg (2008)
4. Ishiguro, H., Kanda, T., Kimoto, K., Ishida, T.: A Robot Architecture Based on Situated Modules. In: Proc. of IEEE/RSJ International Conference on Intelligent Robots and Systems (IROS), Kyongju, Korea, pp. 1617–1623 (1999)
5. Amdahl, G.: Validity of the Single Processor Approach to Achieving Large-Scale Computing Capabilities. In: AFIPS Conference Proceedings, vol. (30), pp. 483–485 (1967)
6. http://www.abovetopsecret.com/forum/thread532721/pg1
7. Guyton, A.C., Hall, J.E.: Medical Physiology, 11th edn. Elsevier Saunders, Amsterdam (2006)
8. Abbott, L.F., Dayan, P.: Theoretical neuroscience: computational and mathe-matical modeling of neural systems. MIT Press, Cambridge (2001) ISBN 0-262-04199-5
9. Haykin, S.: Neural Networks. A Comprehensive Foundation, 2nd edn. Prentice Hall, Englewood Cliffs (1999)
10. Alonso-Nanclares, L., Gonzalez-Soriano, J., Rodriguez, J.R., DeFelipe, J.: Gender differences in human cortical synaptic density. Proc. Nat. Acad. Sci. U.S.A. 105(38), 14615–14619 (2008), doi:10.1073/pnas.0803652105.PMID 18779570
11. Myhrvold, N.: Moore's Law Corollary: Pixel Power. New York Times (June 7, 2006), http://www.nytimes.com/2006/06/07/technology/circuits/07essay.html
12. Image taken from, http://en.wikipedia.org/wiki/Moore's_law
13. http://www.intel.com/support/processors/sb/cs-023143.htm#4
14. http://evolution.mbdojo.com/you_figure_it_out.htm
15. Roth, G., Dicke, U.: Evolution of the Brain and Intelligence. In: TRENDS in Cognitive Sciences, vol. 9(5). Elsevier, Amsterdam (May 2005)
16. http://www.sciencedaily.com/releases/2009/11/091117124009.htm
17. http://weber.ucsd.edu/~jmoore/courses/allometry/allometry.html

The Real-Time Embedded System for a Humanoid: Betty

Meng Cheng Lau and Jacky Baltes

Autonomous Agent Lab,
University of Manitoba,
Winnipeg, Manitoba R3T 2N2, Canada
{laumc,jacky}@cs.umanitoba.ca
http://aalab.cs.umanitoba.ca/

Abstract. This paper investigates the efficiency of the real-time embedded system in our humanoid robot, Betty. In this paper we only discuss the upper body of Betty. Based on several experiments of different queue data structures, communication protocols and PID controller implementations, we measured and analysed the latencies and jitters of Betty's responses. The experimental results indicate the best configuration to optimise the performance of Betty's Control Program.

Keywords: Communication protocol, queue data structure, Preemptive real-time kernel, jitter and latency.

1 Introduction

Designing and implementing a real-time embedded system is one of the most challenging tasks in robot development. It is crucial to ensure that a robot responses in the expected period of time. We tested and analysed our real-time embedded system to perform in real-time by considering the latency and the jitter performances. Latency is the time between starting to send a command and receiving a response from the servos where the command has been executed. Jitter is the time variation of the latencies.

2 Hardware Configuration

Upper body of Betty consists of ten DOFs (degrees of freedom). The head has four DOFs which give pan, tilt, roll and jaw. Each of the hands provides three DOFs, a shoulder allows lateral and frontal motion, and an elbow gives lateral motion. These joints are constructed by four Dynamixel AX-12 servos to the head and three Dynamixel RX-64 servos to each of the hands as shows in Fig. 1. The main reason we choose RX-64 to construct the hands because it has higher final maximum holding torque, 64.4~77.2 kgf.cm compare to only 12~16.5 kgf.cm for AX-12 [1], [2]. It will allow a RX-64 servo to generate sufficient torque to support the weight of the arm.

P. Vadakkepat et al. (Eds.): FIRA 2010, CCIS 103, pp. 122–129, 2010.

In order to control the servos, we use a Dynamixel's dedicated controller, CM-2+ from Robotis as the central control unit. With its AVR ATmega128 microcontroller, we implement our real-time embedded system in C language which establishes communication with the servos. The connections on the CM-2+ is illustrated in Fig. 2 and Fig. 3 shows the overview of Betty's upper body.

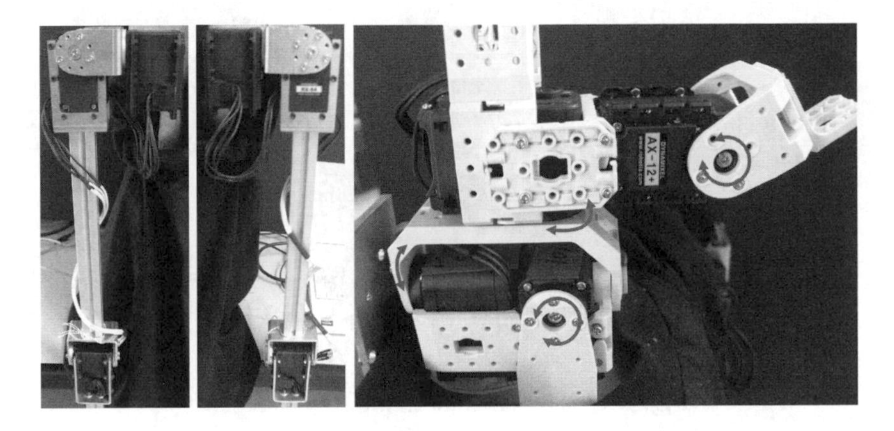

Fig. 1. Betty's upper body construction

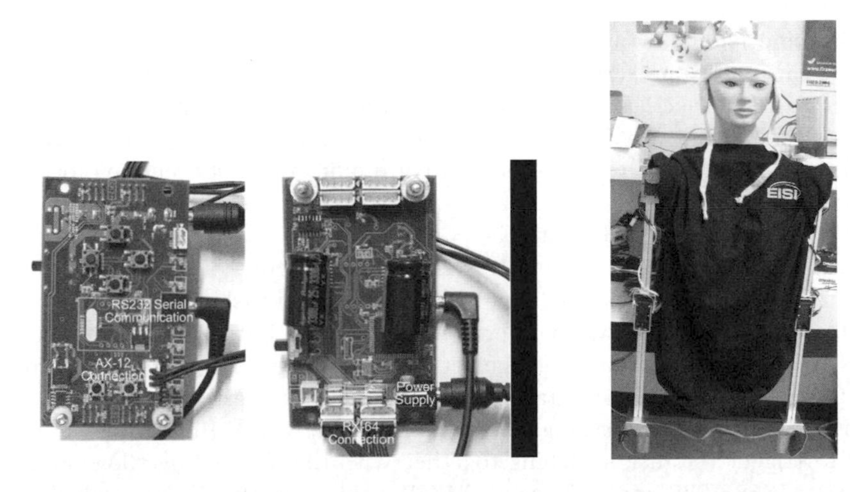

Fig. 2. CM-2+ Connection

Fig. 3. Overview of Betty's upper body

3 Implementation

We program the CM-2+ embedded system in C language as the Control Program to control servos. The Motion Editor program which is written in C++ language is introduced to enable instruction and response communication between the CM-2+ and host computer. Fig. 4 shows the overall framework of the system that explains the connection architecture of the robot's Motion Editor, Control Program, CM-2+ and servos. In order to develop an optimised embedded system, we employed different communication protocols, queue data structures, Preemptive multi-tasking kernel and PID controller in this project.

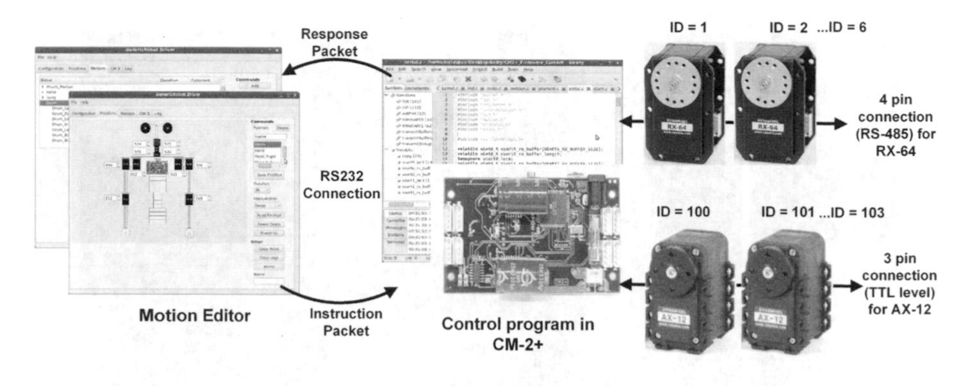

Fig. 4. General program framework of Betty

3.1 Preemptive Multi-tasking Kernel

We use Preemptive multi-tasking kernel in the Control Program to handle task switching. Currently our kernel supports four tasks. First task is an idle task to toggles a LED which is always ready to run to ensure that at least one thread is running and the Control Program is working properly. The second task is the PC thread that handle serial port communication between the Motion Editor program and CM-2+. Third task is the servo thread which will prepare commands to send from CM-2+ to the RX-64 and AX-12 servos. Finally, the last task is the torque thread that reads the current torque of each servo and sends it through the PC thread. The Preemptive multi-tasking kernel uses a timer interrupt to switch between different threads at an initial timer frequency of 10Hz. To execute task switching, the kernel will save the complete task state on the stack and then store the stack pointer in the task control block (TCB).

In our Preemptive multi-tasking kernel, we use *Timer3* ISR (interrupt service routine) to trigger the task switching together with the *Timer3* overflow interrupt vector `TIMER3_OVF_vect`. Then `ISR_NAKED` is added to the program which is responsible to preserve the machine state including the `SREG` register, as well as placing a *reti()* at the end of the interrupt routine and `ISR_BLOCK` where global interrupts are initially disabled by the AVR hardware when entering ISR. So by using the `ISR_NAKED`, *inline asm* can be embedded in our C program to perform context switching in ISR. The correct task ids and stack sizes should be allocated correctly in the TCB. The main advantage of Preemptive scheduling is real-time response on the task level, because the time to tick a task is mainly depends on the interrupt latency [5].

3.2 Communication Protocols

The Control Program uses three different types of communication protocols which are Absolute Position, Sliding Resolution and Difference protocols. Based on these protocols, the Motion Editor program from host computer will send instructions to the CM-2+'s Control Program. For instants, we use `SyncWrite` command to move the servos to specific positions. Table. 1 shows the structures and the differences between these protocols where `SyncWrite` commands were sent to move all servos to move from position 512 to 520 at the speed of 100. We divided the positions and speed by 4 so each of them can be wrapped into one byte in hexadecimal.

According to Table 1, each `SyncWrite` command of the Absolute Position protocol will send 14 bytes of instruction in one packet. On the other hand, Sliding Resolution protocol will send its command in two separate packets which are high and low packets. The high packet consists the most significant four bits of every position whereas the low packet contains the least significant four bits. The advantage of this protocol is to start the servos as soon as a command is received. So while the Control Program receives the high packet, all servos will

Table 1. Communication Protocols

Protocol	Command Structure
Absolute Position	FF011982828282828282828282E5 FF: Header 01: SyncWrite command 19: Speed 82: Position E5: Checksum = lower byte of ~(01+19+82+82+82+82+82+82+82+82+82+82)
Sliding Resolution	FF0101198888888888883C (First frame) FF01021922222222222239 (Second frame) FF: Header 01: SyncWrite command 01: Highbytes or 02: Lowbytes 19: Speed 88: combination of 2 Highbytes 22: Combination of 2 Lowbytes 3C: Checksum = lower byte of ~(01+01+19+88+88+88+88+88)
Difference	FF011922222222223B FF: Header 01: SyncWrite command 19: Speed 22: Each 4-bit represents the difference between the desired and the previous positions. 1 to 7 represent increment and 8 to F represent decrement 3B: Checksum = lower byte of ~(01+19+22+22+22+22+22)

start moving towards their target positions then the final target positions will be determined once the low packet has been received.

In Difference protocol, the Control Program receives two types of instruction packets, Absolute and Relative. The Absolute packet is similar to the one described in the Absolute Position protocol. Then the Relative packet will be sent with only four bits per position which encodes the differences between current and new target positions.

3.3 Data Structure

We tested two different queue data structures in our embedded system which are circular queue and regular queue. For a circular queue, it provides few advantages over regular queue. It only uses a single fixed-size data structure [3] and its array can be accessed at both end. It avoids memory wastage and let the program handles data streaming efficiently [4]. But data in circular queue can be replaced unintentionally if the algorithms in the program is not designed properly. On the other hand, regular queue is easier to implement compare to circular queue because the memory only can be accessed linearly but its downside is more memory has to be allocated if large queue size is needed.

In the Control Program, these interrupt driven queue data structures are adapted into IRS for both transmission and receiving of USART0 and USART1. Data Register Empty Interrupt, UDREI0 and UDREI1 are selected in the Control Program to trigger ISRs on USART0 and USART1 [6],[7]. On the other hand, the Control Program measures the latencies and jitters in every send-receive cycle of CM-2+. Because the UDREIs are implemented in the circular solution, it will introduce latency when ISRs occur. The experimental results are discussed in section 4.

3.4 PID Controller and Optimization

PID (proportional-integral-derivative) controller is the most popular industrial feedback control algorithm. Implementation of the PID controller in our Control Program is based on feedback about the latencies and jitters of the torque responses which will modify the context switching frequency. We need to ensure that our Preemptive kernel is robust enough so that the host program from its Motion Editor can modify its context switching frequency in real-time. To change the context switching frequency, all ISRs are disabled with *cli()* before the assignment of new frequency and they are enabled with *sei()* after its new frequency is updated. Fig. 5 shows the theory of the PID controller.

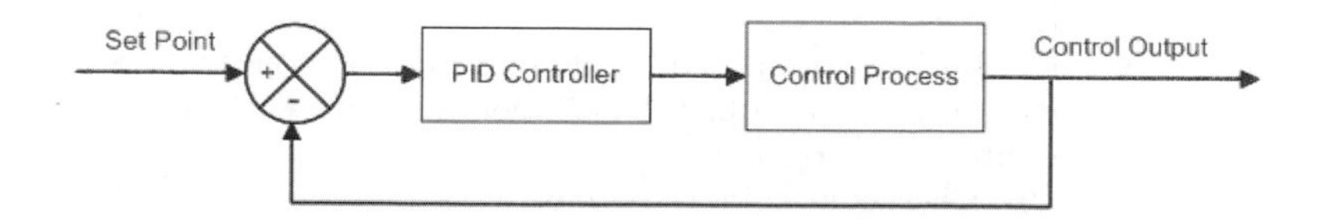

Fig. 5. Block diagram of PID controller

In the PID control loop, the control output (co) is calculated based on the following equations:

$$e(t) = sp - pv \tag{1}$$

$$P_{out} = K_P e(t) \tag{2}$$

$$I_{out} = K_I[e(t) - e(t-1)] \tag{3}$$

$$D_{out} = K_D \sum_{t=0}^{n}(e(t)) \tag{4}$$

$$co = P_{out} + I_{out} + D_{out} \tag{5}$$

Firstly, the error, $e(t)$ is calculated in equation (1) by subtraction of the set point, sp and the process variable, pv which is the latency of current run, t where $t = 0, 1, 2...n$. Then the proportional, integral, and derivative terms of output are determined by equations (2), (3) and (4). Finally, all PID outputs are summed to calculate co as equation (5). K_P, K_I and K_D are the constants gain of proportional, integral and derivative respectively.

Then we extended the PID controller to minimize the latency with the Optimisation Quality Function which returns its control output in the factor of 75% latency and 25% jitter are implemented to control the context switching frequency. The optimization phase is similar to the general PID controller, except it will find the context switching time which has the lowest control output from the quality function. It will be adjusted to reach the desired set point based on the latencies and jitters measurements.

4 Experimental Results

In order to optimise our embedded system, we measured the latencies and jitters of 100 runs to analyse the efficiency of those different implementation discussed in section 3. Baud rate of 1 Mbps is used in USART0 for serial communication. In these experiments, we sent 58 bytes of instruction packet to the Control Program; then we compare the averages of latencies and jitters in different queue data structures.

Table 2. Results of circular and regular queue data structures

	Queue Data Structures	
	Regular	Circular
Latency (μs)	581	1103
Jitter (μs)	2.0	1.4

Based on the experimental results in Table 2, generally both queue data structures produce acceptable latency and jitter. Circular queue needs 1103μs compare to 581μs in regular queue which perform better in this implementation where it takes 10μs to send a byte at 1 Mbps baud rate. However, circular queue provide an advantage over regular queue where it uses less memory because in an embedded system, using minimum amount of memory is as important as real-time transmission.

Performance of the PID controller and Optimisation Quality Function are tested based on three difference protocols introduced in subsection 3.2. Fig. 6 shows the test results of PID controller with 15ms as the set point (SP) of latency. Fig. 7 shows the optimization result of 100 trials at 5000Hz with $\pm$ 50 of context switching frequency in different protocols. The averages of latencies are 15320, 15105 and 15411 for Absolute Position, Difference and Sliding Resolution respectively.

Although the Difference protocol has shown better performance in both experiments but the differences are not significant. Moreover, its advantage only holds if the differences between new and current positions are less then 28 for

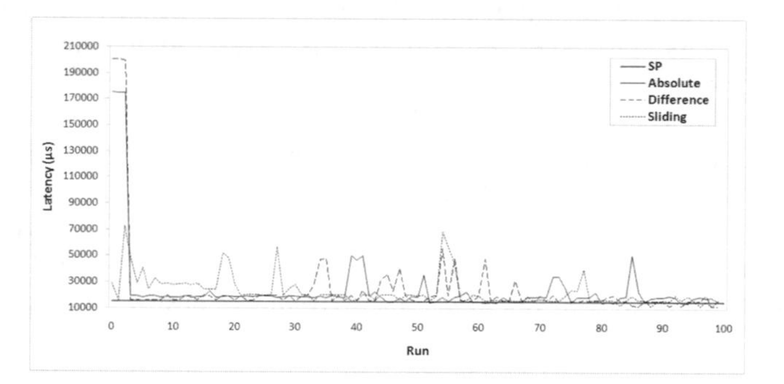

Fig. 6. Performance of PID controller in different communication protocols with $K_P = 0.1$, $K_I = 0$ and $K_D = 0.02$

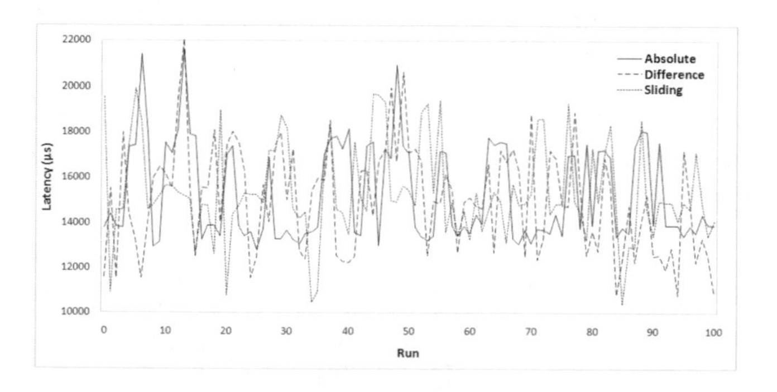

Fig. 7. Performance of Optimisation Quality Function in different communication protocol

each servo movement. In contrast, the Absolute Position protocol is simpler and easier to implement and it has most reliable performance compare to other protocols. Because the latency is fluctuated along the desired set point, therefore we implement control limit approach to acknowledge the steady state after few occurrences of control outputs fall between the control limits are perceived.

5 Conclusion

In this paper we have described the advantages and disadvantages of different implementation of queue data structures, communication protocols and PID controller. Based on the experimental results, we decided to choose the regular queue for lower average latency. On the other hand by choosing the Absolute Position protocol, the communication between the Motion Editor and Control Program becomes more reliable and easier to handle. Generally, Control Program performance has been improved by implementing the Optimisation Quality Function in program PID Controller with our selected configuration.

References

1. Robotis: User's Manual Dynamixel RX-64. Ver 1.10, p. 6 (2007)
2. Robotis: User's Manual Dynamixel AX-12, p. 3 (2006)
3. Circular Buffer, `http://en.wikipedia.org/wiki/Circular_buffer`
4. Wolf, W.H.: Computers as Components: Principles of Embedded Computing System Design. Academic Press, San Diego (2001)
5. On Time RTOS-32 Documentation, `http://www.on-time.com/rtos-32-docs/rtkernel-32/programming-manual/advanced-topics/preemptive-or-cooperative-multitasking.htm`
6. 8-bit AVR Microcontroller with 128K Bytes In-System Programmable Flash, `http://www.atmel.com/dyn/resources/prod_documents/doc2467.pdf`
7. Barnett, R.H., O'Cull, L., Cox, S.: Embedded C Programming ant the Atmel AVR. Cengage Learning, Florence (2007)

Design and Implementation of SOPC Based Motion Control for Human-Sized Biped Robot

Li Tzuu-Hseng S., Su Yu-Te, Chen Wen-Chien, and Hu Jhen-Jia

aiRobots Laboratory, Department of Electrical Engineering
National Cheng Kung University, Tainan 70101, Taiwan, R.O.C.
{thsli,n2895109,n2695420,n2894112}@mail.ncku.edu.tw

Abstract. This paper presents the design of the human-sized biped robot, aiRobot-HBR1 and the implementation of SOPC based motion control system. aiRobot-HBR1 is a human-size biped robot with 110 cm height, 40 Kg weight, and has a total of 12 degree of freedom (D.O.F). After trajectory planning, aiRobot-HBR1 will act according to the motions defined by the operator. Then the SOPC based motion control system is used to command the robot execute the motion trajectory. Designing the motor controller and the graphic user interface along with the NIOS FPGA, we construct a control platform for developing the control strategies of the robot based on SOPC. Finally, the experiment result indicates the validity of the proposed motion control system.

Keywords: Human-sized biped robot, SOPC, motion control.

1 Introduction

In recent years, based on the expect that robot assist human activitities in the daily life, many kinds of robots have been developed. Humanoid robot is one of the popular research fields [1-3]. Humanoid robots are designed that are enable to execute variety functions, such as run [4], entertainment, climbing stair [5], lying down [6], and interact with human beings [7]. The application of humanoid robot is very flexible comparing to other types of robot. Furthermore, Humanoid robot is a highly integrated system which relates to many technical issues, and there are still many tough problems to be solved. Therefore, design a humanoid robot is full of challenge and practicable.

Mechanism design, electrical system design, and motion control system design are the fundamental effects that influence the motion abilities of the humanoid robot. A humanoid robot with flexible motion abilities can execute many tasks in the general environment. For these reasons, this paper devotes to the study of mechanism design, electrical system design, and motion control system design for the human-sized biped robot. We design a human-sized biped robot and establish a motion control structure which can be utilized to develop the control strategies for the biped robot. Thought graphic user interface (GUI), the operator can design the specific motion pattern for the biped robot. According to these motion patterns, the SOPC based [8-10] motion control will command the biped robot in real-time.

P. Vadakkepat et al. (Eds.): FIRA 2010, CCIS 103, pp. 130–137, 2010.

This paper is organized as follow. The hardware design of the human-sized biped robot, aiRobot-HBR1, is presented in Section 2. The biped robot motion control system is depicted in Section 3. In section 4, the experiment result is presented to demonstrate the effectiveness of the proposed hardware design and control method. Conclusions are drawn in Section 5.

2 Mechanism and Hardware of aiRobot-HBR1

The biped robot presented in this paper consists of two legs, and each leg is composed of a thigh, shank, and foot. There are six degrees-of-freedoms (DOFs) in each leg, three of which are located in the hip joint, one in the knee joint, and two in the ankle joint. The arrangement of the motors is shown in Fig. 1 and the 3D design graph is shown in Fig. 2.

Fig. 3. shows the appearance of aiRobot-HBR1. Table 1 lists the specification of aiRobot-HBR1. Fig. 4 depicts the control system structure of aiRobot-HBR1. The central processor unit of the overall control system is Altera Cyclone II EP2C20 FPGA. The core processor of this FPGA is a 32 bits Nios II CPU. FPGA (Field Programmable Gate Array) are now wildly used in the field of digital integrated circuit design and verifying. And there are twelve Maxon RE40 DC motors used as actuators. According to different rated currents, the power system is divided into high power part and low power part. While planning the power supplying system, it is important to distinguish these two parts to avoid destroying of the device. Fig. 5 depicts the planning of the power system.

Fig. 1. Arrangement of motors

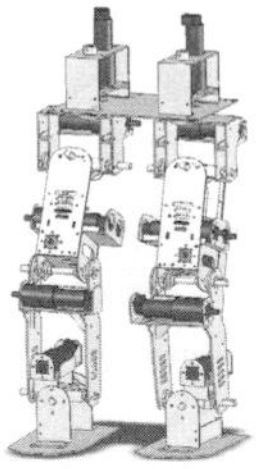

Fig. 2. 3D design graph of aiRobot-HBR1

Table 1. Specification of aiRobot-HBR1

Height		110 cm
Weight		40 kg
DOFs	Hip	3 DOFs
	Knee	1 DOFs
	Ankle	2 DOFs
Controller		Altera NIOS II FPGA
Actuator		DC motor
Material of the structure		A5052 aluminum alloy

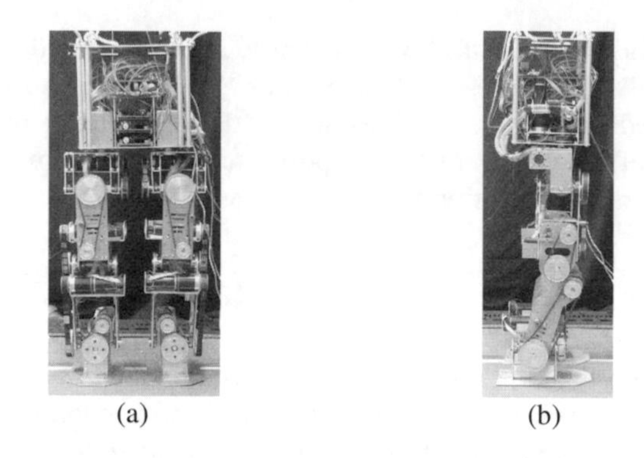

Fig. 3. (a) The front view and (b) the lateral view of aiRobot-HBR1

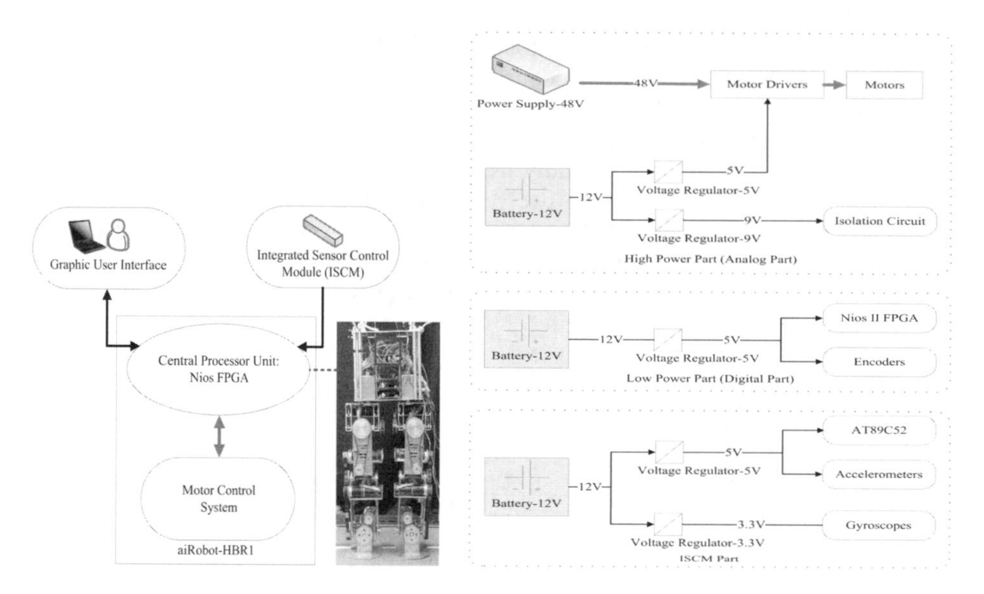

Fig. 4. System structure of aiRobot-HBR1

Fig. 5. Planning of the Power system

3 Motion Control System and Trajectory Planning

This section will first describe the motion control system employed in aiRobot-HBR1. Then introduce the trajectory planning method used in aiRobot-HBR1.

3.1 Motion Control System

The motion control system is to master the robot to achieve the commands which are given by the user or the central processor unit. As depicted in Fig. 6, the motion control system implemented in aiRobot-HBR1 can be partitioned into three blocks, including control interface, central processor unit, and motor control system. The control interface is composed of the integrated sensor control module (ISCM) and the graphic user interface (GUI). The motor control system consists of motor controller, motor driver, motor, and encoder. Most motor controllers are implemented in software in programmable logic controllers. Software programming has the advantages of flexibility and saving time. When system is complicated or needs much computation, software program which executes sequentially becomes not efficient. The excellence of hardware implementation is its high speed and parallel execution. NIOS II FPGA combines the feature of hardware implementation and software programming. One can design the hardware circuit in the FPGA as well as program control strategies in the NIOS II CPU. The mechanism is called "system on a programming chip" (SOPC). Based on the feature of the SOPC, the motion control system can control several motors simultaneously and communicate with computers or other devices easily. Therefore we develop the motor controller in the FPGA. The motor controller can be portioned into four parts: digital noise filter, feedback decoder unit, PD controller, and PWM module. Fig. 7 depicts the control structure of the motor controller for one motor.

The form of the discrete PD algorithm is $u(k) = K_p e(k) + K_d \left[e(k) - e(k-1)\right]$. In real implication, we multiply *u(k)* by a scale factor to get the PWM duty ratio which presents the velocity command of the motor control system. We use VerilogHDL to implement the PD controller in the FPGA, and the control structure of the PD controller is shown in Fig. 8.

The feedback decoder unit utilizes two techniques to implement the function. Finite state machine and digital noise filter are combined to transfer the two incoming quadrature signals of the encoder to the rotation angle, and angular velocity. The encoder outputs two quadrature signals A and B according to the rotation angle of the motor. These two signals have a 90 degrees phase difference. Discerning which signal leads another provides us to realize the rotation direction of the motor. Moreover, the frequency and the width of the signal A and the signal B are proportional to the angular velocity of the motor. Count the number of pulses passing through in a fix interval or calculate the width of the pulse, we can realize the angular velocity of the motor. Generalizing the relationship between the signal A and the signal B, it can be interpreted by a state diagram as shown in Fig. 9.

In the practical application, signals are interfered by noise easily. If we just use the decoder as designed above, it will work badly and miss the rotation information. Hence, we add a finite impulse response (FIR) filter in the front stage of the feedback decoder unit. The digital noise filter is responsible for rejecting noise for the incoming quadrature signals.

The encoder captures the position information of the motor and outputs two quadrature signals. After the signals processed by the feedback unit, the angle of the motor is transmitted to the PD controller. The motor controller will make calculations and subsequently outputs values of PWM duty to the PWM module. The corresponding PWM signal are generated by the PWM module and be transmitted to the outer motor driver. Based on this circuit for one motor, we construct a motion control system which can control twelve motors simultaneously.

3.2 Trajectory Planning

Trajectory planning is to generate motion patterns for the robot to achieve user's planning. Fig. 10 shows a diagram in sagittal plane of the robot that biped robot swing right leg from M_1 to M_4 to avoid an obstacle. Having known these four postures, it is always need some interpolation technique to generate interpolation point among points. Linear interpolation method is a simple interpolation method, and it assumes that the variety among data is linear. The drawback of the method is that the slope of the interpolation line is discontinuous. Discontinuous slope of trajectory always means discontinuous velocity and discontinuous acceleration which bring about vibrations and affect the motion stability of the robot. To generate a smooth trajectory, it is necessary that the first derivative (velocity) terms be differential, the second derivative (acceleration) terms be continuous at all. Therefore, the joint trajectory was obtained by cubic spline interpolation.

The cubic spline interpolation is to fit data with a piecewise continuous curve, passing through each of data. For n points $\theta(t_1)$, $\theta(t_2)$,..., $\theta(t_n)$, $t_1<t_2<\ldots t_n$, there is a cubic spline polynomial $S_i(t)$ for each interval $\left[t_i, t_{i+1}\right]$:

$$S_i(t) = a_i(t-t_i)^3 + b_i(t-t_i)^2 + c_i(t-t_i) + d_i \tag{1}$$

where $i = 1, 2, \ldots, n-1$, $a_i = \left(N_{i+1} - N_i\right)/6h_i$, $b_i = N_i/2$, $c_i = \left(\theta(t_{i+1}) - \theta(t_i)\right)/h_i - h_i\left(\left(N_{i+1} + 2N_i\right)/6\right)$, $d_i = \theta(t_i)$, $h_i = t_{i+1} - t_i$, $N_i = S_i''(t_i)$. The result of the cubic spline interpolation $S(t)$ is that $S(t)$, $S'(t)$, and $S''(t)$ will be continuous on the interval $\left[t_1, t_n\right]$. In our application, the cubic spline interpolation is employed to generate the trajectory of the joint. By the natural of the cubic spline interpolation, the first derivative (velocity) terms, the second derivative (acceleration) terms of the trajectory is continuous.

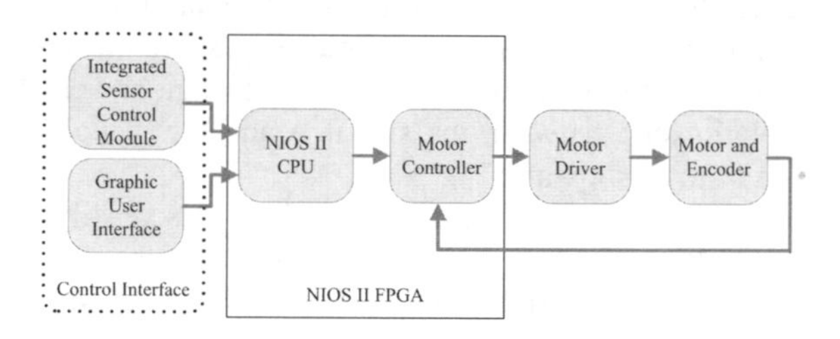

Fig. 6. Structure of the motion control system

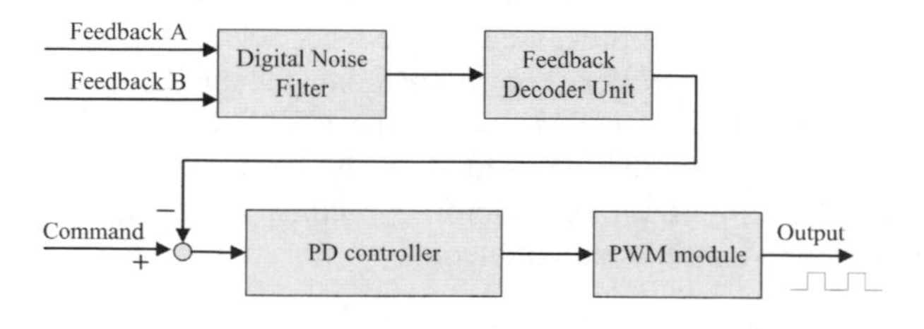

Fig. 7. Control structure of the motor controller

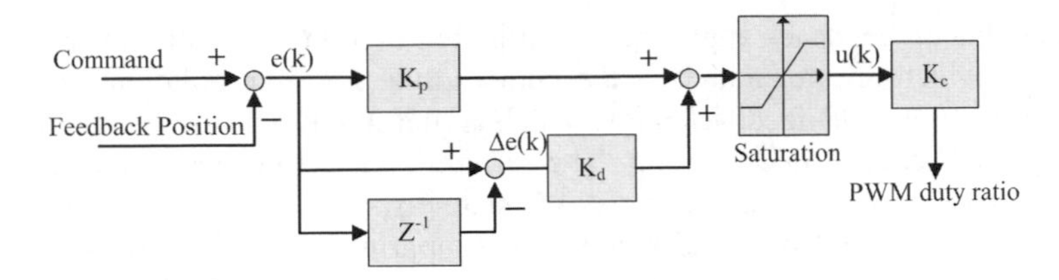

Fig. 8. Control structure of the PD controller

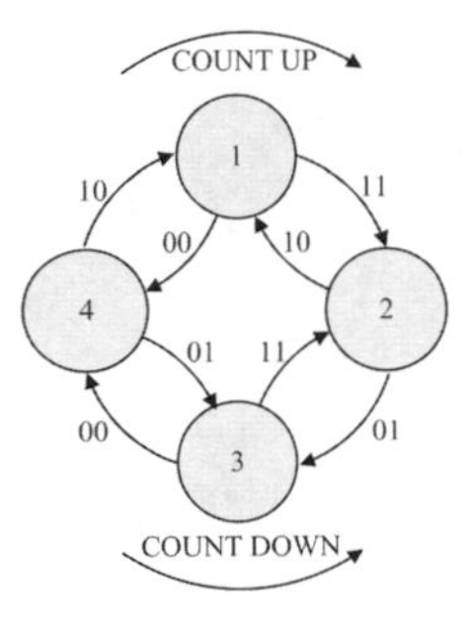

Fig. 9. State diagram of the signal A and the signal B

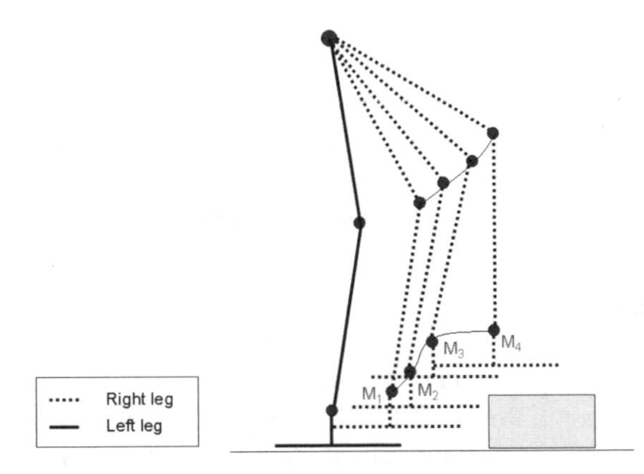

Fig. 10. Swing leg for avoiding obstacle

4 Experiment Result

In this section, an experiment is planned for confirming the performance of the proposed motion control system. The experiment runs the procedure from motion planning to practical motion test, and it is to plan a motion pattern for the robot to make a lifting leg motion. Firstly, recode desired commands for each joint by using GUI. The desired commands for each joint has six desired points as shown in Fig. 11. Second, we use the cubic spline line interpolation method to insert some points among these six points. Finally, the robot makes a lifting leg motion in practice. Fig 11 shows the command and the feedback angles of the lifting leg motion. The solid line presents the commands which are generated by the cubic spline line interpolation method. The dash line represents the feedback angle which is measured from the encoder and processed by the feedback decoder unit. The difference between the command and the feedback angle implies that there is a steady state error existing in the operation of the motor controller. It is trivial and receivable comparing with other errors. Moreover, the ripple or the small squared wave in the command is the result of the sampling and the hold in the central processor unit.

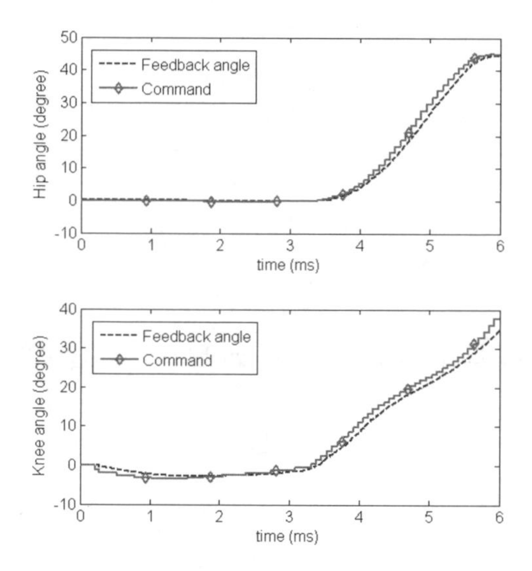

Fig. 11. Angles of the hip and the knee of the robot during lifting leg motion

5 Conclusion

This paper has presented the SOPC based motion control system design for the human-sized biped robot, aiRobot-HBR1. The proposed motion control system design comprises the control interface, the central processor unit, and the motor control system. The graphic user interface can control the robot directly and record the desired commands in the motion pattern planning procedure. The merit of the motor control system is its parallel execution which can control twelve motors simultaneously. Consequently, a motion control system was constructed which provides a platform for developing the control strategies of the human-sized biped robot based on SOPC. The validity of the motion control system has been presented in the practical experiments. In the future, we will improve the mechanism of the robot to be more compact and stronger, thus the robot can execute more tasks. In the motion pattern planning, we do not consider the balance problem of the robot. Thus, we can take the balance problem into account to establish a more reliable motion planning procedure.

Acknowledgment

This work was supported by National Science Council of Taiwan, R.O.C, under Grants NSC97-2221-E006-0-160-MY3 and NSC97-2221-E006-0-172-MY3.

References

1. Chang, C.-M., Lu, M.F.g., Hu, C.-Y., Lai, S.-W., Liu, S.-H., Su, Y.-T., Li, T.-H.: Design and Implementation of Penalty Kick Function for Small-Sized Humanoid Robot by using FPGA. In: Proc. IEEE Workshop on Advanced Robotics and its Social. Impacts (ARSO 2008), Taiwan (August 2008)
2. Su, Y.-T., Li, T.-H., Hsu, C.L., Lu, M.F., Hu, C.Y., Liu, S.H.: Omni-Directional Vision-Based Control Strategy for Humanoid Soccer Robot. In: Proc. of 2007 IEEE IECON, pp. 2950–2955 (November 2007)
3. Su, Y.-T., Li, T.-H.: Design and Implementation of Fuzzy Auto-Balance Control for Humanoid Robot. In: 2007 FIRA RoboWorld Congress, USA (June 2007)
4. Kajita, S., Nagasaki, T., Kaneko, K., Hirukawa, H.: ZMP-Based Biped Running Control. IEEE Robotics and Automation Magazine 4, 63–72 (2007)
5. Fu, C., Chen, K.: Gait Synthesis and Sensory Control of Stair Climbing for a Humanoid Robot. IEEE Transactions on Industrial Electronics 55, 2111–2120 (2008)
6. Kanehiro, F., Kaneko, K., Fujiwara, K., Harada, K., Kajita, S., Yokoi, K., Hirukawa, H., Akachi, K., Isozumi, T.: The first humanoid robot that has the same size as a human and that can lie down and get up. In: IEEE International Conference on Robotics and Automation, ICRA 2003, vol. 2, pp. 1633–1639 (2003)
7. Mitsunaga, N., Smith, C., Kanda, T., Ishiguro, H., Hagita, N.: Adapting Robot Behavior for Human–Robot Interaction. IEEE Transactions on Robotic, 911–916 (August 2008)
8. Su, Y.-T., Hu, C.-Y., Lu, M.F., Chang, C.-M., Lai, S.-W., Liu, S.-H., Li, T.-H.: Design and implementation of SOPC based image and control system for HuroCup. J. of Harbin Institute Tech (New Series) 15, 41–46 (2008)
9. Li, T.-H., Hsu, C.-L., Hu, C.-Y., Su, Y.-T., Lu, M.-F., Liu, S.-H.: SOPC Based Weight Lifting Control Design for Small-Sized Humanoid Robot. In: Proc. 2008 ICCAS, Korea (October 2008)
10. Li, T.-H., Su, Y.-T., Liu, S.-H., Hsiao, M.-Y.: Design and Implementation of a Gait Pattern Generator Based on Fuzzy Control for Small-Sized Humanoid Robot by Using SOPC. In: 2008 CACS Int. Automatic Control Conference, Taiwan (November 2008)

Dynamic Modeling of Energy Efficient Hexapod Robot's Locomotion over Gradient Terrains

Shibendu Shekhar Roy[1], Pranab Sen Choudhury[1], and Dilip Kumar Pratihar[2]

[1] Department of Mechanical Engineering, National Institute of Technology, Durgapur, India
[2] Department of Mechanical Engineering, Indian Institute of Technology, Kharagpur, India
ssroy99@yahoo.com, pranab.nitdgp@gmail.com,
dkpra@mech.iitkgp.ernet.in

Abstract. The minimization of power consumption plays a key role in the locomotion of an autonomous multi-legged robot used for service purpose. This paper presents a detailed dynamic modeling of energy efficient hexapod walking robot during locomotion over gradient terrain. A power consumption model is derived for statically stable wave gaits by considering a minimization of dissipating power for optimal foot force distribution. Two approaches have been developed, such as minimization of norm of foot forces and minimization of norm of joint torques using least squared method. Results of these two approaches have been compared with those of some published literature. The variations of average power consumption and energy per weight per traveled length with the angle of slope and velocity have been compared for the ascending and descending gait generations of the robot.

Keywords: Dynamic modeling, Energy efficiency, Hexapod robot, Gradient terrain.

1 Introduction

Power consumption is one of the main restrictions to get autonomous walking robots. An autonomous walking robot cannot function satisfactorily with low energy efficiency, due to the fact that it has to carry all driving and control units in addition to payload and trunk body [1]. The minimization of power consumption plays an important role in the locomotion of an autonomous multi-legged robot used for service applications. Various approaches are available in the literature to obtain energy efficient gaits of multi-legged robots. Lapshin [2] proposed an energy efficient model of a walking machine and some results had been obtained from the gait-parameter standpoint alone. Orin and Oh [3] tried to resolve the foot force distribution for minimum energy consumption and load balance between several legs. They simplified the friction cone constraint to eliminate the associated nonlinearity by inscribing a pyramid within the desired friction cone. Because of this simplification, the solution was conservative. Nahon and Angeles [4] used quadratic programming to minimize power of robotic systems actuated by DC motors, but permitted power regeneration by motors doing negative work. Marhefka and Orin [5] utilized quadratic programming to solve foot force distribution in hexapod walking robots that minimizes the power

P. Vadakkepat et al. (Eds.): FIRA 2010, CCIS 103, pp. 138–145, 2010.

consumption in DC motors. In this work, gains from power regeneration by the DC motors are not permitted in the optimization problem. Kar et al. [6] performed an analysis of energy efficiency with respect to structural parameters, friction coefficient and duty factor of wave gaits, based on a simplified model of six-legged robot. Kar et al. [6] and Lin and Song [7] took the instantaneous power to be the product of instantaneous joint torques and joint velocities. Such modeling ignores the fact that a considerable amount of power is dissipated on the joints of the supporting legs. In order to eliminate such drawbacks, it is a better approach to consider the integral of the sum of the squares of the joint torques as a criterion of dissipated energy in the actuators. Erden et al. [8] utilized modified simplex method along with Lemke's Complementary Pivoting Algorithm to compute optimum foot force and torque distributions by considering a more practical locomotion performance objective, i.e., minimization of energy dissipation. Nishii [9] used the integral of the weighted sum of the product of instantaneous joint torques and joint velocities and the sum of the squares of the joint torques as the energetic cost and analyzed the energetic cost with respect to the duty factor and velocity in walking of a two joint six-legged robot. Zhoga [10] and Zelinski [11] analyzed energy expenditure and energy efficiency of multi-legged locomotion systems taking into account the leg dynamics and torque, but they failed to consider joint actuator type, although the joint actuator's contribution to energy consumption is decisive. Moreover, they focused their works on walking over level terrain.

In the present study, an attempt has been made to derive a detailed model based on both a dynamic model and an actuator model for hexapod walking robot and also, to minimize the power consumption during ascending and descending over gradient terrain.

2 Modeling of Hexapod Robot

To derive the detailed energy efficient model while moving through various gradient terrains, the following conditions are considered.

(a) The trunk body is held at a constant height with respect to sloping plane and kept parallel to the gradient terrain during locomotion.
(b) The trunk body moves at a constant velocity along a straight line over gradient terrain.
(c) The robot is assumed to describe a wave gaits with duty factor equals to 2/3 [1].
(d) The joint actuators are DC geared motors, which cannot store negative energy. Therefore, any negative energy, i.e., gain in energy supplied by external forces, is lost.

A complete kinematic and dynamic model of a realistic hexapod robot is required to analyze the complex relationships between locomotion parameters and power consumption.

2.1 Kinematic Model of the Hexapod Robot

Figure 1 shows a 3-D model of a six-legged robot considered in the present study. It consists of a trunk of rectangular shape and six legs, which are similar and symmetrically distributed around the trunk body on two sides. Each leg has three degrees of freedom and composed of three links connected by three rotary joints. The lengths of links 1 through 3 are 0.85 m, 0.115 m and 0.1 m, respectively.

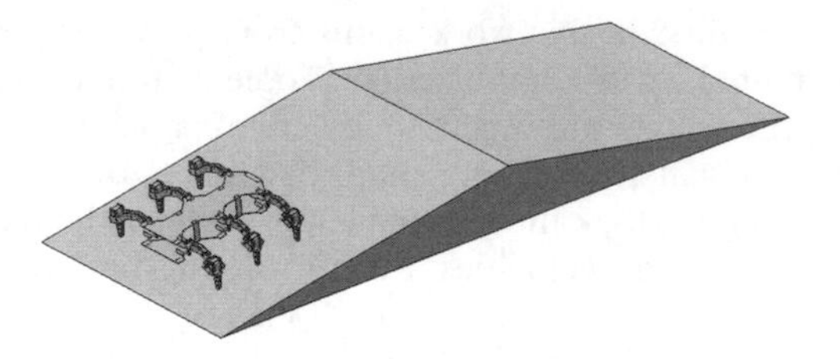

Fig. 1. 3-D model of a hexapod robot

To ensure a smooth functioning, each joint trajectory of the swing legs is assumed to follow a polynomial of fifth degree in time (t). The jth joint of a swing leg, that is, θ_j can be represented in fifth order polynomial as follows:

$$\theta_j = a_{j0}+a_{j1}t+a_{j2}t^2+a_{j3}t^3+a_{j4}t^4+a_{j5}t^5 \text{ ; } j=1, 2, 3, \tag{1}$$

where a_{j0}, a_{j1}, a_{j2}, a_{j3}, a_{j4}, and a_{j5} are coefficients. The boundary conditions of angular displacement and velocity at initial, middle and final point of the trajectory are applied to find these six coefficients for each joint. The velocity and acceleration equations of each joint of a swing leg can be obtained using the following equations:

$$\dot{\theta}_j = a_{j1}+2a_{j2}t+3a_{j3}t^2+4a_{j4}t^3+5a_{j5}t^4 \tag{2}$$

$$\ddot{\theta}_j = 2a_{j2}+6a_{j3}t+12a_{j4}t^2+20a_{j5}t^3 \tag{3}$$

Moreover, the velocity and acceleration equations of for each leg during the support phase can be expressed as follows:

$$\dot{\boldsymbol{\theta}} = \mathbf{J}^{-1}\dot{\mathbf{p}} \text{ and } \ddot{\boldsymbol{\theta}} = \mathbf{J}^{-1}(\ddot{\mathbf{p}} - \dot{\mathbf{J}}\dot{\boldsymbol{\theta}}), \tag{4}$$

where the position vector $\mathbf{p}=[p_x\ p_y\ p_z]^T$ and $\mathbf{J}$ is the Jacobian matrix [12].

2.2 Dynamic Model of the Hexapod Robot

For deriving the dynamic equations and finding joint torques' variations over the locomotion cycle, Lagrange-Euler formulation has been used. A systematic derivation of Lagrange-Euler equations yields a dynamic expression that can be written in the vector-matrix form as given below.

$$\boldsymbol{\tau}_i = [\mathbf{M}(\boldsymbol{\theta})\ddot{\boldsymbol{\theta}} + \mathbf{H}(\boldsymbol{\theta},\dot{\boldsymbol{\theta}}) + \mathbf{G}(\boldsymbol{\theta})]_i - \mathbf{J}_i^T\mathbf{F}_i, \tag{5}$$

where $\mathbf{M(\theta)}$ is the 3×3 mass matrix of the leg, $\mathbf{H}$ is a 3×1 vector of centrifugal and Coriolis terms, $\mathbf{G(\theta)}$ is a 3×1 vector of gravity terms, $\boldsymbol{\tau}_i$ is the 3×1 vector of joint torques and $\mathbf{F}_i$ is the 3×1 vector of ground reaction forces of foot 'i'. During the leg's swing phase, there is no foot-terrain interaction, and $\mathbf{F}_i$ becomes equal to zero. However, during the support phase, ground contact exists and equation (5) becomes undetermined, which has to be solved using an optimization criterion, e.g., optimal foot force distribution.

Optimum foot force distribution

For computing foot-force distributions, the following assumptions are made:

(i) The ground legs are assumed to be supporting the trunk body without any slippage on their tip points.
(ii) The contacts of the tip of the feet with ground can be modeled as hard point contacts with friction.
In the present study, the said problem of foot force distribution has been solved using two approaches as explained below.

Approach 1: Minimization of Norm of Foot Forces

Let us assume that $\mathbf{F}_i=[f_{ix}, f_{iy}, f_{iz}]^T$ is the ground-reaction force vector on foot i. The wrench $\mathbf{W}=[F_x, F_y, F_z, M_x, M_y, M_z]^T$ contains the forces (F_x, F_y, F_z) and moments (M_x, M_y, M_z) acting on the robot's center of gravity and represents the robot's payload, including the effect of surface gradient, any externally applied forces and inertial effects of the robot's body. However, the inertial effects of the legs have been neglected to simplify the study. Under these conditions, six equilibrium equations that balance forces and moments, when n numbers of legs are in their support phase, can be expressed in matrix form as follows:

$$[\mathbf{A}][\mathbf{F}] = -[\mathbf{B}][\mathbf{W}] \tag{6}$$

where

$$[\mathbf{A}] = \begin{bmatrix} \mathbf{I}_3 & \mathbf{I}_3 & \mathbf{I}_3 & \mathbf{I}_3 \\ \mathbf{R}_p & \mathbf{R}_q & \mathbf{R}_r & \mathbf{R}_s \end{bmatrix} \text{ and } \mathbf{B} = \begin{bmatrix} \mathbf{I}_3 & \mathbf{0}_3 \\ \mathbf{R}_c & \mathbf{I}_3 \end{bmatrix}$$

$\mathbf{I}_3$ is the (3×3) identity matrix, $\mathbf{0}_3$ is the (3×3) null matrix and $\mathbf{R}_i$ is the (3×3) skew symmetric matrix of vector $[x_i, y_i, z_i]^T$.

$$\mathbf{R}_i = \begin{bmatrix} 0 & -z_i & y_i \\ z_i & 0 & -x_i \\ -y_i & x_i & 0 \end{bmatrix} \text{ and } \mathbf{I}_3 = \begin{bmatrix} 1 & 0 & 0 \\ 0 & 1 & 0 \\ 0 & 0 & 1 \end{bmatrix}$$

This matrix (R_i) defines the position of tip of a foot i (i=p, q, r, s for wave gait with duty factor 2/3) or that of center of gravity (i=c) with respect to body reference frame. The coordinates of ith foot-ground contact point with respect to body reference frame, located at the body's geometric center, are denoted by (x_i, y_i, z_i). With the known feet positions, the feet forces during a whole locomotion cycle can be computed using equation (6), which is indeterminate, because it consists of six equations and nine unknowns. The solution of equation (6) has been obtained using the least squared method, which gives the minimum norm solution of the indeterminate equilibrium equations. In other words, it is the solution that minimizes the sum of the squares of components of foot forces.

Approach 2: Minimization of Norm of Joint Torques
In this approach, the equation (6) can be reformulated by using the following relations.

$$[\mathbf{F}] = [\mathbf{J}].[\boldsymbol{\tau}] \tag{7}$$

where $[\mathbf{J}]=\begin{bmatrix} {}^{p}\mathbf{J} & \mathbf{0}_3 & \mathbf{0}_3 & \mathbf{0}_3 \\ \mathbf{0}_3 & {}^{q}\mathbf{J} & \mathbf{0}_3 & \mathbf{0}_3 \\ \mathbf{0}_3 & \mathbf{0}_3 & {}^{r}\mathbf{J} & \mathbf{0}_3 \\ \mathbf{0}_3 & \mathbf{0}_3 & \mathbf{0}_3 & {}^{s}\mathbf{J} \end{bmatrix}$; ${}^{i}\mathbf{J}=\left[\mathbf{J}_i^T\right]^{-1}$; i=p, q, r, s and $\mathbf{J}_i$ is the (3×3) Jacobian matrix of leg i. Here, $[\boldsymbol{\tau}]=[\boldsymbol{\tau}_p, \boldsymbol{\tau}_q, \boldsymbol{\tau}_r, \boldsymbol{\tau}_s]^T$ and $\boldsymbol{\tau}_i=[\tau_{i1}, \tau_{i2}, \tau_{i3}]^T$ is the torque vector containing three joint torques at leg i.

The equation (6) can be rewritten as follows:

$$[\mathbf{A}][\mathbf{J}][\boldsymbol{\tau}] = -[\mathbf{B}][\mathbf{W}] \tag{8}$$

$$[\mathbf{A_J}][\boldsymbol{\tau}] = -[\mathbf{B}][\mathbf{W}] \tag{9}$$

The minimum norm solution of the above indeterminate equations has been obtained using a least squared method.

2.3 Power Consumption Model of the Hexapod Robot

The energy consumption in a legged robot is mainly the energy consumed by an actuator in every joint of legs. As joint is driven by a DC motor [9], the consumed energy in motor during a time (T) is given by:

$$E=\int_0^T u_a i\, dt = \int_0^T (u_e + R\, i)\, i\, dt = \int_0^T \tau\dot{\theta} dt + \int_0^T \gamma\tau^2 dt\ ; \tag{10}$$

where u_a is the applied voltage and i is the armature current.

The first term is mechanical energy and the second term is related to energy loss by heat emissions. Although a negative value for the first term, i.e., mechanical energy indicates a gain in energy supplied by external forces, DC motor cannot store this energy. Therefore, the energy consumed by the DC motor during time T is given by

$$E=\int_0^T \left[\Delta\left(\tau\dot{\theta}\right)\right] dt + \int_0^T \left(\gamma\tau^2\right) dt, \tag{11}$$

where $\Delta\left(\tau\dot{\theta}\right) = \begin{cases} \tau\dot{\theta} & \text{if } \tau\dot{\theta} > 0 \\ 0 & \text{if } \tau\dot{\theta} \leq 0 \end{cases}$

Total energy consumed by all motors in six-legged robot becomes

$$E=\int_0^T \sum_{i=1}^{6}\sum_{j=1}^{3}\left[\Delta\left(\tau_{ij}\dot{\theta}_{ij}\right) + \gamma\tau_{ij}^2\right] dt \tag{12}$$

where $\gamma=\frac{RG_s^2}{K_t^2}$; G_s is the speed ratio of the geared motor, K_t is the torque constant, R is the armature resistance, u_e is the induced voltage in the armature windings opposing the applied voltage.

3 Simulation Results and Discussion

In this section, simulation results related to energy-efficient locomotion of a six-legged robot over gradient terrain have been discussed in detail. Table 1 shows the average values of the squares of joint torques of the robot for wave gait with duty factor 2/3 at different gradient terrains. Results indicate that the average value of the squares of joint torques (a close indicator of dissipated power) during one complete locomotion cycle increases with gradient. The average value of the squares of joint torques of the robot as obtained by approach 1 is seen to be higher than that yielded by approach 2.

Table 1. Average values of the squares of joint torques during locomotion over gradient terrains

Terrain Slope	Average of the squares of joint torques $(N\text{-}m)^2$	
	Approach 1	Approach 2
5°	3.8105	1.5608
10°	4.1081	1.8629
15°	4.5941	2.3561
20°	5.2537	3.0255
Stroke = 0.1 m, Velocity = 0.03 m/s, Height of trunk body = 0.13 m		

Table 2. Variations of average power consumption and energy per unit weight per unit traveled length with velocity during ascending through different gradient terrains

Terrain Slope	Velocity (m/s)	Average power consumption (in Watts)		Energy per unit weight per unit traveled length	
		Approach 1	Approach 2	Approach 1	Approach 2
5°	0.020	0.1679	0.1319	0.2445	0.1921
	0.030	0.2042	0.1784	0.1983	0.1732
	0.040	0.2406	0.2249	0.1752	0.1638
	0.050	0.2770	0.2715	0.1613	0.1581
10°	0.020	0.2441	0.2033	0.3555	0.2961
	0.030	0.3149	0.2817	0.3057	0.2735
	0.040	0.3857	0.3602	0.2808	0.2622
	0.050	0.4565	0.4386	0.2659	0.2555
15°	0.020	0.3241	0.2815	0.4720	0.4099
	0.030	0.4288	0.3928	0.4162	0.3814
	0.040	0.5334	0.5042	0.3884	0.3671
	0.050	0.6381	0.6155	0.3717	0.3585
20°	0.020	0.4068	0.3628	0.5924	0.5284
	0.030	0.5446	0.5065	0.5287	0.4917
	0.040	0.6824	0.6501	0.4969	0.4734
	0.050	0.8202	0.7938	0.4778	0 .4624
Stroke = 0.1 m, Height of trunk body = 0.13 m					

Table 3. Variations of average power consumption and energy per unit weight per unit traveled length with velocity during descending through different gradient terrains

Terrain Slope	Velocity (m/s)	Average power consumption (in Watts)		Energy per unit weight per unit traveled length	
		Approach 1	Approach 2	Approach 1	Approach 2
5°	0.020	0.1082	0.0723	0.1576	0.1052
	0.030	0.1147	0.0889	0.1114	0.0863
	0.040	0.1212	0.1056	0.0883	0.0769
	0.050	0.1278	0.1223	0.0744	0.0712
10°	0.020	0.1252	0.0844	0.1824	0.1229
	0.030	0.1365	0.1034	0.1325	0.1003
	0.040	0.1478	0.1223	0.1076	0.0891
	0.050	0.1592	0.1413	0.0927	0.0823
15°	0.020	0.1469	0.1043	0.2139	0.1518
	0.030	0.1629	0.1269	0.1581	0.1232
	0.040	0.1789	0.1497	0.1303	0.1090
	0.050	0.1950	0.1724	0.1136	0.1004
20°	0.020	0.1726	0.1286	0.2514	0.1873
	0.030	0.1933	0.1551	0.1876	0.1506
	0.040	0.2140	0.1817	0.1558	0.1323
	0.050	0.2347	0.2082	0.1367	0.1213
Stroke = 0.1 m, Height of trunk body = 0.13 m					

It is interesting to note that a close similarity has been observed between these results for wave gait and that reported by Erden et al. [8]. Since the average of the squares of joint torques is considered to be proportional to average dissipated power of the joint motor, it can be concluded that approach 2 is more energy efficient than approach 1. Results related to the influences of velocity on average power consumption during ascending and descending through different sloping surfaces with wave gait of the robot are presented in Tables 2 and 3, respectively. The average power consumption is seen to increase with the angle of slope. The variations of average power consumption with the angle of slope have been compared for the ascending and descending cases. The average power consumption in ascending case is found to be more than that of the descending case, as expected. It happens due to the fact that the robot moves against the gravity in the former case. It shows that the velocity should be as low as possible to minimize power consumption. However, traveling with a low velocity takes more time to cover a fixed distance, and consequently, total energy consumption may be increased. The energy required to travel a fixed distance can be quantified using a parameter called specific resistance, that is, the energy consumed per unit weight and per unit traveled length. A close watch on these tables indicates that specific resistance decreases with the increase of velocity. This is in accordance with the observation of Marhefka and Orin [5].

4 Conclusions

In this study, an attempt has been made to minimize power consumption of a six-legged robot during locomotion over gradient terrains. A power consumption model has been derived for statically stable wave gaits by considering a minimization of dissipating power for optimal foot force distribution. Two approaches for optimal foot force distribution have been developed and their performances are tested on gait generation problems of a hexapod robot. The results of these two approaches are found to be in-line with those of some published literatures. It is important to mention that approach 2 is seen to be more energy efficient compared to approach 1. The effects of velocity on average power consumption and specific resistance during locomotion over different gradient terrains with wave gait have been studied. In order to minimize power consumption, the velocity should be as low as possible. However, the velocity is to be as high as possible to minimize the total energy consumption.

Acknowledgments. The first author acknowledges all helps from Mr. Ajay Kr. Singh, Department of Mechanical Engineering, NIT, Durgapur, India.

References

1. Song, S.M., Waldron, K.J.: Machines That Walk: The Adaptive Suspension Vehicle. The MIT Press, Cambridge (1989)
2. Lapshin, V.V.: Energy consumption of a walking machine: model estimations and optimization. In: 7th Conference on Advanced Robotics, pp. 420–425. San Feliu de Guixols (1995)
3. Orin, D.E., Oh, Y.: A mathematical approach to the problem of force distribution in locomotion and manipulation system containing closed kinematic chains. In: 3rd International CISM-IFToMM Symposium, Udine, Italy, pp. 1–23 (1978)
4. Nahon, M.A., Angeles, J.: Minimization of power losses in cooperating manipulators. ASME Journal of Dynamic, Systems, Measurement and Control 114, 213–219 (1992)
5. Marhefka, D.W., Orin, D.E.: Quadratic optimization of force distribution in walking machines. In: IEEE International Conference on Robotics and Automation, Belgium, pp. 477–483 (1998)
6. Kar, D.C., Issac, K.K., Jayarajan, K.: Minimum energy force distribution for a walking robot. Journal of Robotic Systems 18(2), 47–54 (2001)
7. Lin, B.S., Song, S.M.: Dynamic modeling, stability and energy efficiency of a quadrupedal walking machine. Journal of Robotic Systems 18(11), 657–670 (2001)
8. Erden, M.S., Leblebicioglu, K.: Torque distribution in a six-legged robot. IEEE Transactions on Robotics 23(1), 179–186 (2007)
9. Nishii, J.: Gait pattern and energetic cost in hexapods. In: 20th Annual International Conference of the IEEE Engineering in Medicine and Biology Society, vol. 20, pp. 2430–2433 (1998)
10. Zhoga, V.V.: Computation of walking robots movement energy expenditure. In: IEEE International Conference on Robotics and Automation, Belgium, pp. 163–168 (1998)
11. Zelinska, T.: Efficiency analysis in the design of walking machine. J. Theor. Appl. Mech. 38, 693–708 (2000)
12. Roy, S.S., Singh, A.K., Pratihar, D.K.: Analysis of six-legged walking robots. In: 14th National Conference on Machines and Mechanisms, India, pp. 259–265 (2009)

Dynamic Modeling and Optimal Foot Force Distribution of Quadruped Walking Robot

Abhishek Agarwal, Praveen Kumar Gautam, and Shibendu Shekhar Roy

Department of Mechanical Engineering, National Institute of Technology, Durgapur, West Bengal, India – 713209
abhishekagarwalnit@gmail.com, ssroy99@yahoo.com

Abstract. In the present paper, an attempt has been made to carry out kinematic and dynamic analysis of a quadruped walking robot. The direct and inverse kinematic analysis for each leg has been considered in order to develop an overall kinematic model of the robot, when it follows a straight path with two phase discontinuous gait. This study also aims to estimate optimal foot force distributions of quadruped robot, which is necessary for its real-time control. Three different formulations namely, tip-point force formulation, joint torque formulation and joint power formulation, have been developed. Simulation result shows that joint power formulation is more energy efficient foot force formulation than other two formulations.

Keywords: Dynamic modeling, kinematics, quadruped robot, foot force distribution, quadratic programming, discontinuous gait.

1 Introduction

A multi-legged robot possesses a tremendous potential for maneuverability over rough terrain, particularly in comparison to conventional wheeled or tracked mobile robot. It introduces more flexibility and terrain adaptability at the cost of low speed and increased control complexity [1]. In order to develop control algorithm of legged robots, it is important to have good models describing the dynamic behaviour of the complex multi-legged robotic mechanism. Barreto et al. [2] developed the free-body diagram method for kinematic and dynamic modeling of a multi-legged machine. Erden [3] investigated the dynamics of a hexapod walking robot based on Newton-Euler formulation. Koo and Yoon [4] obtained a mathematical model for quadruped walking robot to investigate the dynamics after considering all the inertial effects in the system. A dynamic model of walking machine was derived by Lin and Song [5] to study the dynamic stability and energy efficiency during walking. Pfeiffer et al. [6] investigated the dynamics of a stick insect walking on flat terrain. Freeman and Orin [7] developed an efficient dynamic simulation of a quadruped using a decoupled tree-structure approach. It has been observed that joint torque values largely depend on foot forces during support phase of the robot. For a statically stable quadruped robot, at least three legs should be on the ground at any instant. If a three-dimensional reaction force vector is considered on each ground leg, the foot force distribution problem

P. Vadakkepat et al. (Eds.): FIRA 2010, CCIS 103, pp. 146–153, 2010.

becomes indeterminate during the walking because of the closed chain system. Multiple solutions might exist, which can satisfy the force-moment balance criteria. To control the motion of the robot, the trunk body motion controller calculates the resultant control wrench (i.e., force and moment), that should be applied to the robot's body by its supporting legs. Therefore, one of the important issues of a legged robot's active force control is a successful distribution of its body force to the feet. The pseudo-inverse method had been utilized to obtain the foot force distribution of multi-legged walking robot by Gorinevsky and Shneider [8], Jiang et al. [9], Gonzalez de Santos [10]. This method could provide a solution of the indeterminate equilibrium equations, i.e., force and moment balance equation. However, it did not consider any other practical locomotion performance objectives, such as minimization of joint torques or minimization of energy consumption etc. In the present study, an attempt has been made in the said directions.

2 Kinematic Model of Three Joint Leg

Fig. 1 shows a 3-D model of a quadruped walking robot i.e., SILO4 [10] considered in the present study. Each leg has three degrees of freedom and is composed of three links connected by three rotary joints. The Denavit-Hartenberg (D-H) notations [11] have been used in kinematics of each leg (refer to Fig. 1).

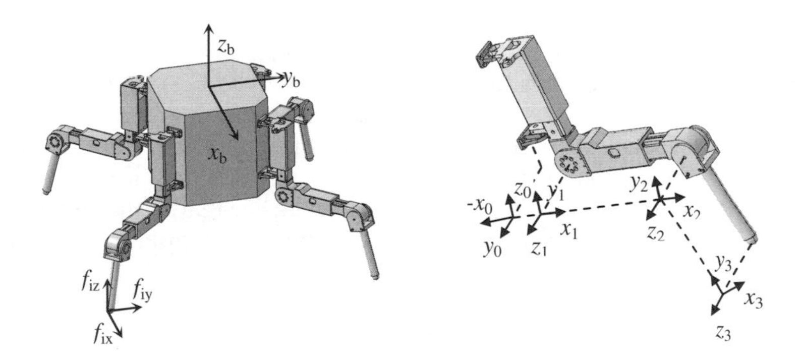

Fig. 1. Model of SILO4 and frame assignment of legs

Table 1 represents four D-H parameters, namely link length (a_i), link twist (α_i), joint distance (d_i), and joint angle (θ_i), required to completely describe the three joint legs.

Table 1. DH parameters for legs

i	d_i	θ_i	a_i	α_i
1	0	θ_1	a_1=0.06	$\pi/2$
2	0	θ_2	a_2=0.24	0
3	0	θ_3	a_3=0.24	0

The transformation matrix between foot tip reference frame {3} and leg or hip reference frame {0} is given as:

$$ {}^{0}\mathbf{T}_3 = {}^{0}\mathbf{T}_1\, {}^{1}\mathbf{T}_2\, {}^{2}\mathbf{T}_3 \tag{1} $$

$$ {}^{0}\mathbf{T}_3 = \begin{bmatrix} c\theta_1 c(\theta_2+\theta_3) & -c\theta_1 s(\theta_2+\theta_3) & s\theta_1 & (a_1+a_2c\theta_2+a_3c(\theta_2+\theta_3))c\theta_1 \\ s\theta_1 c(\theta_2+\theta_3) & -s\theta_1 s(\theta_2+\theta_3) & -c\theta_1 & (a_1+a_2c\theta_2+a_3c(\theta_2+\theta_3))s\theta_1 \\ s(\theta_2+\theta_3) & c(\theta_2+\theta_3) & 0 & a_2s\theta_2+a_3s(\theta_2+\theta_3) \\ 0 & 0 & 0 & 1 \end{bmatrix} $$

The position of the foot is given by the following expressions:

$$ [a_1+a_2\, c\theta_2+a_3\, c(\theta_2+\theta_3)]\, c\theta_1 = p_x \tag{2} $$

$$ [a_1+a_2\, c\theta_2+a_3\, c\, (\theta_2+\theta_3)]\, s\theta_1 = p_y \tag{3} $$

$$ a_2\, s\theta_2+a_3\, s\, (\theta_2+\theta_3) = p_z \tag{4} $$

By solving equations (**2**), (**3**) and (**4**), the joint angles (θ_1, θ_2 and θ_3) have been determined.

3 Dynamic Model of Quadruped Robot

The inverse dynamics based on Newton-Euler formulation [11] is used to calculate the joint torques, which requires foreknowledge of the foot forces. Since, the foot tip-point forces are unknown; the backward equations cannot be used to compute the joint torques. The contact between the foot tip and the ground is assumed to be hard-point contact with friction, which implies that the forces acting at the tip-point are restricted to three components, one normal and two tangential to the surface and no rotational torques act as an even terrain is considered. At any instant of time, considering the static equilibrium of the robot body:

$$ \sum_{i=1}^{n} \mathbf{F}_i + \mathbf{R} = \mathbf{0} \tag{5} $$

$$ \sum_{i=1}^{n} (\mathbf{r}_i \times \mathbf{F}_i) + \mathbf{T} = \mathbf{0} \tag{6} $$

where, **R** and **T** are the resultant external force and moment applied at the CoG of the trunk body. As there are no external forces and moments other than gravity, **R** reduces to $(0, 0, -m.g)^T$ and **T** reduces to null. $\mathbf{F_i}$ and $\mathbf{r_i}$ are foot force vectors and tip-point position vectors with respect to the body frame {*b*} respectively. In equation **(5)** and **(6)**, *n* denotes the number of supporting leg. Here, *n* is 3 when one of the leg is in swing phase and 4 when the robot is in body motion phase.

Equations **(5)** and **(6)** are written in a matrix form as follows:

$$ \mathbf{A.F=W} \tag{7} $$

When the robot legs are in swing, the number of supporting legs, *n*, is 3 such that **A** is 6×9 matrix, **F** is 9×1 column vector of unknown force components and **W** is the wrench column matrix $\mathbf{W}=(0\ 0\ -m.g\ 0\ 0\ 0)^T$. When the robot body is moving, the number of supporting legs, *n*, is 4 such that **A** is 6×12 matrix and **F** is 12×1 column vector. For preventing slippage of the feet another important inequality constraint which restricts the normal and the tangential foot force components, limited by the coefficient of static friction is used, which is, $\sqrt{F_{nx}^2+F_{ny}^2} \le \mu F_{nz}$, which is further linearized such that,

$$F_{nx}-(\mu/\sqrt{2})F_{nz} \le 0 \tag{8}$$

$$F_{ny}-(\mu/\sqrt{2})F_{nz} \le 0 \tag{9}$$

Also, since the foot cannot grasp the ground, the normal component of foot forces cannot be negative, which implies,

$$F_{nz} \ge 0 \tag{10}$$

3.1 Tip Point Force Formulation

The tip point force formulation is the pseudo inverse solution of equation **(7)**, which gives a resultant 9×1 or 12×1 column vector of the foot forces. The resultant force vector is,

$$\mathbf{F} = \mathbf{A}^T(\mathbf{A}\mathbf{A}^T)^{-1}\mathbf{W} \tag{11}$$

3.2 Joint Torque Formulation

In this formulation, the equation **(7)** can be reformulated by using the following relations:

$$\boldsymbol{\tau}_i = \mathbf{J}_i^T\mathbf{F}_i \tag{12}$$

$$\mathbf{F}_i = (\mathbf{J}_i^T)^{-1}\boldsymbol{\tau}_i \tag{13}$$

$$\mathbf{F} = \mathbf{J}.\boldsymbol{\tau} \tag{14}$$

When the legs are in protraction, the foot force component vector, **F** is a 9×1 column vector, **J** is 9×9 diagonal matrix whose diagonals are composed of $(\mathbf{J}_i^T)^{-1} \in 3\times3$ matrix, while $\boldsymbol{\tau}$ is a 9×1 column vector composed of $\boldsymbol{\tau}_i=\left(\tau_{i1} \quad \tau_{i2} \quad \tau_{i3}\right)^T$, where i corresponds to the legs supporting the robot. Similarly, during body motion phase of the robot, **F** is a 12×1 force column vector, **J** is 12×12 diagonal matrix, and, $\boldsymbol{\tau}$ is a 12×1 torque column vector.

The equation (7) can be rewritten as follows:

$$\mathbf{A}.\mathbf{J}.\boldsymbol{\tau}=\mathbf{W} \tag{15}$$

$$\mathbf{A}_t.\boldsymbol{\tau}=\mathbf{W} \tag{16}$$

The minimum norm solution of above in determinant equations has been obtained by using Pseudo-inverse. The solution can be written in matrix form as follows:

$$\boldsymbol{\tau} = \mathbf{A}_t^T(\mathbf{A}_t\mathbf{A}_t^T)^{-1}\mathbf{W} \tag{17}$$

3.3 Joint Power Formulation

The problem is to find the foot force distribution for minimum energy consumption, where the formulation includes: the force and moment balance equations, which are represented by equation **(7)** (equality constraint), the inequality constraints represented by equations **(8)** and **(9)**, the non-negativity constraint by equation **(10)**, and the energy consumption equation (objective function).

The energy consumption model is the one that has been formulated by Orin and Oh, and is given by the equation,

$$P_{inst} = \sum_{\substack{n=1\\j=1}}^{\substack{n=4\\j=3}} P_{nj} = \sum_{\substack{n=1\\j=1}}^{\substack{n=4\\j=3}} (\tau_{nj}\dot{\theta}_{nj} + \gamma_j\tau_{nj}^2) \tag{18}$$

The quadratic equation **(18)** is a function of the unknown foot forces, where the torque τ_{nj} is computed using the dynamic model of the robot system with symbolic unknown foot forces, $\dot{\theta}_{nj}$ is the known angular velocity and γ_j is actuator parameter. The first term is the mechanical power and the second term is the energy dissipation.

The quadratic programming problem for minimum energy consumption model can be stated by the equations,

$$\min\ P_{inst}(F) = c^T F + \frac{1}{2} F^T H F \tag{19}$$

subject to, $AF=W$, and, $A_{ineq}F \le B_{ineq}$ (20)

where,
c = 9×1 or 12×1 column vector of linear weights of P_{inst}
H = 9×9 or 12×12 positive definite weighting matrix of P_{inst}
A_{ineq} = 9×9 or 12×12 inequality constraint matrix,
B_{ineq} = 9×1 or 12×1 inequality constraint vector

Equation **(20)** is formulated using the equations **(8), (9)** and **(10)**.

4 Results and Discussion

In this section, simulation results of the proposed three approaches have been discussed in detail. The cycle time, leg stroke of two phase discontinuous gait, body height, maximum velocity and acceleration of trunk body are assumed to be equal to 24 sec, 0.20 m, 0.41 m, 0.028 m/sec and 0.055 m/sq. sec, respectively.

Fig. 2, 3 and 4 show the distributions of foot forces yielded by tip-point force formulation, joint torque formulation and joint power formulation, respectively over one locomotion cycle. It shows that the front and rear legs complement each other in force, such that the sum of vertical forces of all the ground legs at any given instant of time becomes equal to the weight of the robot. Tip-point force formulation has yielded the forces with either zero or almost zero horizontal components during the phase of constant velocity of the trunk body; therefore, the robot does not make a good use of the friction. However, in joint torque formulation and joint power formulation, horizontal components of the foot forces are found to be significant. These results are quite similar to that reported by Erden and Leblebicioglu [3].

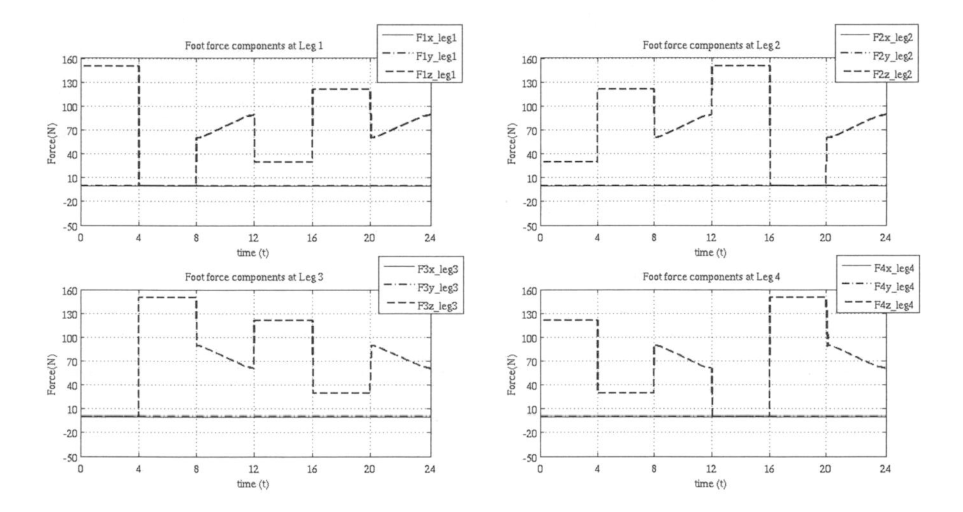

Fig. 2. Foot force distribution for legs 1, 2, 3 and 4 using the tip point force formulation

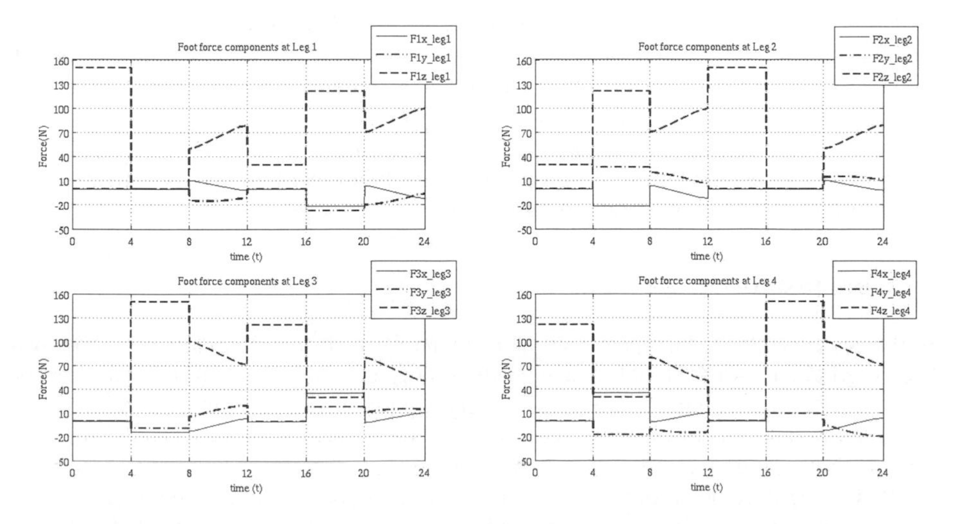

Fig. 3. Foot force distribution for legs 1, 2, 3 and 4 using the joint torque formulation

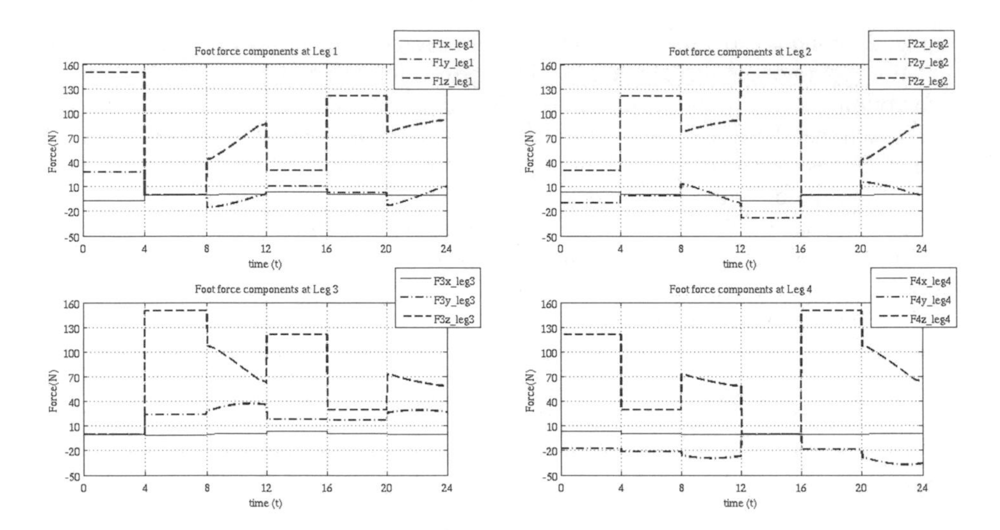

Fig. 4. Foot force distribution for legs 1, 2, 3 and 4 using the joint power formulation

Table 2. Average torque square obtained by three formulations

$\overline{\tau}_j^2$	Tip point Force formulation	Joint Torque formulation	Joint Power formulation
$\overline{\tau}_1^2$	.0245	178.72	5.96
$\overline{\tau}_2^2$	1768.02	2010.63	1117.16
$\overline{\tau}_3^2$	122.71	61.77	178.43
SUM	1890.7545	2251.12	1301.55

Table 2 shows the average values of the squares of joint torques of the quadruped robot obtained by three approaches. The average value of the sum of squares of joint torques of the robot as obtained by joint power formulation is seen to be lesser than that yielded by other two formulations. Since the average of the squares of joint torques is considered to be proportional to average dissipated power of the joint motor, it can be concluded that by joint power formulation is more energy efficient foot force formulation than other two formulations.

5 Conclusion

In this study, both the kinematic as well as dynamic analysis of a quadruped robot has been carried out. An attempt has been made in present study to obtain optimal distributions of foot forces. Three approaches, namely, tip point force formulation, joint torque formulation, joint power formulation have been developed. It is important to mention that quadratic programming is seen to be more energy efficient compared to other two methods. The developed kinematic and dynamic models have been examined for two phase discontinuous gait of the quadruped robot. This work can be extended to tackle the problems related to continuous and non-periodic gait of the walking robot.

Acknowledgments. Authors acknowledge all helps from Mr. Aniket G. Patil, Department of Mechanical Engineering, NIT, Durgapur, India.

References

1. Song, S.M., Waldron, K.J.: Machines That Walk: The Adaptive Suspension Vehicle. The MIT Press, Cambridge (1989)
2. Barreto, J.P., Trigo, A., Menezes, P., Dias, J., de Almeida, A.T.: FBD-The free body diagram method. Kinematic and dynamic modeling of a six leg robot. In: IEEE International Conference on Robotics and Automation, pp. 423–428 (1998)
3. Erden, M.S., Leblebicioglu, K.: Torque distribution in a six-legged robot. IEEE Transaction on Robotics 23(1), 179–186 (2007)
4. Koo, T.W., Yoon, Y.S.: Dynamic instant gait stability measure for quadruped walking. Robotica 17, 59–70 (1999)
5. Lin, B.S., Song, S.M.: Dynamic modeling, stability and energy efficiency of a quadrupedal walking machine. In: IEEE International Conference on Robotics and Automation, pp. 367–373 (1993)
6. Pfeiffer, F., Weidemann, H.J., Danowski, P.: Dynamics of walking stick insect. In: Proceedings of IEEE International Conference on Robotics and Automations, vol. 2, pp. 1458–1463 (1987)
7. Freeman, S., Orin, E.: Efficient dynamic simulation of a quadruped using a decoupled tree-structure approach. The International Journal of Robotics Research 10(6), 619–627 (1991)
8. Gorinevsky, D.M., Shneider, A.Y.: Force control in locomotion of legged vehicles over rigid and soft surfaces. The International Journal of Robotics Research 9(2), 4–23 (1990)
9. Jiang, W.Y., Liu, A.M., Howard, D.: Foot-force distribution in legged robots. In: Proceedings of 4th International Conference in Climbing and Walking Robots, Karlsruhe, Germany, pp. 331–338 (2001)
10. Santos, P.G., Estremera, J., Garcia, E.: Optimizing leg distribution around the body in walking robots. In: Proceedings of IEEE International Conference on Robotics and Automation, Barcelona, Spain, pp. 3207–3212 (2005)
11. Fu, K.S., Gonzalez, R.C., Lee, C.S.G.: Robotics: Control, Sensing, Vision, and, Intelligence. McGraw-Hill, Singapore (1987)

Determination of Optimally Stable Posture for Force Actuator Based Articulated Suspension for Rough Terrain Mobility

Vijay Eathakota, Arun Kumar Singh, Srikant Kolachalam, and K. Madhava Krishna

International Institute of Information Technology, Hyderabad
vijay@research.iiit.ac.in, aks1812@gmail.com,
mkrishna@iiit.ac.in

Abstract. In this paper we develop a novel framework for determining the optimal posture of a rough terrain vehicle. The framework has been applied to a linear force actuator based articulated suspension vehicle. A complete 3D generic framework has been developed for the analysis of the vehicle and the criteria for the optimal posture has been derived from the developed equations of the motion of the vehicle using the zero moment criteria. Extensive simulations are done to show the stability of the vehicle with optimal posture while navigating over rough terrains.

Keywords: Rough Terrain Linear Force Actuator, Optimal Posture, Zero Moment Criteria.

1 Introduction

To improve the mobility of wheeled robots traversing on an uneven terrain having slopes in all three orthogonal directions while maintaining an optimally stable posture is the primary focus of our research. Past research on all Terrain vehicles led to the development of active and passive suspension vehicle. Passive suspension vehicle like SHRIMP [1], Rocky Rover [2] adapts passively by the virtue of the wheel ground contact forces of the underlying terrain. However such systems do not have the ability to control the contact forces at the wheel ground contact points and recover from overturn failure. To overcome these problem active suspension vehicles like HYLOS [3] were developed which uses actuators to control the internal configuration of the vehicle. To govern the choice of the internal configuration, numerous stability metrics have been developed in the past decade. For example vehicle such as HYLOS, SRR [4] uses the force angle stability margin originally proposed by Papadopoulos and Rey [5] to maintain a sub-optimal posture of the vehicle which maintains an equal pre-defined stability margin from all the tip-over axis. Force angle stability measure computes the minimum angle between the resultant force and the tip over axis. This metric in its original form assumes low velocities for the vehicle. An extension of the above metric can be found in [6] which includes the acceleration effects by subjecting the vehicle to a pseudo force. Another measure of stability applied primarily to gait planning of legged robots is the ZMP criterion [7]. The ZMP in its original form is

P. Vadakkepat et al. (Eds.): FIRA 2010, CCIS 103, pp. 154–161, 2010.

insensitive to the height of the centre of mass of the vehicle and thus proves ineffective for a generic wheeled mobile robot in which the height of the chassis can vary.

The posture of the vehicle derived from the above stability margins has significant influence on other parameters of the vehicle like actuator force requirements and velocity of the vehicle. But these effects are completely neglected while computing the stable posture of the vehicle. In this paper we deduce a stable posture of the vehicle considering the all the inertial and gravitational forces acting on the system which optimizes the actuator requirements of the vehicle and hence we refer the derived posture of the vehicle as the optimal posture of the vehicle. We compare the performance in terms of actuator requirements for the vehicle moving with a suboptimal and optimal posture. Extensive uneven terrain simulations are performed to show the efficacy of the proposed mechanism. The rest of the paper is organized as follows. In section 2 we derive the equations of motion for a linear force actuator vehicle. Section 3 describes the framework for the deriving the optimal posture of the vehicle. Section 4 presents the results comparing the performance of the vehicle with the optimal and suboptimal posture.

2 Analysis of the Linear Force Actuator Vehicle (LFA-V)

In this section we develop the mechanical model of the LFA-V. The authors in the previous work [8],[9],[10] have shown that this design can be used to actively control the wheel ground contact forces. As shown in Fig.1 the mechanical structure consists of 4 wheels each pinned to an outer slide link which is connected to an inner slide link through a prismatic joint. The inner slide is fixed to the chassis. To achieve a desired value of the contact force at the ith wheel we propose a force control mechanism in which the prismatic joint is actuated through a linear actuator mounted on the chassis to which a required force $\overline{F}_i^A$ can be commanded. This force acts between the main body and the output slide. Although the input and output slides have finite mass we consider them to be negligible when compared to the mass of the chassis and hence neglect them in our analysis.

Assumption 1: Quasi-static analysis is done on the system assuming the masses of legs and wheels to be negligible when compared to the mass of the chassis.

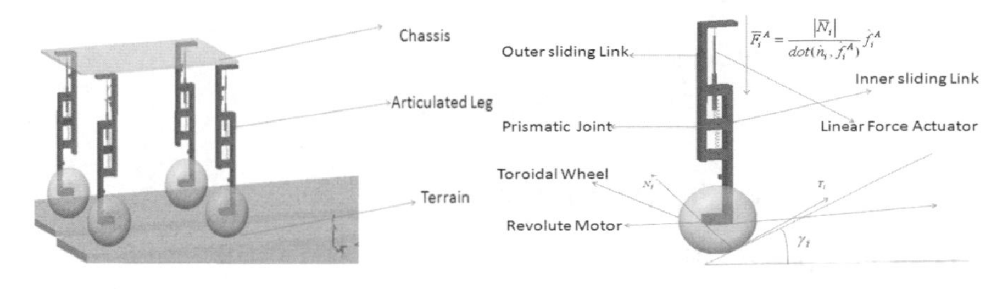

Fig. 1. LFA-V chassis and articulated leg

Force control of the LFA-V is achieved based on D'Alembert's principle of inertial forces. As shown in Figs 1 and 2 the actuator force acting through the prismatic joint causes a force on the chassis and accelerates it upwards. The accelerating chassis can be converted to an equivalent static chassis by adding a pseudo-actor force in the downward direction and this in turn increases the reaction normal forces.

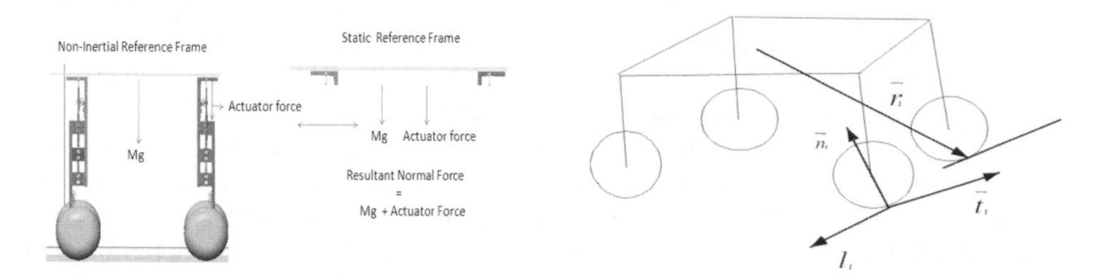

Fig. 2a. D'Alambert's principle of inertial forces **Fig. 2b.** contact of a generic robot

Hence there are three forces acting on the ith wheel. They are $\overline{N}_i, \overline{T}_i$ and the actuator force $\overline{F}_i^{\,A}$.

$\overline{F}_i^{\,A}$ is always perpendicular to the chassis and is given by

$$\overline{F}_i^{\,A} = \frac{|\overline{N}_i|}{dot(\hat{n}_i, \hat{f}_i^{\,A})}\hat{f}_i^{\,A} \quad \text{.Where} \quad \hat{f}_i^{\,A} = R.[0 \;\; 0 \;\; 1]^T \tag{1}$$

Where R is the rotation matrix describing the vehicle reference frame with respect to the chassis reference frame. $\widehat{n_i}$ is the unit force normal vector at the ith wheel as shown in figure 2 b.

The quasi-static equations of motion were derived in [10] and has not been included here for the lack of space. The final expression takes the following form

$$\overline{F} = \sum_{i=1}^{4} (\overline{F}_i^{\,A} + \overline{T}_{neti} + \overline{N}_{neti}) \tag{2}$$

$$\overline{M} = \sum_{i=1}^{4} [(\overline{r}_i \times \overline{F}_i^{\,A}) + \overline{r}_i \times (\overline{T}_{neti} + \overline{N}_{neti})] \tag{3}$$

which can be compactly written as

$$B.C = D \tag{4}$$

Where

$C = [|\overline{T}_1|\,|\overline{N}_1|\;|\overline{T}_2|\,|\overline{N}_2|\;|\overline{T}_3|\,|\overline{N}_3|\;|\overline{T}_4|\,|\overline{N}_4|\,]^T$

$D = [F_x \; F_y \; F_z \; M_x \; M_y \; M_z]^T$

$$B=\begin{bmatrix} t_{netx1} & n_{netx1} & t_{netx2} & n_{netx2} & t_{netx3} & n_{netx3} & t_{netx4} & n_{netx4} \\ t_{nety1} & n_{nety1} & t_{nety2} & n_{nety2} & t_{nety3} & n_{nety4} & t_{nety4} & n_{nety4} \\ t_{netz1} & n_{netz1} & t_{netz2} & n_{netz2} & t_{netz3} & n_{netz3} & t_{netz4} & n_{netz4} \\ m_{tx1} & m_{nx1} & m_{tx2} & m_{nx2} & m_{tx3} & m_{nx3} & m_{tx4} & m_{nx4} \\ m_{ty1} & m_{ny1} & m_{ty2} & m_{ny2} & m_{ty3} & m_{ny3} & m_{ty4} & m_{ny4} \\ m_{tz1} & m_{nz1} & m_{tz2} & m_{nz2} & m_{tz3} & m_{nz3} & m_{tz4} & m_{nz4} \end{bmatrix}$$

3 Determination of Optimally Stable Posture

A suboptimal posture is defined in [3] based on pitch = roll = 0 which cites computational complexity to be a reason for not searching for the angles. In our work here we go beyond pitch, roll = 0 and search for their values which will give lesser motor torques and actuator forces. This is done by the zero moment criteria and for notational simplicity we call it as an optimally stable posture, optimal because it reduces or optimizes forces and still maintains stability. This can be seen by the fact that roll values obtained are still close to zero. Hence the vehicles stability along the roll axis is almost same as [3] and pitch is chosen such that the basic criteria of stability such as energy stability margin [12] and tip-over stability margin are always satisfied. The framework for the optimal posture is derived by equating the moment expression obtained from the quasi-static to zero. These moment equations are highly nonlinear and coupled and hence in our analysis we approximate them to get into cubic polynomial expressions based on Taylor series expansion of trigonometric function. This linearization is justified since the Euler angles all have very low values throughout the simulation as shown later The polynomial expressions are in terms of Euler angles. Solving for the roots of the polynomial gives the Euler angles of the chassis responsible for the optimal posture. This method if computationally efficient and can be applied to real time systems. Let ψ_s, ro_s and yo_s be the Euler angles (pitch, roll, yaw) for the optimally stable posture. Since the vehicle doesn't incorporate steering and lateral slipping is assumed to be negligible we have $T_{xi} \approx 0$ we have $yo_s = 0$ and $|M_z| \approx 0$.

Assumption-2: We assumed $|\psi_s|$ and $|ro_s|$ to be small in magnitude.

Assumption-3: The terrain 1 shown in Fig 3, were modeled such that every point on the surface has a finite and unique gradient in any three orthogonal directions and also the slope at any point i.e. $|\gamma_i| < 1.3^c, \left(1.3^c \approx 75^o\right)$.

By the above assumptions we have $|\psi_s . \tan\gamma_i| < 1$ Therefore we developed an expression for M_x in terms of ψ_s by assuming

$$\sin\psi_s = \psi_s - \frac{\psi^3_s}{3},\ \cos\psi_s = 1 - \frac{\psi^2_s}{2}$$

and $ro_s \approx 0$.

Let $M_x = \sum_{k=0}^{3} A_k.\psi_s^k$, where (5)

$$A_0 = \sum_{i=1}^{4}\left[\left|\overline{T_i}\right|\left(r\cos^2\gamma_i + l_i\cos\gamma_i\right) - \left|\overline{N_i}\right|\left(\frac{r\sin 2\gamma_i}{2} + l_i\sin\gamma_i\right) + \frac{(-1)^{i+1}\left|\overline{N_i}\right|a}{\cos\gamma_i}\right]$$

$$A_1 = \sum_{i=1}^{4}\left[\left|\overline{T_i}\right|\left(r\sin 2\gamma_i + l_i\sin\gamma_i\right) + \left|\overline{N_i}\right|\left(r\left(1-2\sin^2\gamma_i\right) + l_i\cos\gamma_i\right) + \frac{(-1)^{i}\left|\overline{N_i}\right|a\tan\gamma_i}{\cos\gamma_i}\right]$$

$$A_2 = \sum_{i=1}^{4}\left[\left|\overline{T_i}\right|\left(r\sin^2\gamma_i - \frac{l_i\cos\gamma_i}{2}\right) + \left|\overline{N_i}\right|\left(\frac{l_i\sin\gamma_i + r\sin 2\gamma_i}{2}\right) + \frac{(-1)^{i+1}\left|\overline{N_i}\right|a\tan^2\gamma_i}{\cos\gamma_i}\right]$$

$$A_3 = \sum_{i=1}^{4}\left[-\frac{\left(\left|\overline{T_i}\right|l_i\sin\gamma_i + \left|\overline{N_i}\right|l_i\cos\gamma_i\right)}{6} + \frac{(-1)^{i}\left|\overline{N_i}\right|a\tan^3\gamma_i}{\cos\gamma_i}\right]$$ Let $b = \frac{n_{yi}}{n_{zi}}, c = \frac{n_{xi}}{n_{zi}}$

and $D_i = \frac{\left|\overline{N_i}\right|(2.5-i)w}{n_{zi}\left|2.5-i\right|}$.

By *assumption 2* we have $\left|b\psi_s - cro_s\right| < 1$ since $\left|b\right| < 1$ and $\left|c\right| < 1$. Now similarly an expression for M_y was derived in terms of ro_s and ψ_s.

We get $M_y = \sum_{k=0}^{3} B_k.ro_s^k$, where (6)

$$B_0 = \sum_{i=1}^{4} D_i\left(\sum_{j=0}^{3}(b\psi_s)^j\right),\ B_1 = -\sum_{i=1}^{4} D_i\left(3b^2c\psi_s^2 + 2bc\psi_s + b\right),\quad B_2 = \sum_{i=1}^{4} D_i\left(3bc^2\psi_s + c^2\right)$$

and $$B_3 = -\sum_{i=1}^{4} D_i c^3.$$

The higher order terms in the above expressions are neglected due to assumption-2. In order to satisfy condition 1, we solved the equation $M_x = 0$ and obtained ψ_s. This value is inputted in (6) and by solving $M_y = 0$, we obtained ro_s. The above equations are cubic polynomials and may have more than one real root. So the roots which have minimum magnitude were taken to be optimally stable Euler angles in order to satisfy *assumption-2*.

The vehicle is controlled by a PD controller with control law derived in [10].

4 Results and Discussions

Simulations were performed using MATLAB, Simulink and MSC Visual Nastran. The plots of actuator forces, traction forces, and velocity of the chassis in Y direction and Euler angles of the chassis for both the terrains were shown in Fig's 4-9. The plots corresponding to the optimally stable posture determined in section 3 are shown in Fig's 4,6,8 and Fig's 5,7,9 were plotted for $\psi_s = ro_s = yo_s = 0$ .The controllers were applied to maintain V_d =0.5m/s and h_d =0.42m by assuming M =9.31 Kg,

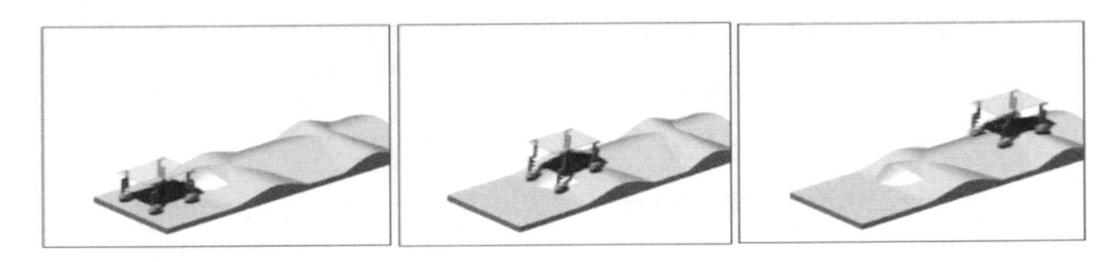

Fig. 3. LFA-V negotiating terrain -1

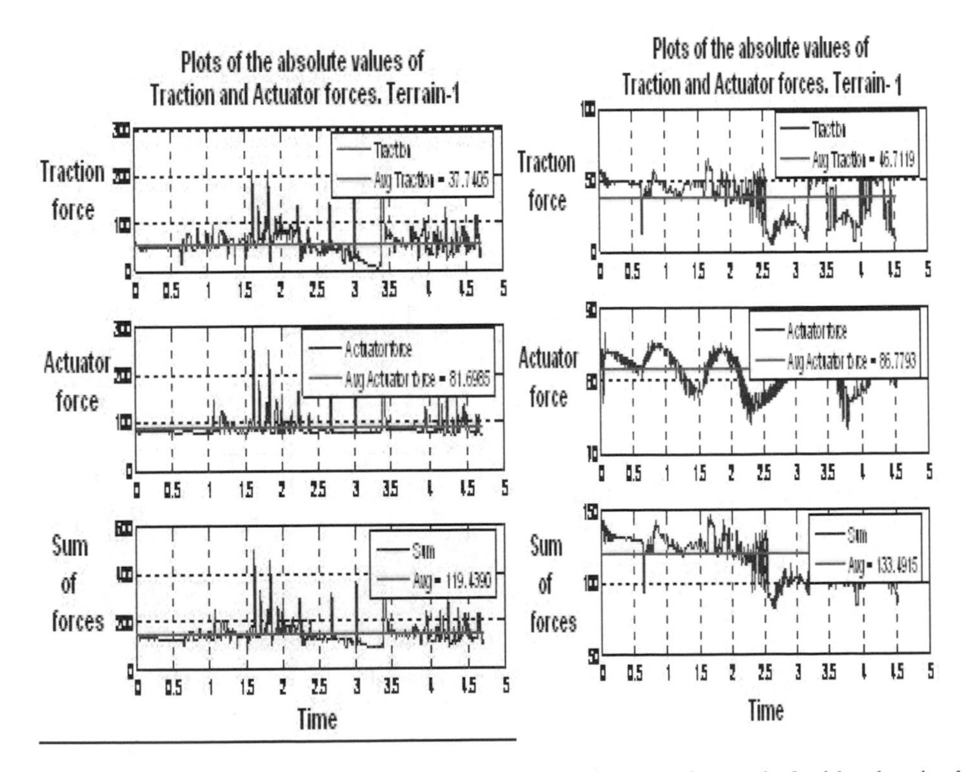

Fig. 4. Forces for terrain-1 with Optimal Posture

Fig. 5. Forces for terrain-2 with suboptimal posture

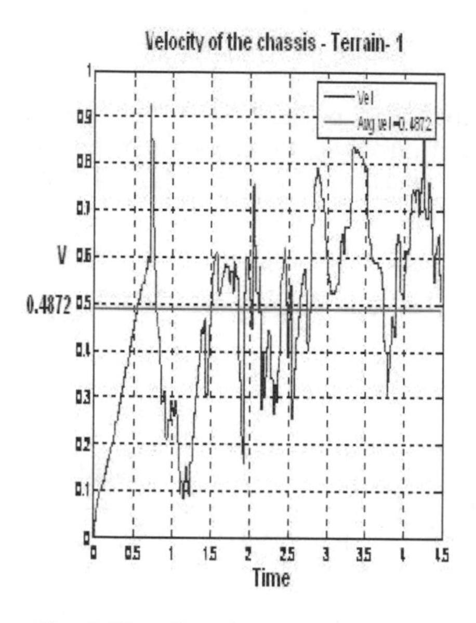

Fig. 6. Plot of V with optimal Posture

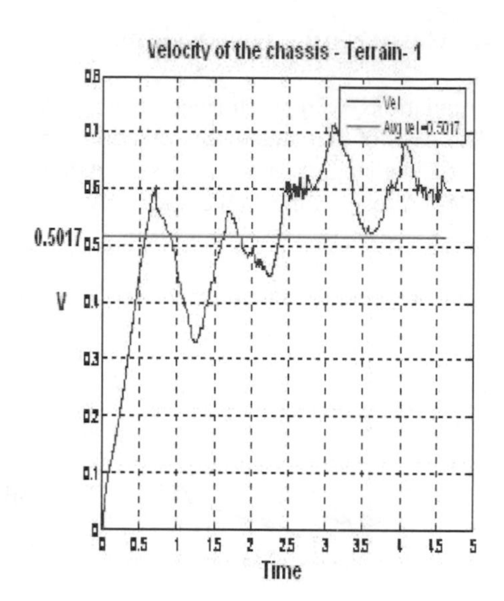

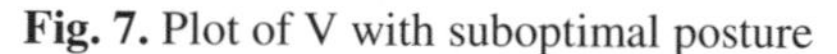

Fig. 7. Plot of V with suboptimal posture

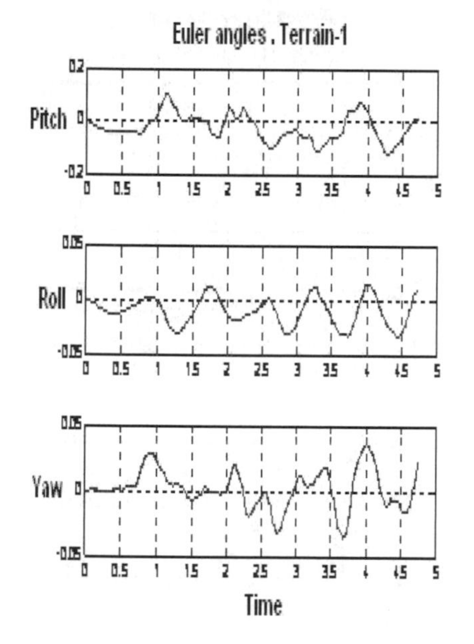

Fig. 8. Plot of optimal Euler angles - terrain-1

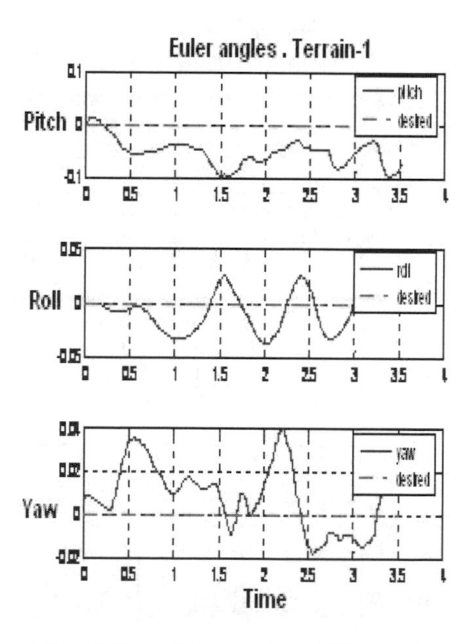

Fig. 9. Plot of sub- optimal Euler angles - terrain-1

a =0.3m, w =0.2m, $r_t = 0.0125$ m and r =0.05m. From figures 4and 5 we can see that the average traction force and linear actuator force requirement in the case of optimal posture are 37.74 N and 81.70 N respectively as compared to the requirement of 45.71 N and 86.77 N for the suboptimal posture. Hence it can be seen that there has been a reduction of around 5N in the actuator force and an average reduction of 9N in the traction force and since $\Gamma_i \approx \left|\overline{T}_i\right| r$ we can say that it will result in the reduction of motor torques as well. But the reduction in the velocity is around .0145m/s which is very less. So we are saving in terms of energy needed to drive the linear actuators and also the wheel motors torques and still being able to maintain almost the same velocity of the vehicle. Similar results were obtained for various other terrains that are not reported for conciseness. The reductions in forces may appear small but are significant considering the light weight of the vehicle.

References

[1] Estier, T., Crausaz, Y., Merminod, B., Lauria, M., Piguet, R., Siegwart, R.: An innovative space rover with extended climbing abilities. In: Proc. Int. Conf. Robotics in Challenging Environments, Albuquerque, USA (2000)

[2] Volpe, R., Balaram, J., Ohm, T., Ivlev, R.: Rocky 7: A next generation mars rover prototype. J. Advanced Robotics 11(4), 341–358 (1997)

[3] Grand, C.H., BenAmar, F., Plumet, F., Bidaud, P.: Stability and traction optimization of a reconfigurable wheel-legged robot. Int. J. Robotics Research (October 2004)

[4] Iagnemma, K., Rzepniewski, A., Dubowsky, S., Schenker, P.: Control of Robotic Vehicles with Actively Articulated Suspensions in Rough Terrain. Autonomous Robots 14(1), 5–16 (2003)

[5] Papadopoulos, E.G., Rey, D.A.: A new measure of tipover stability margin for mobile manipulators. In: Proc. of the IEEE Int. Conf. on Robotics and Automation, pp. 3111–3116 (1996)

[6] Peters, S.C., Iagnemma, K.: An analysis of roll over stability measurement for high speed mobile robots. In: Proc. of International Conference on Robotics and Automation (ICRA 2006), Orlando Florida, pp. 3711–3716 (2006)

[7] Sugano, S., Huang, Q., Kato, I.: Stability criteria in controlling mobile robotic systems. In: Proc. of the IEEE/RSJ Int. Conf. on Intelligent Robots and Systems, pp. 832–838 (1993)

[8] Eathakota, V.P., Kolachalama, S., Singh, A.K., Madhava Krishna, K.: Force actuator based suspension vehicle for rough terrain mobility. In: IEEE- International Symposium on Measurements and Control in Robotics (ISMCR 2008), Bangalore, India (2008)

[9] Sanan, S., Singh, S., Madhava Krishna, K.: Controlling an Actively Articulated Suspension Vehicle for Mobility on rough terrain. In: CLAWAR 2007 (2007)

[10] Sanan, S., Rao, N., Madhava Krishna, K., Singh, S.: On improving the Mobility of Vehicles on uneven terrain. In: Proceedings of ICAR 2007 (2007)

[11] Craig, J.J.: Introduction to Robotics – Mechanics and Control, 3rd edn. Prentice-Hall, Englewood Cliffs

[12] Messuri, D.A., Klien, C.A.: Automatic body regulation for maintaining stability of legged vehicle during rough terrain locomotion. IEEE International Journal of Robotics and Automation

Multi-Agent Rendezvous Algorithm with Rectilinear Decision Domain

Kaushik Das* and Debasish Ghose

GCDSL, Dept. of AE,
Indian Institute of Science,
Bangalore, India
{kaushikdas,dghose}@aero.iisc.ernet.in

Abstract. The aim of this paper is to develop a computationally efficient decentralized rendezvous algorithm for a group of autonomous agents. The algorithm generalizes the notion of sensor domain and decision domain of agents to enable implementation of simple computational algorithms. Specifically, the algorithm proposed in this paper uses a rectilinear decision domain (RDD) as against the circular decision domain assumed in earlier work. Because of this, the computational complexity of the algorithm reduces considerably and, when compared to the standard Ando's algorithm available in the literature, the RDD algorithm shows very significant improvement in convergence time performance. Analytical results to prove convergence and supporting simulation results are presented in the paper.

Keywords: Multi-agent, Rendezvous, Decision domain, Consensus.

1 Introduction

Research on multiple agents in the context of robotics is motivated by the fact that instead of using a highly sophisticated and expensive automated agent, it may be advantageous to use a group of small, simple, and relatively cheap autonomous agents (mobile robots or UAVs).

The autonomous control of multi agents has emerged as a challenging problem. The agents are assumed to have limited sensor and communication range and execute some local rule-based strategy depending on the information collected by each agent from the environment and from neighboring agents. One of the generic tasks that such a system of agent is often called upon to perform is to physically bring all the agents to a common point. This is called a multi-agent rendezvous problem [1]. This problem is important because if rendezvous is feasible, then more general formations are also achievable [2]. Previous work in this area are by Ando et al. [3] and Lin et al. [4] where each agent moves toward the rendezvous point by performing a sequence of "stop-and-go" moves. The *stop* mode is basically the *sensing period* and is an interval of fixed length. In the *go* mode the agents will maneuver in an interval of variable length and will move from its current position to a new position. In [3], as also in our paper, the sensing period

* The authors gratefully acknowledge the AOARD/AFOSR for their grant to this project.

P. Vadakkepat et al. (Eds.): FIRA 2010, CCIS 103, pp. 162–169, 2010.

is assumed to be zero. Both [3] and [4] use algorithms that require determination of the smallest circle that contains a given set of agent positions. These algorithms are called "circumcenter algorithms". Although the complexity of this algorithm is proved to be subexponential of order $O(ne^{(2+o(1))\sqrt{2\ln n}})$ [5], the number of actual computations is fairly high. In our paper we generalize the notion of sensing domain and decision domain and show that by using a rectilinear decision domain the computations can be simplified considerably, thus bringing down the convergence time.

We show that our algorithm is far superior in terms of computational time than Ando's algorithm [3] which is the standard algorithm in the literature. we consider point robots with simple kinematics and instantaneous directional motion as in [3], [4].

2 Preliminaries

Let $R = \{a_1, a_2, \ldots, a_n\}$ be the set of robots or agents. The positions of agent a_i is given by $p_i = (x_i, y_i) \in \mathbb{R}^2$. The sensor domain of a agent a_i is denoted as S_i and its decision domain is denoted by D_i, where $D_i \subset S_i$. Information sensed from the decision domain is used to implement the algorithm. Essentially, we introduce the concept that information from the whole of the sensor domain need not be used for decision-making. In Fig. 1a, we give a schematic of these concepts. Note that, in general, p_i need not be inside S_i. An agent determines its set of neighbouring agents based on D_i. In this paper we assume that the sensing domain (S_i) of all the agents is circular with radius r. The decision domain (D_i) is a square of side $2d$, with $d < \frac{r}{\sqrt{2}}$, aligned with a pre-specified global (X,Y) reference frame. This is shown in Fig. 1b. The set of neighbors of agent a_i is defined as $N_i = \{a_j \mid (|x_i - x_j|) \leq d \text{ and } (|y_i - y_j|) \leq d\}$.

Note that an agent is also its own neighbor, so $a_i \in N_i$. Also, if $a_j \in N_i$ then $a_i \in N_j$. In Ando et al. [3] S_i and D_i are the same and are circles.

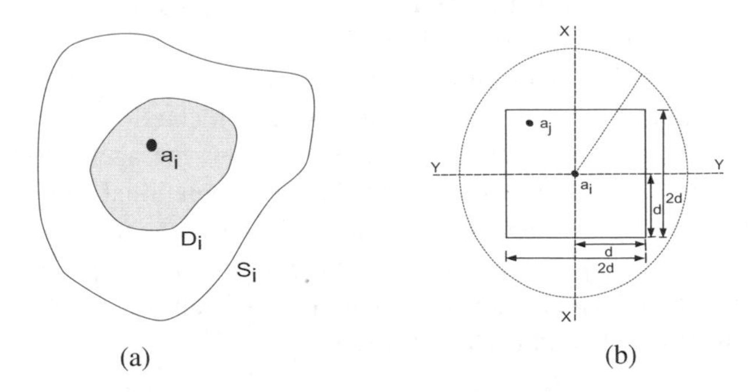

Fig. 1. (a) General sensor domain and decision domain (b) The sensor domain is circular and the decision domain is a square aligned with the global $X-Y$ coordinates

3 Rectilinear Decision Domain (RDD) Algorithm

Convergence to a rendezvous point in [3] is proved through two properties: (i) Agents who are neighbours remain as neighbours (ii) Agents come closer with each other in

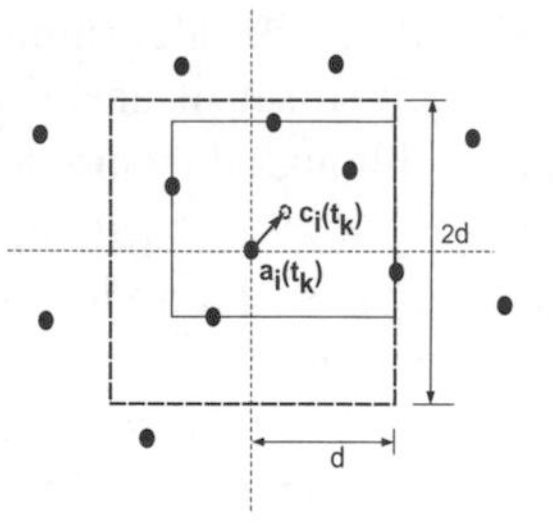

Fig. 2. Agent a_i will move to the centroid of the rectangle. The square with broken lines is the decision domain of agent a_i and agents inside it belong to N_i. The solid rectangle is the smallest rectangle that contains the agents in N_i.

some sense till they meet at a point. The RDD (rectilinear decision domain) algorithm retains these two ideas. Again as in [3] the RDD algorithm has the assumption that the initial graph is connected.

Algorithm RDD *(Rectilinear Decision Domain)*

Step 1: Each agent a_i determines its neighbour set N_i using its decision domain D_i.
Step 2: Each agent constructs the smallest rectangle, aligned with the global coordinate axes, that contains all the members of its neighbour set.
Step 3: Each agent computes the centroid of the rectangle and moves to it.

Fig. 2 illustrates these steps where $c_i(t_k)$ is the centroid of the rectangle at the time instant t_k. These steps are similar to Ando's algorithm [3], but for a few significant differences. In Step 1, Ando's algorithm determines neighbours using the sensor domain S_i. In Steps 2 and 3, Ando's algorithm computes the circumcenter of the neighbours and moves *toward* it subject to a constraint. Unlike RDD which allows the agents to move directly to the centroid, Anod's algorithm may not allow the agents to reach the centroid. These two important differences lead to high computational complexity, and thus higher convergence time, in Ando's algorithm. In RDD, an agent a_i uses the information $P_i = \{(x_j, y_j) | a_j \in N_i\}$ where $a_j \in N_i$, and computes $\max\{x_j\}$, $\min\{x_j\}$, $\max\{y_j\}$ and $\min\{y_j\}$ to obtain the rectangle. The computational complexity of this operation is $O(n)$. We will now state two important theorems.

Theorem 1. *An agent's movement will be confined to a square of side d centered at the agent's current position and aligned with the global reference frame.*

Proof. Consider the maximum deviation of the centroid of the rectangular area along the X-axis from the agent's current position. This will be less than $d/2$ because the maximum deviation of a neighbor's position along the X-axis is d. Similar arguments hold for the Y-axis. So the agent movement will be confined within a square of side d, centered at the agent's current position. □

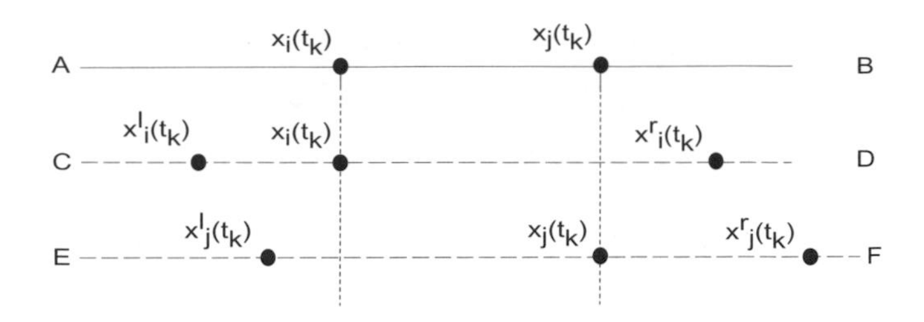

Fig. 3. Agent a_i,a_j and its neighbor projection along X-axis

Theorem 2. *If at any time t_k, agents a_i and a_j are neighbors, then they will be neighbors for all time $t > t_k$.*

Proof. Let the projections of agents a_i and a_j on the X axis at time t_k be denoted by $X_i(t_k)$ and $X_j(t_k)$, respectively, where $X_j(t_k) > X_i(t_k)$. Let the left most and right most projections of the neighbors of agent a_i be $X_i^l(t_k)$ and $X_i^r(t_k)$, respectively. The position of agents a_i and a_j on the X axis at time instant t_{k+1} is then $X_p(t_{k+1}) = \frac{X_p^l(t_k)+X_p^r(t_k)}{2}$; where $p = i, j$. The distance between the agents a_i and a_j along the X axis at time instant t_{k+1} is $\mid X_j(t_{k+1}) - X_i(t_{k+1}) \mid = \frac{1}{2} \mid (X_j^l(t_k) - X_i^l(t_k)) + (X_j^r(t_k) - X_i^r(t_k)) \mid$.

We can show that $\mid X_j^l(t_k) - X_i^l(t_k) \mid \leq d$, since when $X_i^l(t_k) \leq X_j^l(t_k) \leq X_i(t_k)$, we can write $0 \leq X_j^l(t_k) - X_i^l(t_k) \leq X_i(t_k) - X_i^l(t_k) \leq d$. Again, when $X_j^l(t_k) \leq X_i^l(t_k) \leq X_i(t_k)$, we can write $X_j^l(t_k) - X_i^l(t_k) \leq 0 \leq X_i(t_k) - X_i^l(t_k) \leq d$. Similarly, we can prove $\mid X_i^r(t_k) - X_j^r(t_k) \mid \leq d$. This implies that $X_j(t_{k+1}) - X_i(t_{k+1}) \leq d$. We can prove the same along the Y axis too. □

The theorem proves that connectivity between agents, which plays an important role in the performance of the algorithm, is preserved.

4 Analysis of the RDD Algorithm

Let the global convex hull made by the positions of the agents at the time instant t_k be denoted by $Co(t_k)$. We can define the diameter of the convex hull at the time instant t_k as $dia(Co(t_k)) = max\{\| p_i(t_k) - p_j(t_k) \|\},\ \ i, j \in \{1, 2, \ldots, n\}$. When rendezvous is achieved the diameter of the global convex hull is zero. We will first show that the diameter of the global convex hull will reduce at each step.

Let us consider $Co_i(t_k)$ as the convex hull made by the neighbor set of the agent a_i at the time instant t_k. Let the smallest rectangle containing the neighbor set of agent a_i, and aligned along the global coordinate axes, at time instant t_k be $R_i(t_k)$. It is obvious that $Co_i(t_k) \subset R_i(t_k)$. Let $y^i_{min} = \min\{y_j\}$, where y_j is the $y-$th coordinate of $a_j \in N_i$. Other variables are similarly defined. Then we have the following result.

$$mid(R_i(t_k)) = (1/2)((x^i_{min} + x^i_{max}), (y^i_{min} + y^i_{max})) \tag{1}$$

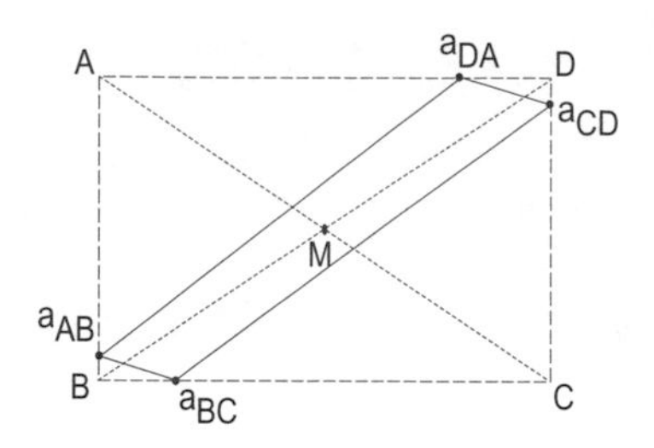

Fig. 4. Rectangle made by the neighbor of agent a_i

Theorem 3. *mid* $(R_i(t_k)) \in Co_i(t_k)$ *and mid* $(R_i(t_k))$ *is not an corner point of* $Co_i(t_k)$

Proof. Consider Fig. 4. Here $R_i(t_k)$ is the rectangle $ABCD$ and mid point of it is M. Each side must have an agent. Let N_{AB} be the set of agents on side AB. Similarly, N_{BC}, N_{CD} and N_{DA} are the set of agents on the side BC, CD and DA, respectively. It is clear that $N_{AB} \cap N_{CD} = \emptyset$ and $N_{BC} \cap N_{DA} = \emptyset$. Choose agents a_{AB}, a_{BC}, a_{CD} and a_{AB} in such a way that $a_{AB} \in N_{AB}$, $a_{BC} \in N_{BC}$, $a_{CD} \in N_{CD}$ and $a_{DA} \in N_{DA}$. The line joining a_{AB} and a_{BC} will separate B and M. Similarly, the line joining a_{BC} and a_{CD}will separate B and M, the line joining a_{CD} and a_{DA} will separate D and M and line joining a_{DA} and a_{AB} will separate A and M. So, we can write $M \in Co\langle a_{AB}, a_{BC}, a_{CD}, a_{DA}\rangle$. As $Co\langle a_{AB}, a_{BC}, a_{CD}, a_{DA}\rangle \subset Co_i(t_k)$, then we can write $M \in Co_i(t_k)$. Let, the coordinate of the mid $(R_i(t_k))$ is (x^i_{mid}, y^i_{mid}). If the rectangle is not a point then it is evident $x^i_{mid} \neq x^i_{min}$, $x^i_{mid} \neq x^i_{max}$, $y^i_{mid} \neq y^i_{min}$ and $y^i_{mid} \neq y^i_{max}$. This observation and $M \in Co_i(t_k)$implies that M can not be a corner point of $Co_i(t_k)$. □

Now consider any agent a_i. Let the maximum distance along the X-axis on the right side between agent a_i and its neighbors be d^i_{rx} and on the left side is d^i_{lx}. Similarly, along the Y-axis the maximum distance above a_i is d^i_{ay} and below of a_i is d^i_{by}.The position of the agent a_i at time instant (t_{k+1}) will be

$$x_i(t_{k+1}) = x_i(t_k) - (d^i_{lx} - d^i_{rx})/2; \;\; y_i(t_{k+1}) = y_i(t_k) - (d^i_{by} - d^i_{ay})/2 \tag{2}$$

The movement of the agent a_i will depends upon $\{d^i_{lx}, d^i_{rx}, d^i_{ay}, d^i_{by}\}$. The agent a_i will be stationary if $d^i_{lx} = d^i_{rx}$ and $d^i_{ay} = d^i_{by}$. Next we will state a theorem that agents at the corner points global convex hull $Co(t_k)$ cannot remain stationary.

Theorem 4. *For any agent* a_i *which is at the corner of* $Co(t_k)$ *and has at least one non-located neighbor, both* $d^i_{lx} = d^i_{rx}$ *and* $d^i_{ay} = d^i_{by}$ *cannot be satisfied.*

Proof. We will prove this by contradiction. Let $d^i_{lx} = d^i_{rx}$ and $d^i_{ay} = d^i_{by}$ for an agent a_i which is at a corner point. Let the agent a_i be at corner point A in Fig. 5. The neighbors a_j and a_k for which $d^i_{lx} = d^i_{rx}$ and $d^i_{ay} = d^i_{by}$ is at B and C, respectively. From Fig. 5, we can write $d^i_{rx} = d_1 \sin\theta_1$, $d^i_{lx} = d_2 \sin\theta_2$, $d^i_{by} = d_1 \cos\theta_1$ and $d^i_{ay} = d_2 \cos\theta_2$. If $d^i_{rx} = d^i_{lx}$ and $d^i_{ay} = d^i_{ay}$,

$$d_1/d_2 = \sin\theta_2/\sin\theta_1 = \cos\theta_2/\cos\theta_1 \tag{3}$$

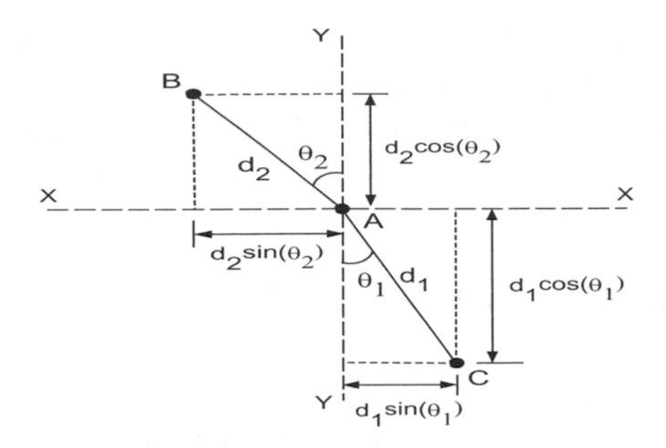

Fig. 5. Agent a_i is at the corner point of $Co(t_k)$

Equating the above two equations $\theta_2 = \theta_1$. Thus, A will be on the line BC. So, A is not a corner point. This leads to the contradiction. □

From Theorem 4 and Theorem 3, it is clear that agents at the corner of the convex will move either inside of the convex hull or move along the edges of the convex hull.

Theorem 5. *No agent can be at a corner point of the convex hull $Co(t_k)$ at time t_{k+1}.*

Proof. We prove this by contradiction. Suppose an agent a_i reaches the corner point c_1 of the convex hull $Co(t_k)$ at time t_{k+1}. This implies that c_1 is the mid point of the rectangle made by the neighbors of agent a_i at time t_k. Then, according to Theorem 3, c_1 is inside the convex hull $Co_i(t_k)$. Again $Co_i(t_k) \subseteq Co(t_k)$. This implies that c_1 is the convex combination of some points of $Co(t_k)$. This leads to the contradiction. □

Theorem 6. $Co(t_{k+1}) \subset Co(t_k)$

Proof. Let there be m corner points of the convex hull $Co(t_k)$ given by $P_c(t_k) = \{p_1, \dots, p_m\}$ and m_1 corner points of the convex hull $Co(t_{k+1})$ given by $P_c(t_{k+1}) = \{\widehat{p}_1, \dots, \widehat{p}_{m_1}\}$. Now, $P_c(t_k) \cap P_c(t_{k+1}) = \emptyset$, because no agent can reach at the corner of the convex hull $Co(t_k)$ according to lemma 5 at time instant t_{k+1}. The corner point agents of $Co(t_k)$ must move (according to Theorem 4). Since $\widehat{p}_i \in Co(t_k)$ and $p_i \notin Co(t_{k+1})$, for all $p_i \in P_c(t_k)$, we have $Co(t_{k+1}) \subset Co(t_k)$. □

Next we will show that the diameter of the global convex hull will reduce at each step.

Theorem 7. $dia(Co(t_{k+1}) < dia(Co(t_k))$

Proof. As before, let $P_c(t_k) = \{p_1, \dots, p_m\}$ be the set of corner points of $Co(t_k)$ and $P_c(t_{k+1}) = \{\widehat{p}_1, \dots, \widehat{p}_{m_1}\}$ be the set of corner points of $Co(t_{k+1})$. According to the definition of the diameter of a convex hull[7]

$$dia(Co(t_k)) = \max\{\|p_i - p_j\|, \forall p_i, p_j \in P_c(t_k)\} \tag{4}$$

$$dia(Co(t_{k+1})) = \max\{\|\widehat{p}_i - \widehat{p}_j\|, \forall \widehat{p}_i, \widehat{p}_j \in P_c(t_{k+1})\} \tag{5}$$

Again $\|p_i - p_j\| < dia(Co(t_k))$ for all $p_i, p_j \in Co(t_k)$ and $p_i, p_j \notin P_c(t_k)$. Here $P_c(t_{k+1}) \in Co(t_k)$ and $P_c(t_{k+1}) \cap P_c(t_k) = \emptyset$. From this, we get $dia(Co(t_{k+1})) < dia(Co(t_k))$. □

The above theorem tells us that the sequence of convex hulls, generated by the positions of the agents will make a descending chain of convex sets. Under similar arguments as Ando et al.[3] it can be shown that the convex hull will become a point as all the agents converge to a point. We omit the proof due to space limitations.

5 Simulation Results and Implementation Issues

In Fig. 6, four snapshots for 10 agents have been shown. One can see that all the agents eventually converge to a point. The system converges when the maximum distance between the agents along the X axis and Y axis is less than the decision domain distance (d), since the agents would converge to a single point in the very next step.

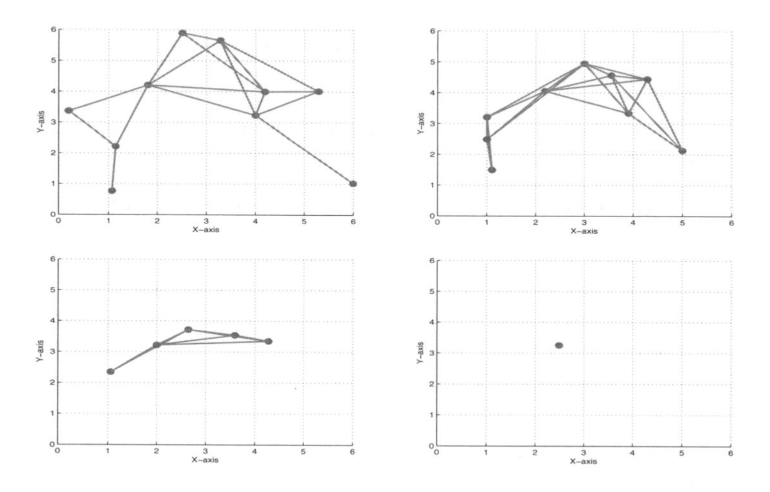

Fig. 6. 10 agents placed randomly converge to a point

The comparative study between Ando's algorithm and the RDD algorithm is executed. In Table 1, the comparison of computational time to converge is given. The results shows that RDD algorithm is superior in terms of computational time.

In Table 2 comparison of the number of iterations to converge is given. The iteration number in case of RDD algorithm is a bit higher than Ando's algorithm. The reason behind for this is the decision domain in case of RDD algorithm is smaller than Ando's algorithm.

The algorithm can be implemented by any agent (robot) that can gather information about and from its neighbors (positions and orientations) using sonar sensors, vision sensors or laser scanners. After generating the control command the agents can broadcast the control command with positions and orientation of the centroid information by a short range communication mechanism. An agent receiving a broadcast control command from another agent which is not in its decision domain will ignore the command.

Table 1. Comparison of Computational Time

Time to Converge (sec)			
Number of Agents	Ando	RDD	Ando/RDD
20	0.5856s	0.0329	17.8065
50	3.0677s	0.1758	17.4544
100	12.1486	0.5439	22.3360
150	40.7101	1.7395	23.4031

Table 2. Comparison of Number of Iterations

Number of Iterations to Converge				
Number of Agents	Ando	RDD	Ando (Time/Iteration)	RDD (Time/Iteration)
20	3.8000	5.2667	0.1541	0.0062
50	6.8667	9.3333	0.4468	0.0188
100	8.4667	11.8667	1.4349	0.0458
150	11.3333	15.9333	3.5921	0.1092

6 Conclusions

We presented and analyzed a rendezvous algorithm considering a rectilinear decision domain. The computational complexity of RDD algorithm is low compared to the well established Ando's algorithm in the literature. The RDD algorithm is simpler in terms of few computation needed and in relaxing the restriction on the movement of the agents to the centroid of the rectangle. In this sense, it is a purely centroidal algorithm. The simulation results support our claim.

References

1. Kranakis, E., Krizanc, D., Rajsbaum, S.: Mobile Agent Rendezvous: A Survey. In: Flocchini, P., Gąsieniec, L. (eds.) SIROCCO 2006. LNCS, vol. 4056, pp. 1–9. Springer, Heidelberg (2006)
2. Lin, Z., Broucke, M., Francis, B.: Local Control Strategies for Groups of Mobile Autonomous Agents. IEEE Transactions on Automatic Control 49, 622–628 (2004)
3. Ando, H., Oasa, Y., Suzuki, I., Yamashita, M.: Distributed Memoryless Point Convergence for Mobile Robots with Limited Visibility. IEEE Transactions on Robotics and Automation 15, 818–828 (1999)
4. Lin, J., Morse, A.S., Anderson, B.D.O.: The Multi-Agent Rendezvous problem. In: Conference on Decesion & Control, Maui, Hawaii, USA, pp. 1508–1513 (2003)
5. Gartner, G.: A subexponential algorithm for abstarct optimization problems. SIAM Journal Computers 24(5), 1018–1035 (1995)
6. Das, K., Ghose, D.: Positional Consensus in Multi-Agent Systems using a Broadcast Control Mechanism. In: American Control Conference, St. louis, Missouri, USA, pp. 5731–5736 (2009)
7. Rockafellar, R.T.: Convex Analysis. Princeton University Press, Princeton (1972)

Task and Role Selection Strategy for Multi-robot Cooperation in Robot Soccer

Dong-Hyun Lee, Ki-In Na, and Jong-Hwan Kim

Department of Electrical Engineering, KAIST, 355 Gwahangno, Yuseong-gu, Daejeon, Republic of Korea

Abstract. Robot soccer is played in a dynamic environment. All robots in a team should select their task dynamically and adaptively in a cooperative manner. This paper proposes a framework for task and role selection strategy for distributed multi-robot coordination and its application to robot soccer. To deal with such a dynamic environment, context-based cooperation method is provided. The robot team has a task strategy and a task strategy ratio that can change the weight of the tasks, based on environmental changes. Using information on task strategy ratio and environmental changes, each robot in a team can effectively select its task and role. Computer simulations are carried out to demonstrate the effectiveness of the proposed framework for task and role selection for multi-robot cooperation.

1 Introduction

There are many complex problems such as such reconnaissance, hazardous cleanup, automated construction, exploration in outdoor environments, etc., which may not be solved by a single robot. Multi-robot coordination is one of solutions to solve such problems. Robot team can solve such problems more effectively and robustly with available redundant resources. For example, robot team can complete given tasks, even if one of the robots in the team malfunctions or goes out of order. Robot team can perform tasks more efficiently by utilizing the resources of each robot to meet specific mission capabilities.

To perform complex tasks, it is required for robot team to have robustness, scalability, information and resource sharing strategy and efficient task allocation method. The robot soccer is one of dynamical and adversarial applications in real world. The ball moves unpredictably and the opponent team robots act aggressively to catch and shoot the ball. In this adversarial and rapidly changing environment, planning and scheduling are less effective and they might lead the team to a worse situation. Therefore, in this domain, tasks and roles should be selected and performed dynamically in real time.

Considering this problem, in this paper, a framework for task and role selection strategy is proposed for multi-robot coordination. In this framework, there are two key modules such as multi-robot context module and environment context module. The former is used for sharing the information among robots in a team and the latter is provided to collect information on environmental changes. The

P. Vadakkepat et al. (Eds.): FIRA 2010, CCIS 103, pp. 170–177, 2010.

task strategy and its ratio are used to change the allocated number of robots for each task. The effectiveness of the proposed framework is demonstrated by computer simulations.

The remainder of this paper is organized as follows. Section 2 introduces the research related to multi-robot coordination. Section 3 defines the terms used in the multi-robot coordination and describes the perception, context module and task strategy. Section 4 presents the simulation results. Finally, conclusions follow in Section 5.

2 Related Research

There have been many researches about multi-robot coordination such as multi-robot architecture, resource sharing, task allocation, etc. ALLIANCE architecture is a behavior-based multi-robot architecture [1]. Each robot in the architecture has two internal states, impatience and acquiescence. If one robot performs a certain task poorly, the impatience of the other robots increases and if it goes over a certain threshold, the one which has the highest impatience takes over the task. In the same manner, if the robot recognizes that it can not perform a task well, its acquiescence increases and if it goes over a certain threshold, the robot abandons its task. Each behavior has behavior motivation and the robot selects the behavior which has the highest motivation. PAB (Port-Arbitrated Behavior) is extended versions of subsumption architecture for multi-robot coordination system [2]. PAB is a behavior-based architecture and consists of cross inhibition and cross suppression for behavior selection.

As research on multi-robot resource sharing, ASyMTRe (Automated Synthesis of Multi-robot Task solutions through software Reconfiguration) and the market-based approach were proposed. In ASyMTRe, each robot consists of different kind of schemas such as communication schema, motor schema and perceptual schema [3][4]. It coordinates those schemas for robots to perform a task with a tightly-coupled manner. In the market-based approach, any robot can call an auction for selling tasks and bid on the task [5]-[7]. Any robot can call an auction for selling tasks and bid on the task. Each robot calculates its bidding value based on its resources and state.

MCMRA (Motivation and Context-based Multi-Robot Architecture) is a multi-robot coordination architecture for dynamic task, role and behavior selections [8]. It employs the motivation of task, the utility of role, a probabilistic behavior selection and a team strategy for efficient multi-robot coordination. The proposed algorithm in this paper is based on MCMRA and applied to FIRA robot soccer [9].

3 Task and Role Selection Strategy

3.1 Definition

This section introduces the terms used in this paper. Robot team can perform different kinds of tasks, where each task consists of different types of roles. In robot soccer, robot can select either an offense task ($task_1$) or a defense task ($task_2$). The roles for $task_1$ and the role for $task_2$ are defined as

$$\begin{aligned} Roles\ for\ task_1 &= \{Striker, Fwd, Centerwing, \\ &\quad Leftwing, Rightwing\} \\ Roles\ for\ task_2 &= \{Goalkeeper, Sweeper, Centerback, \\ &\quad Leftback, Rightback\}. \end{aligned}$$

Resources represent hardware features of robots. Based on the resources, each robot has different preferences for tasks. In robot soccer, a robot team is comprised of offensive and defensive robots, which have high preference on offense and defense, respectively. The resources of the robots are defined as

$$Res = [res_1, res_2, res_3], \qquad (0 \le res_i \le 1)$$

where res_1, res_2 and res_3 represent the normalized maximum velocity, maximum torque and frame size, respectively. The preference of robots is defined as

$$Pref = [pref_1, pref_2], \qquad (0 \le pref_i \le 1)$$

where $pref_1$ and $pref_2$ represent the preference of the robot on $task_1$ and $task_2$, respectively. For example, the resources of the offensive and defensive robots are respectively defined as

$$\begin{aligned} \text{Resource of the offensive robot} : Res &= [0.7, 0.4, 0.4] \\ \text{Resource of the defensive robot} : Res &= [0.4, 0.7, 0.7]. \end{aligned}$$

Using this, preference for each task is described as follow:

$$\begin{gathered} \text{Preference of the offensive robot for tasks :} \\ Pref = [0.7, 0.3] \\ \text{Preference of the offensive robot for tasks :} \\ Pref = [0.3, 0.7]. \end{gathered}$$

3.2 Environment Context Module

Environment context module is used to detect environmental events or particular conditions. In robot soccer, fifteen environment contexts are defined as in Table 1. For example, $e_{c8}(t)$ is defined as score difference between the home team and the opponent team is divided by maximum allowable score difference, as follows:

$$e_{c8}(t) = \frac{S_D}{S_{DMax}} \tag{1}$$

with

$$S_D = \begin{cases} S_H - S_O & \text{if } |S_H - S_O| < S_{DMax} \\ S_{DMax} & \text{if } |S_H - S_O| < S_{DMax} \\ -S_{DMax} & \text{otherwise.} \end{cases} \tag{2}$$

where S_H and S_O are scores of the home team and the opponent team, and S_{DMax} is the maximum allowable score difference which is set to 3 in this paper. If the home team is winning, $e_{c8}(t)$ is a positive value and otherwise, it is a negative value or zero.

Table 1. Environmental context

$E_C(t)$	Meaning
$e_{c1}(t)$	The home team possesses the ball
$e_{c2}(t)$	The opponent team possesses the ball
$e_{c3}(t)$	The robot possesses the ball oneself
$e_{c4}(t)$	There is a home team robot aside
$e_{c5}(t)$	There is a opponent team robot aside
$e_{c6}(t)$	Approaching goal post of the opponent team
$e_{c7}(t)$	Time ratio $(0 \leq e_{c7}(t) \leq 1)$
$e_{c8}(t)$	Score difference ratio $(-1 \leq e_{c8}(t) \leq 1)$
$e_{c9}(t)$	$e_{c5}(t) = 1 \cap e_{c6}(t) = 0$
$e_{c10}(t)$	$e_{c1}(t) = 0 \cap e_{c2}(t) = 0$
$e_{c11}(t)$	$e_{c1}(t) = 1 \cap e_{c3}(t) = 0$
$e_{c12}(t)$	$e_{c3}(t) = 1 \cap e_{c4}(t) = 0 \cap e_{c5}(t) = 1$
$e_{c13}(t)$	$e_{c3}(t) = 1 \cap e_{c4}(t) = -1 \cap e_{c5}(t) = 1$
$e_{c14}(t)$	$e_{c3}(t) = 1 \cap e_{c5}(t) = 0$
$e_{c15}(t)$	$e_{c9}(t) = 1$ is maintained for a certain period of time

3.3 Multi-robot Context Module

Each robot broadcasts its preference, position and conditions to the others. Therefore, all robots in a team have identical information in their multi-robot context module. The multi-robot context module of i-th robot for robot soccer is defined in Fig. 1. In the figure, 'Ball distance' is the distance between the robot and the ball and 'Ball possession' represents which team possesses the ball. 'Abandon' is broadcast when the robot cannot perform task any more and 'Help' is broadcast when the robot needs a help.

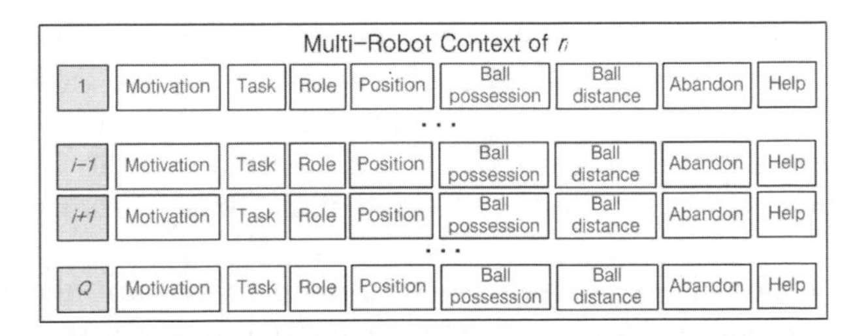

Fig. 1. Multi-robot context module

3.4 Task Strategy and Task Strategy Ratio

Task strategy represents relative importance of each task and can be changed by user's decision when robot team is performing a task. Task strategy ratio adjusts the required number of robots in each task according to external conditions and task strategy. In other words, when tasks are performed at the same time, team members select each task according to the task strategy ratio. The i-th task strategy, s_i is defined as

$$s_i = [w_1, w_2], \qquad w_1 + w_2 = 1 \tag{3}$$

where w_1 and w_2 are the weights of offense and defense tasks, respectively. In robot soccer, three task strategies are defined as

$$\begin{aligned} s_1 &= [0.5, 0.5], \\ s_2 &= [0.7, 0.3], \\ s_3 &= [0.3, 0.7] \end{aligned} \tag{4}$$

where s_1, s_2 and s_3 represent normal strategy, offensive strategy and defensive strategy, respectively. Informations from task strategy and environmental context module are considered to determine the task strategy ratio which is defined as

$$\begin{aligned} v_1 &= \begin{cases} 0.5 & \text{if } s_1 \cup (e_{c1} = 0 \cap e_{c2} = 0) \\ w - \frac{e_{c7}(t) \cdot e_{c8}(t)}{2} & \text{otherwise,} \end{cases} \\ v_2 &= 1 - v_1 \end{aligned} \tag{5}$$

where v_1 and v_2 represent the task strategy ratio of offense and defense tasks, respectively, and w is defined in Table 2.

Table 2. Setting of w according to conditions

The team possessing the ball	The normal strategy	The offensive strategy	The defensive strategy
The home team	0.5	0.55	0.45
The opponent team	0.3	0.45	0.15

As e_{c7} increases, e_{c8} effects more on task strategy ratio. For example, one score difference in the second half has a bigger effect than one score difference in the first half. In other words, the strength of offensive attribute increases with the increase of negative score difference and the decrease of the remaining time. Since w is dependent on strategy, the degree of attribute is changed depending on it.

On task selection, if v_1 is less than 0.5, more robots are allocated to the defensive task. In normal strategy and free ball situation, v_1 is 0.5 and the same number of robots are allocated into each task. This is shown in Table 3, where N_R is the total number of robots.

To allocate the same number of robots into each task except *Goalkeeper* which is always in goal area, one more robot is allocated to the defensive task when v_1 is 0.5. Moreover, if the number of total robots is even number, $\frac{N_R-1}{2}$ value is rounded down. In this case, one remained robot is allocated to a task according to its preference. For example. when there are three offensive robots and three defensive robots in normal strategy, two robots are allocated to $task_1$,

Table 3. Robot allocation according to offensive task strategy ratio

Offense task strategy ratio	Number of allocated robots on $task_1$	Number of allocated robots on $task_2$
$v_1 < 0.5$	1	$N_R - 1$
$v_1 = 0.5$	$\frac{N_R-1}{2} - 1$	$\frac{N_R-1}{2} + 1$
$v_1 > 0.5$	$N_R - 1$	1

three robots are allocated to $task_2$ and one remained robot is allocated to $task_1$ because of offensive attribute.

4 Experiments

This experiment is to demonstrate the effectiveness of task and role allocations according to the normal strategy, the offensive strategy and the defensive strategy. The normal strategy allocates the same number of robots to both the offense task and the defense task. The offensive strategy allocates more robots to the offensive task than the defensive task. On the contrary, the defensive strategy allocates more robots into defensive task. Moreover, every strategy adjusts weight of each task depending on task strategy ratio changed by the game situation such as score difference and remaining time. In this experiment, robot team was composed of three offensive robots and three defensive robots. Changing predefined conditions of each robot such as strategy, ball possession and score difference during the game, task and role of each robot were compared with those of the others.

$e_{c7}(t)$, time ratio, was increased from 0 to 1 during the game. Condition of the ball was changed as Free ball $\Rightarrow$ Home team ball $\Rightarrow$ Free ball $\Rightarrow$ Opponent team ball, periodically. Score difference was also changed as 0:0 $\Rightarrow$ 0:1 $\Rightarrow$ 0:2 $\Rightarrow$ 1:2 $\Rightarrow$ $2:2$ $\Rightarrow$ 3:2.

In the experiment, robot team consisted of six robots including *Striker* and *Goalkeeper*. $Robot_2$ and $Robot_3$ were the defensive robots and $Robot_4$ and $Robot_5$ were the offensive robots. Since $Robot_2$ and $Robot_3$ had a strong defensive attribute, they mainly carried out defensive tasks and occasionally carried out the offensive task according to the score difference and the remaining time. In the same manner, $Robot_4$ and $Robot_5$ performed the offensive tasks and occasionally performed the defensive task according to the situation.

As shown in the graphs of Fig. 2 and Fig. 3, the offensive attribute became stronger as the remaining time decreased because the home team was losing the game. This is clearly shown in the offensive strategy of Fig. 2. Fig. 3 was also shown that the defensive robot was performing the offensive task when the team was losing the game for about 20 seconds. After about 30 seconds, since the robot team turns the game around and there was not much time left, the defensive attribute became stronger.

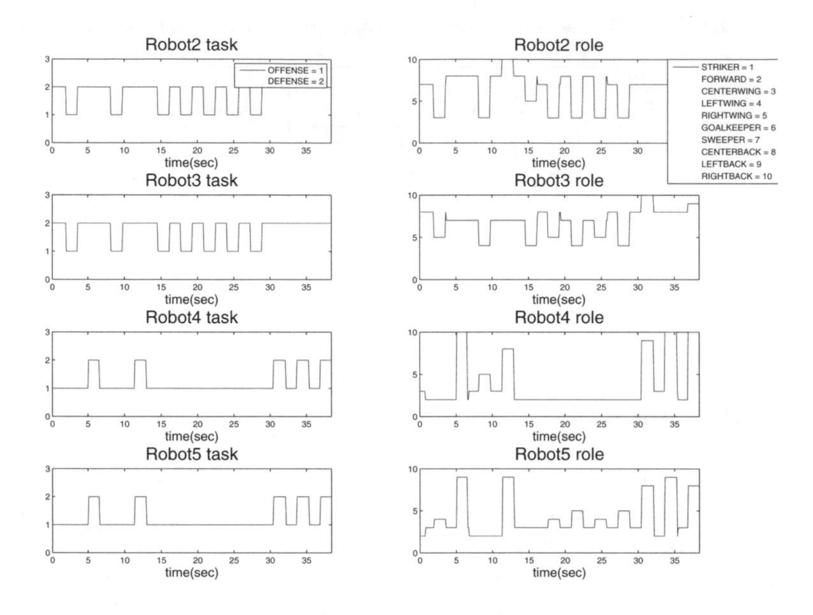

Fig. 2. Task allocation according to the offensive strategy

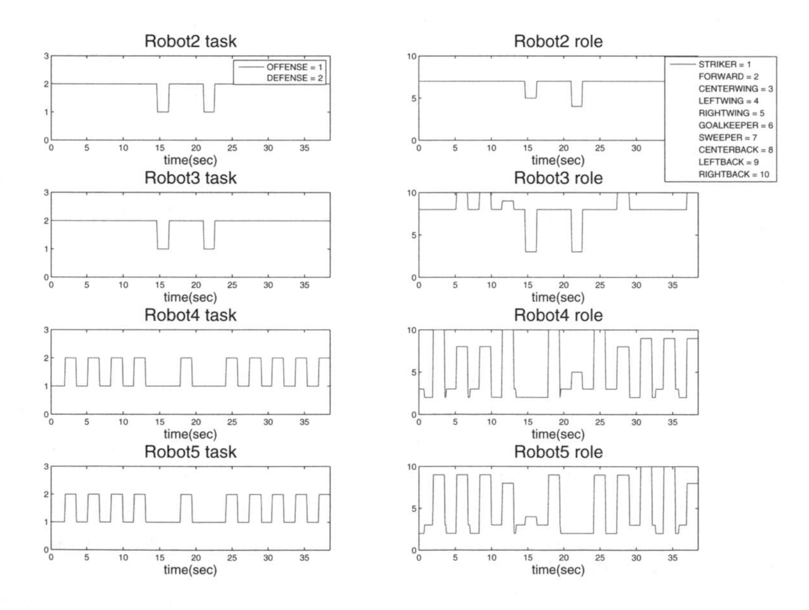

Fig. 3. Task allocation according to the defensive strategy

5 Conclusions

This paper proposed task and role selection strategy for multi-robot coordination in robot soccer. Each robot had preference for different tasks depending on its own resources. The number of robots for each task was allocated considering their preferences on tasks, environment context, multi-robot context, and task strategy. The simulation results showed that robots were able to select task and role considering their preference, score difference, remaining time, task strategy and task strategy ratio. When offensive strategy was used, they tend to select offense task and roles in the offense task. When defensive strategy was used, on the other hand, they showed the tendency of defense rather than offense.

Acknowledgement

This work was supported (National Robotics Research Center for Robot Intelligence Technology, KAIST) by Ministry of Knowledge Economy under Human Resources Development Program for Convergence Robot Specialists.

References

1. Parker, L.E.: ALLIANCE: An architecture for fault-tolerant multirobot cooperation. IEEE Trans. on Robotics and Automation 14, 220–240 (1998)
2. Werger, B., Mataric, M.J.: Broadcast of Local Eligibility for Multi-Target Observation. In: Proceedings, 5th International Symposium on Distributed Autonomous Robotic Systems, pp. 347–356 (2000)
3. Fang, T., Parker, L.E.: ASyMTRe: Automated Synthesis of Multi-Robot Task Solutions through Software Reconfiguration. In: Proceedings, The 2005 IEEE International Conference on Robotics and Automation, pp. 1513–1520 (2005)
4. Parker, L.E., Schneider, F.E., Schultz, A.C.: Enabling Autonomous Sensor-Sharing for Tightly-Coupled Cooperative Tasks. In: Proceedings, The 2005 International Workshop on Multi-Robot Systems, pp. 119–130 (2005)
5. Gerkey, B.P., Mataric, M.J.: Mataric: Sold!: Auction Methods for Multirobot Coordination. IEEE Trans. on Robotics and Automation 18, 758–768 (2002)
6. Dias, M.B.: Traderbots: A new paradigm for robust and efficient multirobot coordination in dynamic environments. Ph.D. dissertation. Robotics Institute, Carnegie Mellon University, Pittsburgh (2004)
7. Zlot, R.: An Auction-Based Approach to Complex Task Allocation for Multirobot Teams. Ph.D. dissertation, Robotics Institute, Carnegie Mellon University, Pittsburgh (2006)
8. Lee, D.-H., Kim, J.-H.: Motivation and Context-Based Multi-Robot Architecture for Dynamic Task, Role and Behavior Selections. In: Proceeding, FIRA RoboWorld Congress, pp. 161–170 (2009)
9. Kim, J.-H., Kim, D.-H., Kim, Y.-J., Seow, K.-T.: Soccer Robotics (Springer Tracts in Advanced Robotics), vol. 326. Springer, Heidelberg (2004)

Positional Consensus of Multi-Agent Systems Using Linear Programming Based Decentralized Control with Rectilinear Decision Domain

Kaushik Das* and Debasish Ghose

GCDSL, Dept. of AE,
Indian Institute of Science,
Bangalore, India
{kaushikdas,dghose}@aero.iisc.ernet.in

Abstract. In this paper we develop a Linear Programming (LP) based decentralized algorithm for a group of multiple autonomous agents to achieve positional consensus. Each agent is capable of exchanging information about its position and orientation with other agents within their sensing region. The method is computationally feasible and easy to implement. Analytical results are presented. The effectiveness of the approach is illustrated with simulation results.

Keywords: Multi-agent, Linear programming, Consensus, Decentralize.

1 Introduction

In many applications teams of robots or autonomous agents are being used to achieve a mission that was earlier entrusted to a single autonomous robot. Such systems have the obvious advantage of redundancy, architectural simplicity, and low cost. But they are more difficult to control as they have to be controlled autonomously through a decentralized control strategy. The main goal of control strategies for groups of autonomous agents is to achieve a coordinated objective while using local information. One of the problems that is of great importance in this field is that of achieving consensus, that is, achieving identical values for some specified subset of the states of the agents. For instance, the agents may try to converge to the same direction of movement [1] after some time or they might want to converge to a point. Both are problems in achieving consensus. If we have a centralized system with perfect information then achieving consensus is a trivial matter, since the central controller can instruct each agent suitably to reach a common consensus point. The challenging task is to achieve consensus using a distributed and decentralized strategy. Several classes of problems in the area has been addressed in the literature in [2] - [5].

In [6] a *Linear Programming (LP)* based centralized formulation was proposed to achieve positional consensus of multiple agents. The central controller observes the position and orientation of the agents and, using this information, it computes a common control for all the agents based on a *Linear Programming (LP)* technique. The central

* The authors gratefully acknowledge the AOARD/AFOSR for their grant to this project.

P. Vadakkepat et al. (Eds.): FIRA 2010, CCIS 103, pp. 178–185, 2010.

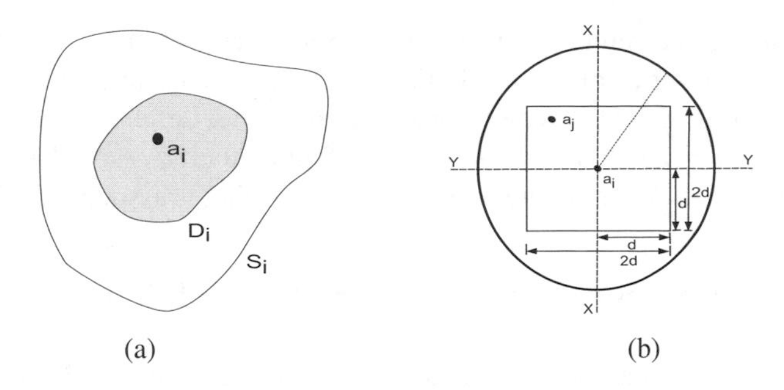

Fig. 1. (a) General sensor domain and decision domain (b) The sensor domain is circular and the decision domain is a square aligned with the global $X-Y$ coordinates

controller broadcasts identical control command to all the agents. The agents implement the common control command in their own local reference frame to move toward consensus. It was shown that perfect consensus is not possible by this method when the number of agents is more than two. So, a post-command randomization was introduced for the agents. Using this technique it was shown that consensus for larger number of agents could be achieved. However, the computation of the command was done in a centralized framework.

In this paper we show that it is possible to achieve positional consensus in a *decentralized* framework, using a similar LP based formulation. We assume the agents' decision domain in $\mathbb{R}^2$ space agent is a square. We assume that agents are able to communicate with each other within the sensing zone and they are also able to exchange information about their position and orientation. Using these local information, each agent will run the LP based algorithm and will compute a control for itself. After computing the control command, an agent can directly implement it or it can obtain information from its neighbors about their control commands and then implement an average control. We show that the decentralized strategy leads to consensus.

2 Preliminary Definitions

Let $R = \{a_1, a_2, \ldots, a_n\}$ be the set of robots or agents. The positions of agent a_i is $x_i = (p_{i1}, p_{i2}) \in \mathbb{R}^2$ and the orientations is $\theta_i \in (-\pi, \pi]$. The sensor domain of an agent a_i is denoted as S_i and its decision domain is denoted by D_i, where $D_i \subset S_i$. Information sensed from the decision domain is used to implement the algorithm. Essentially, we introduce the concept that information from the whole of the sensor domain need not be used for decision-making. In fact, we can select decision domains that make computations simpler. In Fig. 1a, we give a schematic of these concepts. Note that, in general, p_i need not be inside S_i. An agent determines its set of neighbouring agents based on D_i. In this paper we assume that the sensing domain (S_i) of all the agents is circular with radius r. The decision domain (D_i) is a square of side $2d$, with $d < \frac{r}{\sqrt{2}}$, aligned with a pre-specified global (X,Y) reference frame. This is shown in Fig. 1b. The set of neighbors of agent a_i is defined as $N_i = \{a_j \mid (|x_i - x_j|) \leq d; \;\; (|y_i - y_j|) \leq d; \;\; i \neq j\}$.

This system can be represented by a proximity graph $G(t) = (V, E(t))$, where V denotes the node or vertex of the graph. Each node represents an agent or a robot. $E(t)$ represents the set of edges between the nodes at any time. Let $a_i \sim a_j$ denote the edges between the agents a_i and a_j and $(a_i\, a_j) \in E(t)$ if and only if $a_j \in N_i$. The degree of any node represents the number of neighbors of the agents that is associated with that node.

3 Algorithm: A Linear Programming Formulation

The LP based centralized consensus algorithm is described in [6]. In which a central controller computes the control command using the position and orientation of the agents and the control is broadcast to all the agents. The computed control will be in $(d, \Delta\theta)$ form. After receiving this control command, each agent will execute a rotation of angle $\Delta\theta$ and translation d. Here, we present an algorithm which is *decentralized* in nature.

Let the set of positions and orientation of the neighbors of agent a_i be $\{\{x_j\}, \{\theta_j\}\}$, where $a_j \in N_i$. As in [6], we define the control command computed by agent a_i as $(\Delta\theta_i, d_i)$ using information from its neighbors only. We define our performance measure as the half length, denoted by $r_i > 0$, of the side of a square oriented along the global coordinate frame, and containing all the final positions of the agent and its neighbors. Let this square be centered at $z_i = (z_{1_i}, z_{2_i}) \in \mathbb{R}^2$. Since all neighbors of agent a_i execute the same command $(\Delta\theta_i, d_i)$, their final positions, given by $x_{jf} = [q_{i1} \;\; q_{i2}] \in \mathbb{R}^2$, are

$$\begin{bmatrix} q_{i1} \\ q_{i2} \end{bmatrix} = \begin{bmatrix} p_{i1} \\ p_{i2} \end{bmatrix} + \begin{bmatrix} \cos\theta_{i0} & -\sin\theta_{i0} \\ \sin\theta_{i0} & \cos\theta_{i0} \end{bmatrix} \begin{bmatrix} u_{1_i} \\ u_{2_i} \end{bmatrix} \tag{1}$$

where, $u_{1_i} = d_i \cos(\Delta\theta_i)$ and $u_{2_i} = d_i \sin(\Delta\theta_i)$ are the control variables used. Note that Eqn. (1) is a set of linear equations. Now, we formulate the linear programming problem for agent a_i as,

Minimize r_i
Subject to

$$-r_i \le p_{j1} + u_{1_i}\cos\theta_j - u_{2_i}\sin\theta_j - z_{1_i} \le r_i \tag{2}$$

$$-r_i \le p_{j2} + u_{1_i}\sin\theta_j + u_{2_i}\cos\theta_j - z_{2_i} \le r_i \tag{3}$$

$$r_i \ge 0; \;\; j \in N_i \tag{4}$$

This is a linear programming problem with the decision vector as $(r_i, z_{1_i}, z_{2_i}, u_{1_i}, u_{2_i})$. Note that the decision vector remains the same irrespective of the number of neighbors. Only the number of inequality constraint increases with the number of agents. Also, note that z_{1_i},z_{2_i},u_{1_i} and u_{2_i} are free variables and can take both +ve or -ve.

4 Decentralized Control

In this section we will give a formal description of the algorithm. It is obvious that the outcome of this algorithm depends upon the positions and orientations of the neighbors of an agent as well as of its own. The positions of the agents a_i at time instant t_{k+1} can be written as

$$x_i(t_{k+1}) = x_i(t_k) + U(\Delta\theta_j(t_k), d_j(t_k)) \tag{5}$$

where, $U(\Delta\theta_j, d_j) = f(x_j t_k, \theta_j t_k)$.

While computing the control command, agent a_i may or may not include its own position and orientation. Also, after computing the control command it may or may not share its control with its neighbor. This gives four possible operation for implementing this algorithm. The four cases are shown in the Table 1. When agent will share control command with its neighbors, we will say that it uses a broadcast mechanisms (which is restricted to its neighbors) to communicate with its neighbor. When there is no sharing of control command then we will say that there is no broadcast communication. Note that broadcast based algorithm needs some effort in implementing the algorithm as the broadcast is restricted.

Table 1. Four different cases

	No Broadcast	Broadcast
Self Included $j \in N_i \cup \{i\}$	Case A	Case B
Self not Included $j \in N_i$	Case C	Case D

4.1 No Broadcast: Case A and Case C

The agents will share its positions and orientation but it will not share its control with its neighbor. It may or may not include its position and orientation while computing the control command. Under this paradigm there are two cases

1. Case A: Compute control using the information of the neighbor as well as of its own. Then, $U(\Delta\theta_j, d_j) = f(x_i(t_k), \theta_i(t_k))_{i \in N_j \cup \{j\}}$.
2. Case C: Compute control using the information of the neighbor only. $U(\Delta\theta_j, d_j) = f(x_i(t_k), \theta_i(t_k))_{i \in N_j}$

4.2 Broadcasting – Averaging Rule: Case B and Case D

The agents will send or broadcast its control to its neighbors. So a neighbor will have more than one control input. The agents will compute an average of all its control and will move to a new position. Here also two case can arise

1. Case B: Compute the control using information from neighbors as well as of its own and broadcast to the neighbors. Then, $U_{a_i} = \frac{1}{m_i+1}\left((\Delta\theta_i(t_k), d_i(t_k)) + \Sigma_{j \in N_i}(\Delta\theta_j, d_j)\right)$.

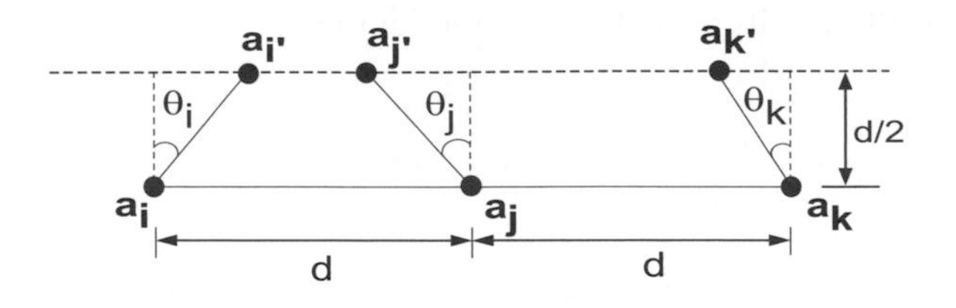

Fig. 2. An illusration of three agents

2. Case D: Compute control using only the neighbor information and broadcast to neighbors. Then, $U_{a_i} = \frac{1}{m_i}\left((\Delta\theta_i(t_k), d_i(t_k)) + \Sigma_{j \in N_i}(\Delta\theta_j, d_j)\right)$.

5 Modifications: Preserving Connectivity

Applying the algorithm has the problem that since the connectivity between agents may not be preserved, as will be seen from the following example.

Consider only three agents which are collinearly placed at intervals of distance d, i.e., the maximum sensing distance (see Fig. 2) along the $X-$axis. Without loss of generality we put a boundary at $d/2$ distance in the Y-axis. This will ensure that the connectivity along the Y-direction will not be violated. From Fig. 2 we can see that the initial graph is connected. The positions of the agents at the time instant t_{k+1} are a'_i,a'_j and a'_k. The graph will not be connected if $\theta_k \geq \theta_j$ because $(d/2)\sin(\theta_k) \leq (d/2)\sin(\theta_j)$. We can see from the above result that the graph will not be connected using the above algorithms and positional consensus may not always be possible. To ensure positional consensus, we have to ensure the connectivity of the graph always.

In order to achieve connectivity, the movement of the agents should be restricted. We call this as pairwise motion constraint. This is shown in Fig.3(a). In this figure agent a_i has one neighbor a_j. The movement of agent a_i is restricted within the square. The center of the square is C_{ij} which is the mid point of the line joining agents a_i and a_j. The side length is d. This constraint ensures that if two agents connect at time t_k, then they remain connected for all time $t > t_k$.

In Fig.3(b) we can see that agent a_i has two neighbors a_j and a_k, g_i is the goal point of the agents, computed by the algorithm. Now the shaded region is the common overlapping zone between the two squares. For agent a_j, it can move towards the goal by a distance d_{ij} and for agent a_k it can move a distance d_{ik}. Eventually, the agent a_i will move the distance $d = \min\{d_{ij}, d_{ik}\}$ towards the goal. So, in general,an agent will move towards the goal $d = \min\{d_{ij}\}, a_j \in N_i$.

The common overlap zone is shown by the region $ABCD$ in Fig.4. The agent position at time instant t_k is $a_i(t_k)$. The goal positions is at $g_i(t_k)$ which is outside of the common overlap zone. The agent can move up to the point a_f which is at the border of the

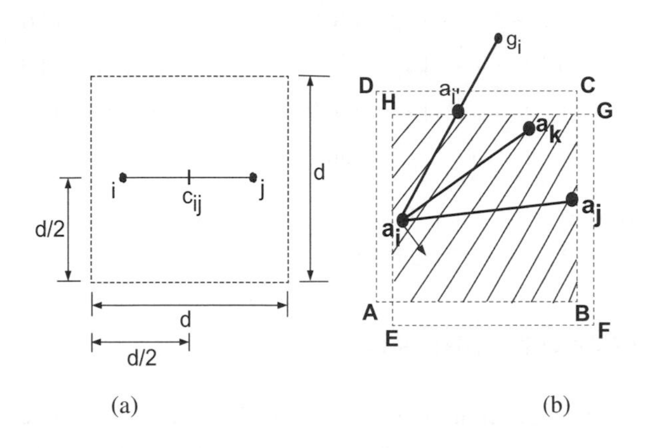

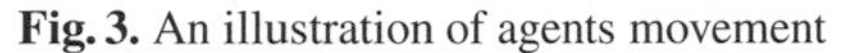

Fig. 3. An illustration of agents movement

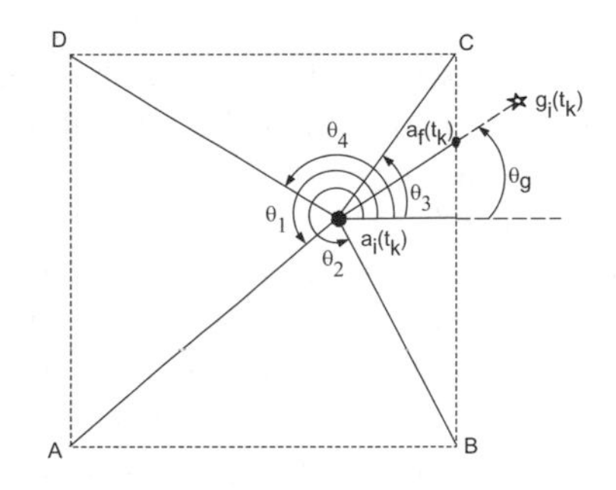

Fig. 4. Agent movement in common overlap Region

common overlap region. To compute the final positions of the agent a_i, we have divided the whole overlap zone into four parts.

1. $In_1 : \theta_g \in [0, \theta_1) \cup [\theta_4, 2\pi)$
2. $In_2 : \theta_g \in [\theta_1, \theta_2)$
3. $In_3 : \theta_g \in [\theta_2, \theta_3)$
4. $In_4 : \theta_g \in [\theta_3, \theta_4)$

Now we define four index function I_1, I_2, I_3 and I_4. The index I_1 can take values from $\{0,1\}$. We can write

$$I_i = \begin{cases} 1 \ , \theta_g \in In_i \\ 0 \ , \theta_g \notin In_i \end{cases} ; \quad i \in \{1, \ldots, 4\}$$

In the four different cases the movement of the agents will be different. The final positions coordinate of the agents are,

$$p_{i1_f} = p_{i1} + (p_{2_{c_3}} - p_{i2}) * I_1 * \cot\theta_3 + (p_{2_{c_3}} - p_{i2}) * I_2 * \cot\theta_g$$
$$+ (p_{2_{c_1}} - p_{i2}) * I_3 * \cot\theta_1 + (p_{2_{c_1}} - p_{i2}) * I_4 * \cot\theta_g \quad (6)$$
$$p_{i2_f} = p_{i2} + (p_{1_{c_3}} - p_{i1}) * I_1 * \tan\theta_g + (p_{1_{c_3}} - p_{i1}) * I_2 * \tan\theta_3$$
$$+ (p_{1_{c_1}} - p_{i1}) * I_3 * \tan\theta_g + (p_{1_{c_1}} - p_{i1}) * I_4 * \tan\theta_1 \quad (7)$$

The above equations are valid when the goal point is outside common overlap region or on the boundary the common overlap region.

6 Achieving Perfect Consensus

The solution of the linear programming (LP) based algorithm will yield control instructions that can be broadcast to all the neighbors using which the agents will move to a new position. It can be shown that no further improvement of the performance (that is, the agent will be stationary) can be achieved by repeated use of the algorithm. The graph will not change if the set of neighbors do not change. In other words, repeated application of the LP based decentralized algorithm with the new final positions, will not yield any solution. Suppose we represent the LP algorithm as an operation L on the initial conditions that yields the solution as, $L(x_{i0}, \theta_{i0} | i = 1, \ldots, n) = (u_1^*, u_2^*, r*, x_{if}, \theta_{if})$.

If the neighbor set remains, then, $L(x_{if}, \theta_{if} | i = 1, \ldots, n) = (0, 0, r*, x_{if}, \theta_{if})$. This was the case in the centralized strategy [6]. However, in the decentralized case, it is possible that the neighbors set of an agent can change and will yield non-zero central commands for the agents at each successive applications of the strategy. But, subsequently when agents come sufficiently close the graph becomes completely connected and the control command becomes zero. In order to continuously get non zero control commands, we introduce a perturbation strategy given below.

In the *perturbed case* where the agents randomly perturb final orientation angle after each LP solution is now implemented as follows [6].

$$\hat{\theta}_{i,k+1} = \theta_{i,k+1} + \nu_{i,k+1} \quad (8)$$

In (8) $\nu_{i,k+1}$ represents the perturbation angle.

7 Simulations Results and Implementation Issues

In the first set of simulations we have considered 5 agents which are placed randomly at the $\mathbb{R}^2$ plane. We have set their orientation randomly. The result is shown in Fig. 5. We can see that using this algorithm all the agents eventually come to consensus.

The algorithm can be implemented by any agent (robot) that can gather information about and from its neighbors (positions and orientations) using sonar sensors, vision sensors or laser scanners. After generating the control command the agents can broadcast the control command with positions and orientation of the centroid information by a short range communication mechanism. An agent receiving a broadcast control command from another agent which is not in its decision domain will ignore the command.

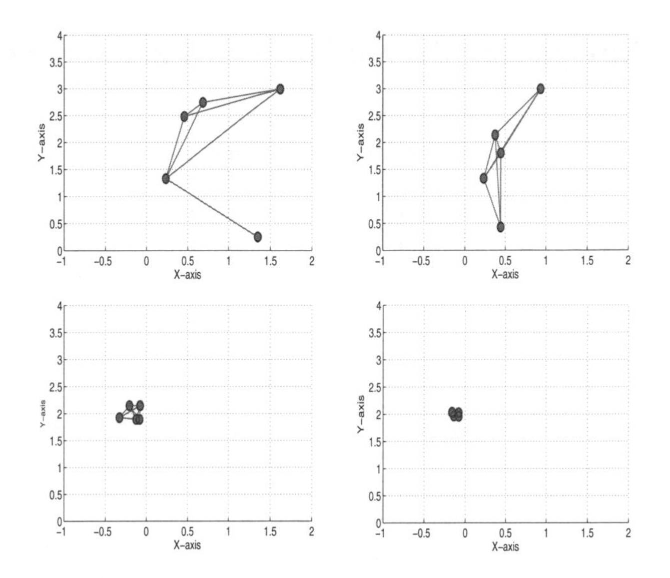

Fig. 5. Modified Algorithm: 5 agent preserves connectivity at each instant of time and come closer to a consensus

8 Conclusions

We presented a decentralize LP based algorithm. This work extends previous work on broadcast based centralized control of agents to a decentralized framework. The flexibility, that the algorithm gives to the individual agents allows us to design effective consensus algorithms. In future, we will focus on computational time and implementation issues.

References

1. Vicsek, T., Czirok, A., Jacob, E.B., Cohen, I., Schochet, O.: Novel type of phase transistions in a system of self-driven particles. Phys. Rev. Lett. 75, 1226–1229 (1995)
2. Jadababaie, A., Lin, J., Morse, A.S.: Coordination of groups of mobile autonomous agents using nearest neighbour rules. IEEE Trans. Automat. Contr. 48(6), 988–1001 (2003)
3. Saber, R.O.: Flocking in multiagent dynamic systems: Algorithms and theory. IEEE Trans. Automat. Contr. 51 (2006)
4. Ando, H., Oasa, Y., Suzuki, I., Yamashita, M.: Distributed Memoryless Point Convergence for Mobile Robots with Limited Visibility. IEEE Transactions on Robotics and Automation 15, 818–828 (1999)
5. Cortes, J., Martínez, S., Bullo, F.: Robust Rendezvous for Mobile Autonomous Agents via Proximity Graphs in Arbitrary Dimensions. IEEE Trans. Automat. Contr. 51, 1289–1298 (2006)
6. Das, K., Ghose, D.: Positional Consensus in Multi-Agent Systems using a Broadcast Control Mechanism. In: American Control Conference, St. louis, Missouri, USA, pp. 5731–5736 (2009)
7. Godsil, C., Royle, G.: Algebric Graph Theory. Springer, Heidelberg (2001)

Tacit Navigation Method for Multi-agent System*

J.W. Kim, Y.H. Kim, B.C. Min, and D.H. Kim

The Department of Electrical Engineering, Kyung Hee University
1 Seocheon-dong, Giheung-gu, Yongin-si, Gyeonggi-do, 446-701, Korea
kjw810313@khu.ac.kr, gemblerz@khu.ac.kr, minbc@khu.ac.kr,
donghani@khu.ac.kr

Abstract. The goal of this paper is to make robots move effectively according to the navigation method and reach the destination in a short time. We propose a new navigation method is based on the standard rules of airplane traffic: If both airplanes approach each other from opposite sides, they are supposed to give way by turning right away from each other to avoid a collision, and if flying airplanes come into conflicting paths side by side, the left airplane turns right to yield. The robot that moves according to this navigation method would either stop or use the limit-cycle to avoid a collision. As a result, robots remain safe and reach their destinations faster.

Keywords: Multi-Agent, tacit, the rule of airplane, Limit-cycle.

1 Introduction

According to development of the robot industry, the navigation method has been studied actively for efficiency of robot movement. Examples of the studied navigation methods are described below:

D-H. Kim and J-H. Kim [1] have designed a real-time limit-cycle navigation method for fast mobile robots and its application to robot soccer. B.C. Min et al. [2] have presented Fuzzy Logic Path Planner and Motion Controller by Evolutionary Programming for Mobile Robots. Shingo Shimoda et al. [3] have dealt with High-speed navigation of unmanned ground vehicles on uneven terrain using potential fields. These navigation methods make robot to avoid the nearest way and reach the goal in a short time, but it doesn't reduce the chance of a collision. So this paper describes that by using the collision inspection [4], the limit-cycle navigation method, and rules of airplane traffic [5], tacit navigation method for multi-agent system can translocate robots effectively. Using this method, UGV (Unmanned Ground Vehicle) reduces the chance of a collision and reaches the goal effectively in a short time.

"Rules of the air" of the International Civil Aviation Organization (ICAO) annex 2 and "Right of way" of Federal Aviation Regulation (FAR) 91.113 require the following: "If both airplanes approach from opposite sides, they are supposed to give way

* This research was carried out under the General R/D Program of the Daegu Gyeongbuk Institute of Science and Technology(DGIST), funded by the Ministry of Education, Science and Technology(MEST) of the Republic of Korea.

P. Vadakkepat et al. (Eds.): FIRA 2010, CCIS 103, pp. 186–193, 2010.

by turning right away from each other to avoid a collision, and if flying airplanes come into conflicting paths side by side, the left airplane turns right to yield." Tacit navigation method for multi-agent system applies them to the UGV. "If two UGVs approach each other from opposite sides, they are supposed to give way by turning right away from each other to avoid a collision, and if moving UGVs come into conflicting paths side by side, the left UGV waits for a while to yield."

This paper is organized as follows: first, collision inspection process between moving objects such as UGV robot and opponent robot is stated (Section 2). Second, descriptions are stated that by using the limit-cycle navigation method and rules of airplane traffic, tacit navigation method for multi-agent system is completed (Section 3). Third, the superiority of tacit navigation method is demonstrated through the simulation and experiment (Section 4, 5). Finally, conclusions will be drawn in Section 6.

2 Collision Inspection

Fig. 1 simply draws collision inspection process between moving objects such as UGV robot and opponent robot. We suppose two objects move from t=0 to t=1 at a certain speed in a straight line and cannot cross or overlap.

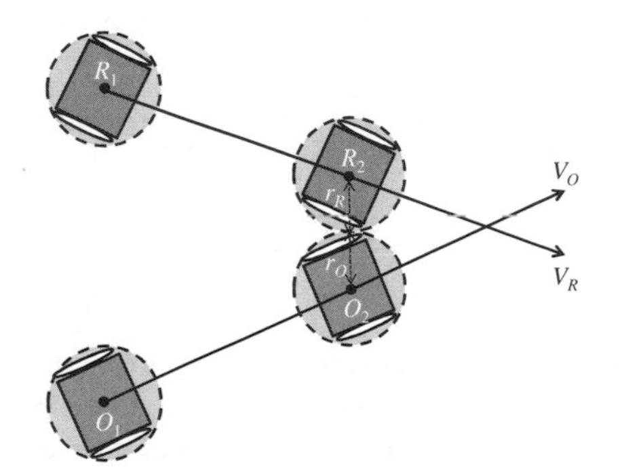

Fig. 1. Collision inspection between two moving objects

The initial and current positions of the UGV robot are R_1, R_2, and the initial and current positions of opponent robot are O_1, O_2. In this case V_R and V_O, equals to each object's velocity vector, are as follows.

$$\begin{aligned} V_R &= R_1 - R_2 \\ V_O &= O_1 - O_2 \end{aligned} \tag{1}$$

The position $R(t)$ of radius (r_R) of robot and the position $O(t)$ of radius (r_O) of opponent robot are as follows.

$$\begin{aligned} R(t) &= R_1 - tV_R \\ O(t) &= O_1 - tV_O \end{aligned} \tag{2}$$

At this point, when the distance between $R(t)$ and $O(t)$ equals to $r_R + r_O$, then robot and opponent robot will collide in time t. First of all, square of the distance between $R(t)$ and $O(t)$ is as follows,

$$d^2 = \|R(t) - Q(t)\|^2 \tag{3}$$

Eq. (2) can be substituted into Eq. (3),

$$d^2 = \|R_1 + tV_R - O_1 - tV_O\|^2 \tag{4}$$

Assuming A and B as,

$$\begin{aligned} A &= R_1 - O_1 \\ B &= V_R - V_O \end{aligned} \tag{5}$$

Eq. (4) can be rewritten as follows,

$$\begin{aligned} d^2 &= \|A + tB\|^2 \\ &= A^2 + 2t(A \cdot B) + t^2 B^2 \end{aligned} \tag{6}$$

When the Eq. (6) is two quadratic equations about t, t is as follows.

$$\begin{aligned} t_1 &= \frac{-(A \cdot B) - \sqrt{(A \cdot B)^2 - B^2(A^2 - d^2)}}{B^2} \\ t_2 &= \frac{-(A \cdot B) + \sqrt{(A \cdot B)^2 - B^2(A^2 - d^2)}}{B^2} \end{aligned} \tag{7}$$

In case the value in square root is negative, the robot and opponent robot are not bumped against each other. In another case B^2 equals to zero, the moving robot and opponent robot are hauled up or two objects move the same direction and at the same velocity, that is, the robot and opponent robot are not bumped against each other either.

At this point, the collision time of robot and opponent robot is when d equals to $r_R + r_O$. And, the contact time t_1, t_2 of two objects can figure out by Eq. (7). At the beginning of the paper, because we supposed two things cannot cross and overlap, the collision time t_1 of two objects is only needed to calculate. Finally, the collision time of opponent robot is Eq. (8).

$$t = \frac{-(A \cdot B) - \sqrt{(A \cdot B)^2 - B^2(A^2 - r_R + r_O)}}{B^2} \tag{8}$$

3 Collision Avoidance

This paper uses limit-cycle navigation method, only turning right in order to effectively avoid a collision. Also, the protocol has a fixed rule modification of “Rules of the air” of ICAO annex 2 and “right of way” of FAR 91.113.

3.1 Limit-Cycle Navigation Method of Only Turning Right

The limit-cycle navigation method generates a circular trajectory when distinguishing the stability of a system using a phase portrait, graphical analysis methods of second-order nonlinear system.

Originally, this method was used to decide an avoidance direction by detecting obstacles after drawing a straight line from the current position to the goal, but this paper uses the limit-cycle only for the purpose of turning right.

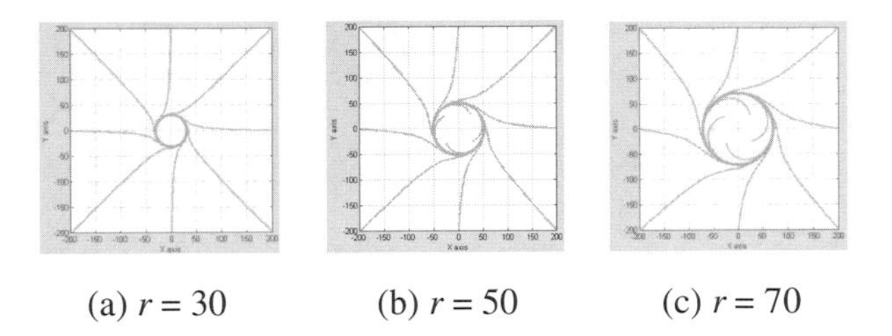

(a) $r = 30$ (b) $r = 50$ (c) $r = 70$

Fig. 2. Phase portrait of Limit-cycle turning right

If the 2nd-order nonlinear system is given as follows,

$$\begin{aligned} \dot{x}_1 &= -x_2 + x_1(r - x_1^2 - x_2^2) \\ \dot{x}_2 &= x_1 + x_2(r - x_1^2 - x_2^2) \end{aligned} \tag{9}$$

Then, all the trajectories will be moving inward counter-clockwise, as depicted in Fig. 2. To utilize this theory as a path planning method, the variable r in Eq. (9) should be changed according to the radius of the obstacle.

If $x_1$and x_2 are matched with x and y in the global coordinate Σ_{OXY}, the form of Eq. (9) can be transformed as follows,

$$\begin{aligned} \dot{x} &= -y + x(r^2 - x^2 - y^2) \\ \dot{y} &= x + y(r^2 - x^2 - y^2) \end{aligned} \tag{10}$$

where x and $y$are relative values to the obstacle.

Consequently, we can adjust the radius, and the Limit-cycle method constitutes a local navigation plan. This selects an efficient way by which the UGV robot can avoid obstacles rather than moving far away from them. These limit-cycle characteristics are applied to the navigation plan.

3.2 Tacit Navigation Method for Multi-agent System

According to the standard rules of airplane traffic, “If both airplanes approach each other from opposite sides, they are supposed to give way by turning right away from each other to avoid a collision, and if flying airplanes come into conflicting paths side by side, the left airplane turns right to yield." Tacit navigation method for multi-agent system converts these rules into as follows.

(1) Consider a situation where two UGVs are approaching each other from opposite directions (i.e., more than 90). They would turn right, away from each other, using limit-cycle navigation method to avoid the collision as in Fig. 3.

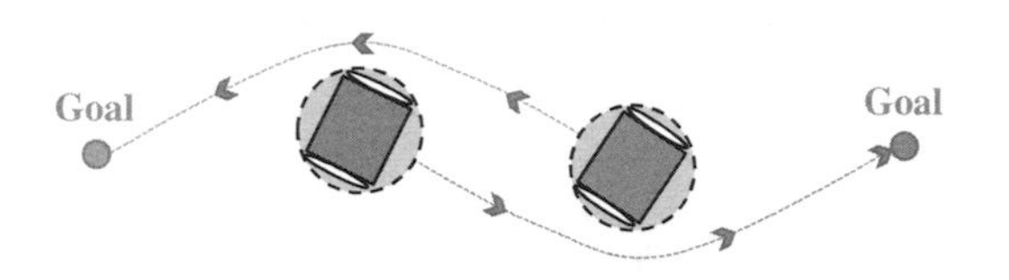

Fig. 3. UGV turning right only

(2) Consider if moving UGVs come into a conflict of paths side by side (less than 90), left UGV move after waiting for a while and right UGV go straight as in Fig. 4.

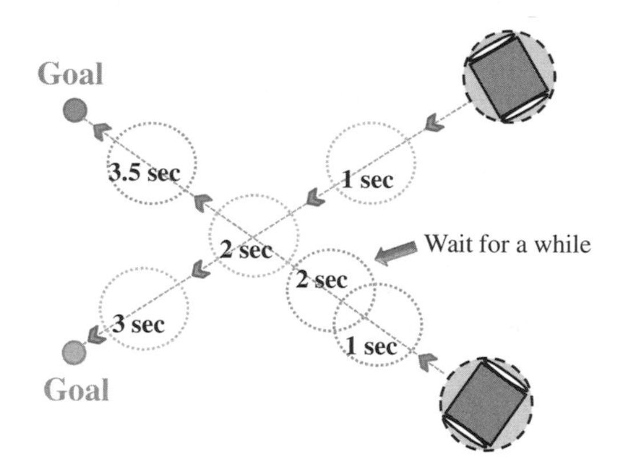

Fig. 4. Collision avoidance by waiting for a while

Two navigation methods are compared in Fig. 5. UGVs using limit-cycle navigation method and collision inspection avoid collisions in Fig. 5 (a). The first UGV avoids the chance of three collisions, and the others also avoid the chance of three collisions. So, navigation paths aren't clear and reach the goal with difficulty.

When we use tacit navigation method, the first UGV in Fig. 5 (b) seems to take longer than the UGV using conventional navigation method in Fig. 5 (a). But, numerous UGVs keep to the right as in Fig. 5 (b) and minimize the chance of collision while arriving at the goal in a short time. Finally, the second UGV in Fig. 5 (b) can arrive the goal effectively in short time by going straight.

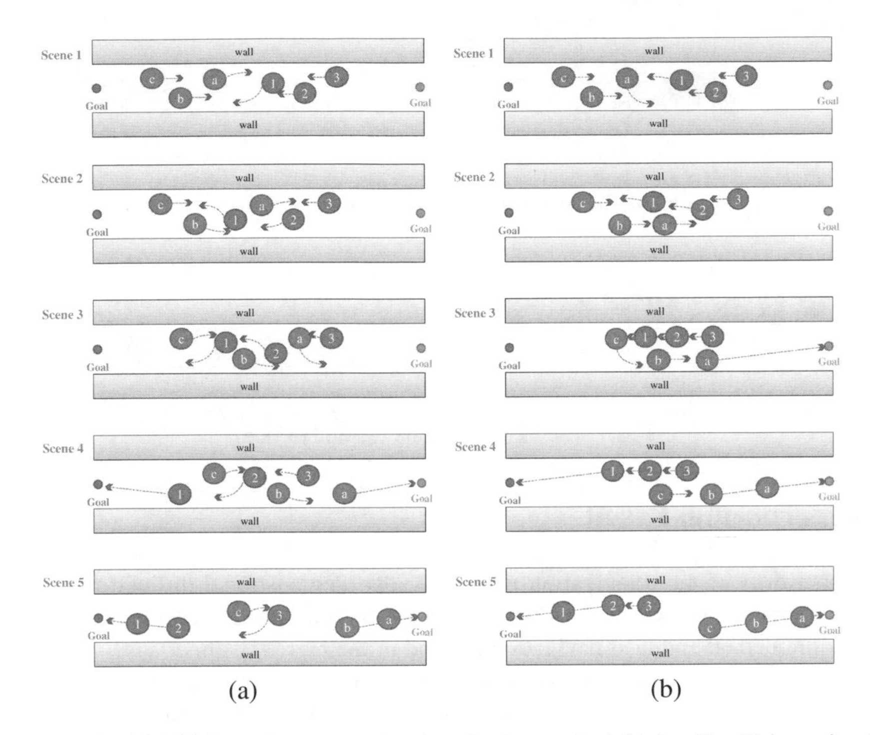

Fig. 5. (a) Avoid collision using conventional navigation method (b) Avoid collision using tacit navigation method

4 Simulation Result

The superiority of tacit navigation method is demonstrated through the simulation. To perform the simulation, we compared two navigation methods in narrow path. The first navigation is conventional navigation method mentioned at the intro. The conventional navigation method is made by using the collision inspection and the limit-cycle navigation method. The other navigation is tacit navigation method provided in this paper. Using the tacit navigation method, the UGV minimizes the chance of a collision and arrives at its destination more quickly than in the other navigation method.

Fig. 6 and 7 show several UGVs moving towards a goal. Fig. 6 shows collision avoidance using conventional navigation method; and Fig. 7 shows that according to tacit navigation method, several UGVs turn right using the limit-cycle. Add to that, when two UGVs moving side by side, the left UGV waits for a while and the right UGV continues going straight.

Simulation results shows that the more UGVs using tacit navigation method minimize the chance of a collision and arrive faster than in other situations. Actually, Fig. 7 arrives 1sec faster than Fig. 6 in the same situation and condition. Also, Fig. 9 shows that a simple path is generated by UGVs using tacit navigation method and minimizes the chance of a collision.

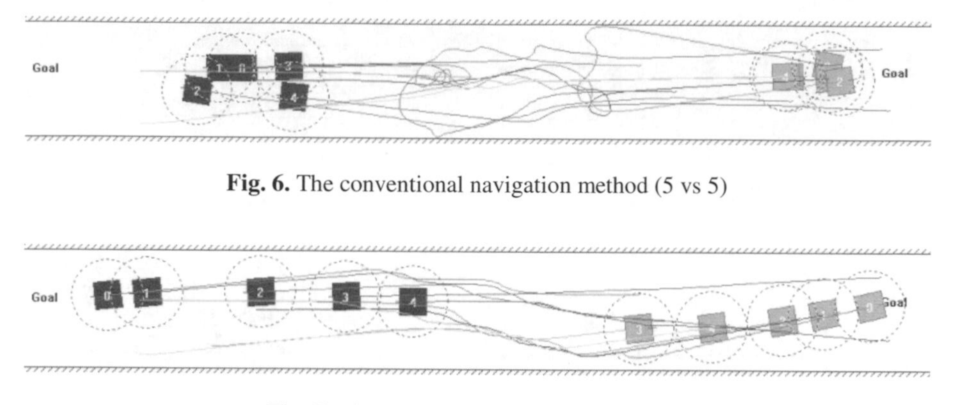

Fig. 6. The conventional navigation method (5 vs 5)

Fig. 7. The tacit navigation method (5 vs 5)

5 Experimental Result

The superiority of tacit navigation method is also demonstrated through the experiment with robot soccer system. The experiment is also conducted the conventional navigation method and tacit navigation method in the narrow path.

Fig. 8 shows collision avoidance using conventional navigation method. This method is difficult to avoid all the walls and UGVs in the narrow path. So we cannot measure the moving time exactly. Fig. 9 shows that according to tacit navigation method, several UGVs smoothly turn right using the limit-cycle. Add to that, when two UGVs moving side by side, the left UGV waits for a while and the right UGV continues going straight. Finally, UGVs reduces the chance of a collision and reaches the goal effectively in a short time.

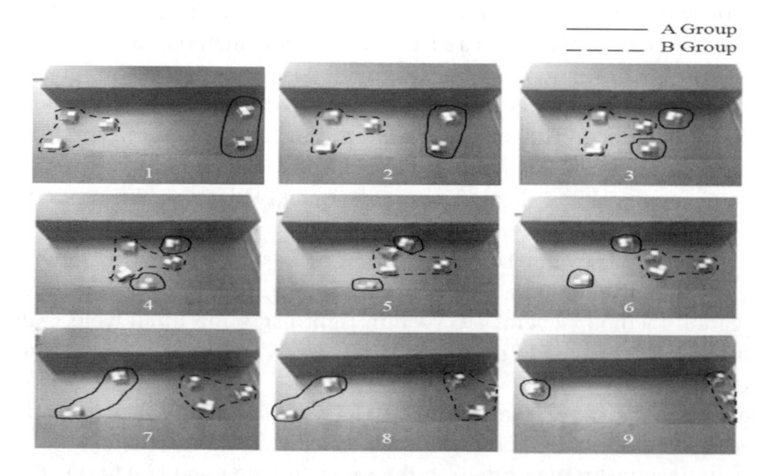

Fig. 8. The conventional navigation method

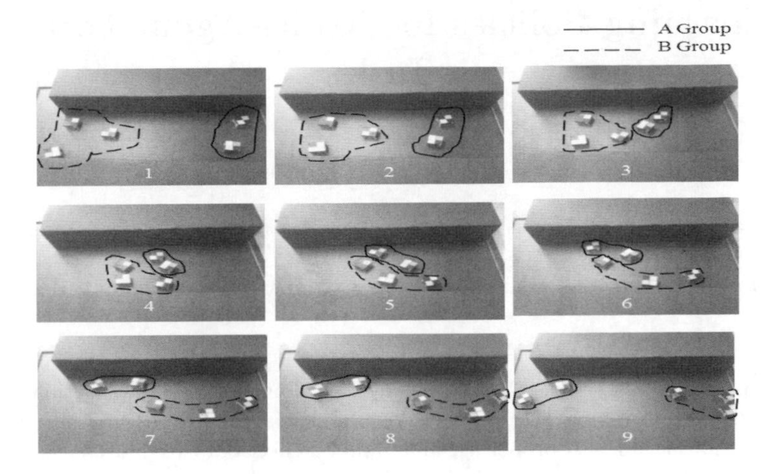

Fig. 9. The tacit navigation method

6 Conclusion

If robots become so popular that we have a robot for each person, then robots will need social conventions just like people. In order to live with numerous UGV robots, this paper makes tacit navigation method by applying the simple formulas of the collision inspection, the limit-cycle and the rules of aircraft. The simulation results showed that UGVs arrive at their goal correctly and in a short time when tacit navigation method is used. Especially, because tacit navigation method makes robot to keep to the right, it minimizes the chance of a collision and reduces moving time. In future work, we will develop an advanced tacit navigation method to enable moving regularly in the wide place.

References

1. Kim, D.-H., Kim, J.-H.: A real-time limit-cycle navigation method for fast mobile robots and its application to robot soccer. Robotics and Autonomous Systems 42, 17–30 (2003)
2. Min, B.C., Kim, M., Kim, D.: Fuzzy Logic Path Planner and Motion Controller by Evolutionary Programming for Mobile Robots. International Journal of Fuzzy Systems 11(3) (September 2009)
3. Shimoda, S., Kuroda, Y., Iagnemma, K.: High-speed navigation of unmanned ground vehicles on uneven terrain using potential fields. Robotica 25, 409–424 (2007)
4. Cho, C.: Obstacle Avoidance of an Omni-directional Robot using Limit-cycle Navigation Method, M.S. thesis, Kyung Hee University (Febrary 2010)
5. Jang, D.-S., Cho, S.-J., Tahk, M.-J., Koo, H.-J., Kim, J.-S.: Fuzzy based Collision Avoidance against Multiple Threats for Unmanned Aerial Vehicles. In: Proceedings of the Aircraft Symposium, vol. 43(26). FIG.9, TBL.4, REF.4 (2005)

Q-Learning Policies for Multi-Agent Foraging Task

Yogeswaran M. and Ponnambalam S.G.

School of Engineering, Monash University, Sunway campus, 46150 Petaling Jaya, Selangor, Malaysia
yogeswaran.mohan@eng.monash.edu.my, sgponnambalam@eng.monash.edu.my

Abstract. The trade-off issue between exploitation and exploration in multi-agent systems learning have been a crucial area of research for the past few decades. A proper learning policy is necessary to address the issue for the agents to react rapidly and adapt in a dynamic environment. A family of core learning policies were identified in the open literature that are suitable for non-stationary multi-agent foraging task modeled in this paper. The model is used to compare and contrast between the identified learning policies namely greedy, ϵ-greedy and Boltzmann distribution. A simple random search is also included to justify the convergence of q-learning. A number of simulation-based experiments was conducted and based on the numerical results that was obtained, the performances of the learning policies are discussed.

1 Introduction

Consider the problem of an agent that is operating in an environment of incomplete information. The agent is frequently faced with the choice between maximizing its expected profits based on its current knowledge of its environment and trying to learn more about the environment in order to improve the quality of its decisions. This situation is widely known as the exploitation-exploration trade-off [1].

The exploitation-exploration trade-off issue in multi-agent system's learning have been a crucial area of research for the past few decades. A proper learning policy is necessary to address the issue for the agents to react rapidly and adapt in a dynamic environment. Policies identified in literatures are namely greedy, ϵ-greedy and Boltzmann distribution. A simple random search is also included to justify the convergence of q-learning.

Foraging is an act of searching for food, which is a widely used test application in multi-agent domain. The task consists in searching for objects spread in the environment and in bringing them to a specific place called home. In this paper,a multi-robot foraging task was modeled consisting of four mobile robots (Khepera 2 model) and 10 pucks which are enveloped by an environment. The modeling and the simulation was carried out using [2]. In this version of foraging task, the agents have to learn to search and retrieve pucks which are scattered in the environment back to the home location.

P. Vadakkepat et al. (Eds.): FIRA 2010, CCIS 103, pp. 194–201, 2010.

A policy defines the learning agent's way of behaving at a given time. It is a mapping from perceived states of the environment to actions to be taken when in those states. In this paper, policies namely greedy, ϵ-greedy and Boltzmann distribution have been selected and the performance of the policies are reported in section 6. Several attempts have been made to study the impact of the policies in q-learning. One of most recent work was by [1] where the paper examines multi-armed bandit tasks to solve the balancing problem of exploitation and exploration using policies such as ϵ-greedy, Boltzmann distribution, probability matching method and adaptive pursuit method. In [3] an improved immune mechanism in q-learning and compared the results against ϵ-greedy and Boltzmann distribution.

In [4], ϵ-greedy policy was used to model the q-learning agents and their behavior was analyzed on a game called "The Prisoners Dilemma". [5] used the same game theory together with Boltzmann distribution to show the convergence of q-learning in multi agent systems. Boltzmann distribution was also used in [6] where a new framework for self-organized flocking of agents using q-learning was modeled. [7] applied a simple q-learning algorithm to solve the problem of the displacement of an agent within a 3 by 5 matrix. Boltzmann distribution was used as the policy in this paper. Some detailed analysis was done on ϵ-greedy, Boltzmann distribution and interval estimation policy in [8]. The authors have applied the policies in 2 different tasks specifically mountain car task and server job scheduling. Implementation of greedy and ϵ-greedy can be seen in [9], where they described a formal and principled approach to imitation called implicit imitation between two agents.

Based on the review conducted, it is clear that none of the researchers have studied the effects of the Q-learning policies in a sophisticated task such as foraging. Most of the researchers implemented the policies in game theories for instance the prisoner's dilemma , rock-paper-scissors, pennies game, tic-tac-toe and soccer . The experiments reported in the papers are sufficient to distinguish the differences in those policies but does not carry enough evidence that the policies will converge favorably towards a more sophisticated and challenging task such as foraging. Therefore, a foraging environment is modeled and a critical study on the reported policies was conducted.

2 Q-Learning

Q-learning is a form of model-free learning, where an agent does not need to have any model of the environment, instead the agent only needs to know the available states and possible actions in those states. The agent observes the environment and determines its current state $s \in S$ and chooses an action $a \in A$. Once action $a \in A$ have been executed, a reward $r \in R$ is generated which is used to evaluate the quality of the action taken by updating the corresponding $Q_{(s,a)}$ values. The $Q_{(s,a)}$ values represents the sum of immediate reward r_i obtained by executing action a_i at state s_i and the total discounted expected future rewards obtained from new available states. Equation (1) used to update the Q-values, where $0 \leq \alpha \leq 1$ is the learning rate and $0 \leq \gamma \leq 1$ is the discount rate.

$$Q_{(s',a')} \leftarrow Q_{(s,a)} + \alpha \left[r + \gamma max_a Q_{(s',a')} - Q_{(s,a)} \right] \quad (1)$$

The learning rate weighs the influence of the received rewards $r \in R$ in the learning process. A rate of 0 will make the agent not learn anything, while a rate of 1 would make the agent consider only the most recent reward. The discount factor weighs the influence of the future rewards. A rate of 0 will make the agent opportunistic by only considering current rewards, while a rate approaching 1 will make it venture for a long-term high reward. The learning rate and the discount rate can only be set by the means of trial and error. In this paper, α and γ are adjusted to 0.8 and 0.2 respectively to obtain a marginally satisfactory result.

3 Policy

3.1 Random Search

$$a_i = rand(a_n) \quad (2)$$

An agent under random search policy will select its action $a \in A$ without any influence of the rewards that are observed from the environment. This makes the agent to explore around in the environment with no proper directions or goals. On the other hand, continuous exploration gives the agent to react rapidly to changes that takes place in the environment.

3.2 Greedy

$$a_i = max_a Q_{(s,a)} \quad (3)$$

Following the greedy policy, the agent selects action $a \in A$ based on the highest $Q_{(s,a)}$ estimate of the available actions. The action considered by the greedy policy may depend on actions made so far but not on future actions. Thus, the greedy approach will select the action with the greatest mean, and would not attempt to perform exploration.

3.3 ϵ-Greedy

$$a_i = \begin{cases} rand(a_n) & rand(0,1) \leq \epsilon \\ max_a Q_{(s,a)} & otherwise \end{cases} \quad (4)$$

The ϵ-greedy policy is an extension of the greedy policy. The only difference is ϵ-greedy allows a certain degree of random exploration with a probability of ϵ. The ϵ is a small positive value, $0 < \epsilon < 1$ which is determined by trial and error. High values of ϵ will force the agent to explore more frequently and as a result will prevent the agent from concentrating its choices to the optimal action, while giving the agent the ability to react rapidly to changes that takes place in the environment. Low value of ϵ will drive the agent to choose optimal actions. For the studied case, ϵ=0.2 improved the quality of the solution marginally out of the set of tested ϵ values.

3.4 Boltzmann Distribution

$$P = \frac{e^{(max_a Q_{(s,a)}/T)}}{\sum e^{(Q_{(s,a)}/T)}} \tag{5}$$

$$a_i = \begin{cases} max_a Q_{(s,a)} & P > x \\ rand\,(a_n) & otherwise \end{cases} \tag{6}$$

$$T_{new} = e^{(-dj)} T_{max} + 1 \tag{7}$$

An agent under Boltzmann distribution will choose an action based on the probability P shown in (5). T is a positive parameter called the temperature which is set through trial and error. Parameter x is also to be set through trial and error. High temperature causes the available actions to be equally preferable. As the temperature decays based on (7) when the iteration j increases, Boltzmann distribution becomes closer to greedy policy. For the foraging task in this paper, temperature T_{max} set to 500, decay rate d is set 0.009 and the threshold value x is set to 0.26.

4 Problem Environment

In this section, an integrated model representing the multi-agent foraging task and Q-learning is proposed and the influences of various policies mentioned in the previous sections are studied. The foraging task consists of four agents, an environment which is divided into few rooms by walls, a home which is located exactly in the middle of environment and ten pucks which are placed randomly in the environment. The simulation environment was modeled completely in [2].

4.1 The Agent

The agent is modeled strictly based on Khepera II mobile robot. The maximum sensing range of the proximity sensors are set to 50 mm and the minimum range to 20 mm. When an obstacle enters the minimum sensing range of the agent, the agent is considered to have collided with an obstacle.A proximity sensor is used to sense the presence of puck in the gripper module. Based on the input from the sensor, the agent will react accordingly by gripping or releasing the puck from or to the desired location. The simulation model can be directly downloaded into a real Khepera II mobile robot from Webots for real time experiments. The width and length of the environment each are 2000mm. The environment is separated into grid of 100 by 100 mm each resulting in 441 states for the agent to explore. Walls are located in the environment to divide the environment into four rooms and also act as a boundary for the environment. The wall also acts as obstacles for the agent. The home location is located exactly in the middle of the environment. The agent will deposit the collected pucks at the home position. Pucks are placed in the environment at predetermined locations. There are ten pucks distributed evenly in the environment.

5 Methodology

5.1 Objective Functions

$$F_1 = min \frac{\sum_{i=1}^{n} E_i}{\sum_{i=1}^{n} A_i} \quad (8)$$

$$F_2 = max \frac{\sum_{i=1}^{m} P_i x_i}{P} \quad (9)$$

$$F_3 = min \frac{S_p}{S} \quad (10)$$

$$F_{all} = max \left[w_1 (1 - F_1) + w_2 F_2 + w_3 (1 - F_3) \right] \quad (11)$$

The less number of collisions made over the total sum of actions taken throughout the simulation characterizes a healthier performance for the agent which can defined via (8). Higher number of pucks collected throughout the entire simulation period characterizes a healthier performance for the agent which can be defined via (9). Less amount of time taken to retrieve all the pucks characterizes a healthier performance for the agent which is defined via (10). The agent's aim is to minimize (8), maximize (9) and minimize (10). This can be achieved by using equation derived in (11). A weight factor of w_1, w_2 and w_3 were used to introduce selective preferences on the objective functions. In this paper, w_1 is set to 0.2, w_2 is set to 0.5 and w_3 is set to 0.3. In an ideal case, (11) will produce the value of 1. Total number of pucks in the environment, P is set to 10, while the sum of actions throughout the simulation period, A and the sum of collision made, E varies depending on the agent's reaction based on the simulation's conditions. Available simulation time S is set to 36000 milliseconds and S_t is the time taken to retrieve all the pucks in the environment.

5.2 States and Actions

Set of states s_i observed by an agent are coordinate values in the form of $\{x_i, z_i\}$. At a given time, the agent has four available states $s = \{(x_1, y_1), (x_2, y_2), (x_3, y_3), (x_4, y_4)\}$ and four available actions $a = \{\text{front; back; left; right}\}$. Based on the policy adopted the agent will choose the desirable action a_i at the current available states s_i. The set of actions changes relatively to the agent. The x-axis and z-axis has a coordinate value ranging from ± 1000 to 0 which results in 441 states overall for the agent to explore and exploit. Q-values that are updated or reinforced are stored into a table called Q-table.

5.3 Rewards

Reward shows the desirability of the action taken by the agent towards the perceived states. If the action taken is attractive, a positive reward is given to the agent. If the action taken is not attractive then a negative reward is given to the agent. For the foraging task tested in this paper, a set of rewards are defined. A +1.0 reward is given to the agent for picking up a puck from the environment. This triggers the agent to visit the same location again to look for more rewards. A reward of +1.0 is also given to the agent for dropping the puck at home location. The agent is given a punishment of -0.1 if the agent wanders around the home without a puck in hand. This will lead the agent to escape the home location and perform search in the environment. A punishment of -1.0 is given if the agent experiences any collision with the obstacles in the environment.

5.4 Experimental Procedures

The main sub-tasks are wander and homing. The Q-table for wandering and homing are separated. This is to avoid the agent from confusing the rewards received from both of the sub-tasks. The Q-table$_{wander}$ will prompt the agent to leave the home location and wander into the environment to search for pucks. Therefore the Q-table$_{wander}$ will have negative rewards congregated at the home location. Once the agent encounters with a puck in the environment, the agent picks up the puck, updates the $Q_{(s,a)}$ value in Q-table$_{wander}$ and that point forward the Q-table$_{homing}$ will used to lookup on the $Q_{(s,a)}$ values. The Q-table$_{homing}$ will have positive rewards congregated at the home location and therefore the agent will pursue the home location to deposit the puck. The simulation starts by setting the episode count and Q$_{(s,a)}$ values in the both Q-tables are set to 0. In the initialization stage, the timer count is set to 0 followed by the reseting of agent's and the puck's locations to their respective coordinates in the environment. The agent and the pucks are located to their relative coordinates in the environment. Then the condition to verify whether the agent is holding a puck in its gripper is checked. Since the episode count is 0 at the beginning of the simulation, the condition will be false and therefore the agent goes into a wandering mode. All the relevant $Q_{(s,a)}$ values are updated and referred from the Q-table$_{wander}$ .Then the agent will observe the available states in the environment. The agent will then select an action based on the policy given. Rewards for the action taken is calculated. The relevant $Q_{(s,a)}$ value in the Q-table$_{wander}$ is updated using (1). If ten hours of the simulation time is reached, the episode counter will be incremented by one indicating the completion of one episode and the condition is checked for the completion of 30 episodes. Else the agent will continue to map the state-action pairs until it reaches ten hours of simulation time and proceed into checking for the completion of 30 episodes. If the 30 episodes have been completed, the simulation will be terminated. Otherwise the same process will commence until the termination condition is met. If the agent found a puck, it will pick up the puck and a reward of +1 will be given to the agent. This reward will be updated in the Q-table$_{wander}$. After this state, the agent will be using Q-table$_{homing}$ to update the relevant Q$_{(s,a)}$ values until the puck is deposited at the home location. This process continues until all the pucks are collected or the termination condition is met.

6 Results and Discussion

The performances of the learning policies in the experiment conducted are reported in this section. Each of the policies is tested for 30 runs and the results are presented below. Figure 1 shows the overall performance that is calculated using (11). The greedy policy doesn't allow any exploration of the states once the learning process starts converging. So the agent will be forced to select the best action in every state and through that the states that are not desirable to the agent will not be explored again. This prevents the agent from experiencing the same collision again. Therefore greedy policy is well suited for collision avoidance. The ϵ-greedy policy on the other hand allows the agent to explore the undesirable states with a probability of ϵ. So at some point, even if the agent has found the optimal solution, the agent will still choose a random action ϵ of the time. Therefore there is a probability that the collision may occur in the environment again. Boltzmann distribution allows exploration in the beginning of the episodes and slowly diverges to greedy policy's characteristics as the temperature drops. A low temperature means that the selection will be greedier while a high temperature means that it will be more random. The major advantage of this selection method is that the $Q_{(s,a)}$ value is taken into account when selecting an action, meaning that more exploration will be done when the $Q_{(s,a)}$ values are similar, and less when there is one action that has a significantly higher $Q_{(s,a)}$ value than the others. The down side of the Boltzmann distribution is that the policy generally requires more time to converge compared to other policies. The random search performed poorly compared to the policies modeled in this paper. Based on the plot, it is obvious that the multi-agent system can perform better in a foraging task by adopting ϵ-greedy policy.

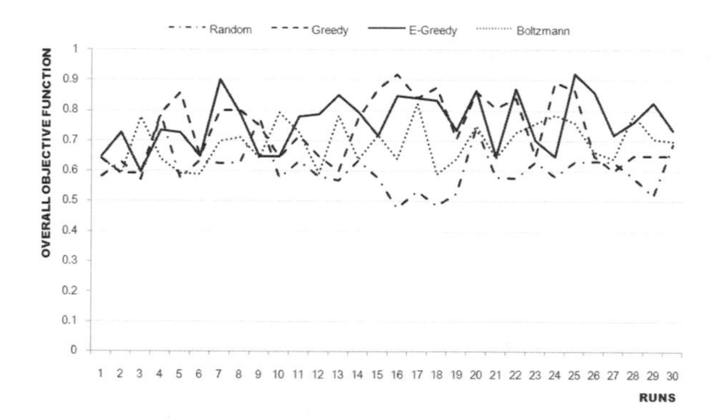

Fig. 1. Overall performance

7 Conclusions

In this paper, the behavior of a q-learning agent using the reported policies namely greedy policy, ϵ-greedy policy, Boltzmann distribution and random search have been studied and the results are reported. Through the experiment conducted, although all the policies carry their own advantages and disadvantages, it is clear that ϵ-greedy policy is much more practical and effective compared to the other policies studied. It is necessary to identify the suitable policies in q-learning for different types of tasks as the suitable policies helps to accelerate the convergence of the agent's learning process. There are more directed policies available in the open literature which can be also considered for the study.

Acknowledgment

This research is funded by e-Science grant provided by Ministry of Science, Technology and Innovation (MOSTI), Project Number: MONASH0000186.

References

1. Koulouriotis, D.E., Xanthopoulos, A.: Reinforcement Learning and Evolutionary Algorithms for Non-stationary Multi-armed Bandit Problems. Applied Mathematics and Computation 196, 913–922 (2008)
2. Webots: Commercial Mobile Robot Simulation Software, `http://www.cyberbotics.com`
3. Ji, Z., Wu, Q., Sid-Ahmed, M.: An Improved Immune Q-learning Algorithm. In: IEEE International Conference on Systems, Man and Cybernetics, pp. 1636–1641 (2007)
4. Gomes, E., Kowalczyk, R.: Dynamic Analysis of Multiagent Q-learning with E-greedy Exploration. In: Proceedings of the 26th International Conference on Machine Learning, vol. 382, pp. 369–376 (2009)
5. Tuyls, K., Verbeeck, K., Lenaerts, T.: A Selection-Mutation Model for Q-learning in Multi-agent Systems. In: Proceedings of the Second International Joint Conference on Autonomous Agents and Multiagent Systems, pp. 693–700 (2003)
6. Morihiro, K., Isokawa, T., Nishimura, H., Matsui, N.: Emergence of Flocking Behavior Based on Reinforcement Learning. In: Gabrys, B., Howlett, R.J., Jain, L.C. (eds.) KES 2006. LNCS (LNAI), vol. 4253, pp. 699–706. Springer, Heidelberg (2006)
7. Dahmani, Y., Benyettou, A.: Seek of an Optimal Way by Q-Learning. Journal of Computer Science 1(1), 28–30 (2005)
8. Whiteson, S., Taylor, M., Stone, P.: Empirical Studies in Action Selection with Reinforcement Learning. Adaptive Behavior 15(1), 33–50 (2007)
9. Price, B., Boutilier, C.: Accelerating Reinforcement Learning Through Implicit Imitation. Journal of Artificial Intelligence Research 19, 569–629 (2003)

An Intelligent Algorithm for the Path Planning of Autonomous Mobile Robot for Dynamic Environment

Soumic Sarkar, Sankar Nath Shome, and S. Nandy

Robotics and Automation
Central Mechanical Engineering Research Institute, Durgapur, CSIR, India
Soumic4it@gmail.com

Abstract. Collision avoidance is an important aspect of path planning for mobile robotic systems, and has remained an issue of intensive research for researchers all over the world. Among the several techniques adopted for the same purpose, potential field force (PFF) is the most widely accepted and reliable strategy. However, the technique is impregnated with the drawback concerning "Local Minima". The aim of this paper is to present an algorithm that obviously survives the limitations posed by PFF which has been taken care efficiently with the help of reinforcement of a virtual attractive force that makes the resultant force parallel to the boundary of the obstacle. The algorithm also includes the speed control mechanism of the robot. Computational complexity of the given algorithm is also provided. Simulations results are presented that prove the effectiveness of the proposed method. The algorithm has been tested on an indoor mobile robotic platform.

Keywords: Potential Field Force, Obstacle Avoidance, Virtual Field Force, Path Panning, Wall Follow-Through Method, Speed Control.

1 Introduction

Obstacle avoidance is one of the key issues to Intelligent Path Planning of autonomous mobile robots. All autonomous mobile robots exhibit some kind of collision avoidance feature, ranging from primitive algorithms [1-6] that detect an obstacle and stop the robot short of it in order to avoid a collision, through sophisticated algorithms, that enable the robot to detour obstacles. The latter algorithms using Fuzzy Logic or Genetic Algorithm are much more complex, since they involve not only the detection of an obstacle, but also some kind of quantitative measurements concerning the obstacle's dimensions. Once these have been determined, the obstacle avoidance algorithm needs to steer the robot around the obstacle and move toward the original target.

Autonomous navigation, in general, assumes an environment with known and unknown obstacles, and it includes global path planning algorithms to plan the robot's path among the known obstacles, as well as local path planning for avoiding unknown obstacles. This article, however, assumes motion in the presence of unknown obstacles, and therefore concentrates only on the local obstacle avoidance aspect so that the algorithm can be used in untried territory like in undersea applications.

P. Vadakkepat et al. (Eds.): FIRA 2010, CCIS 103, pp. 202–209, 2010.

One approach to autonomous navigation is the wall-following method (WFM). Here the robot navigation is based on moving alongside walls at a predefined distance. If an obstacle is encountered, the robot regards the obstacle as just another wall, following the obstacle's contour until it may resume its original course. This kind of navigation is technologically less demanding, since one major problem of mobile robots (the determination of their own position) is largely facilitated. But the problem with "local minima" of potential field force can be eliminated if it's integrated with WFM. Naturally, robot navigation by the WFM along with potential field method is versatile and is suitable for many applications.

2 Potential Field Approach

Potential field path planning creates a field, or gradient, across the robots map that directs the robot to the goal position from multiple prior positions. This approach was originally invented for robot manipulator path planning [1] and is used often and under many variants in the mobile robotics community. The potential field method treats the robot as a point under the influence of an artificial potential field $U(q)$. The basic idea behind all potential field approaches is that the robot is attracted towards the goal while being repulsed by the obstacles that are known in advance.

If new obstacles appear during robot motion, one could update the potential field in order to integrate this new information. In the simplest case, we assume that the robot is a point thus the robot's orientation θ is neglected and the resulting potential field is only 2D (x, y). If we assume a differentiable potential field function $U(q)$, we can find the related artificial force $F(q)$ acting at the position $q=(x,y)$.

$$F(q) = -\nabla U(q)$$

Where, $\nabla U(q)$ denotes the gradient vector of U at position q.

$$\nabla U = \begin{bmatrix} \frac{\partial U}{\partial x} \\ \frac{\partial U}{\partial y} \end{bmatrix}$$

The potential acting on the robot is the computed as the sum of attractive field of the goal and the repulsive field of the obstacles:

$$U(q) = U_{att}(q) + U_{rep}(q)$$

Similarly, the forces can also be separated in an attracting and a repulsing part:

$$\begin{aligned} F(q) &= F_{attr}(q) - F_{rep}(q) \\ &= -\nabla U_{attr(q)} - \nabla U_{rep}(q) \cdot \end{aligned}$$

3 Attractive Potential

An attractive potential [1, 4] is defined as a parabolic function

$$U_{attr}(q) = \frac{1}{2} k_{attr} . \rho^2_{goal}(q)$$

where, k_{attr} is a positive scaling factor and $\rho_{goal}(q)$ denotes the Euclidean distance $\left\| q - q_{goal} \right\|$. This attractive potential is differentiable, leading to the attractive force F_{attr}.

$$\begin{aligned} F_{attr}(q) &= -\nabla U_{attr}(q) \\ &= -k_{attr} .(q - q_{goal}) \\ &= k_{attr} .\rho_{goal}(q) \nabla \rho_{goal}(q) \end{aligned}$$

The attractive force converges linearly toward 0 as the robot reaches the goal.

4 Repulsive Potential

The idea behind the repulsive potential [1,4] is to generate a force away from all known obstacles. This repulsive potential should be very strong when the robot is closed to the object but shouldn't influence its movement when the robot is far from the object. One example of such a repulsive field is

$$U_{rep}(q) = \begin{cases} \frac{1}{2} k_{rep} \left(\frac{1}{\rho(q)} - \frac{1}{\rho_0} \right)^2 & if \;\; \rho(q) \leq \rho_0 \\ 0 & if \;\; \rho(q) \geq \rho_0 \end{cases}$$

Where k_{rep} is again a scaling factor $\rho(q)$ is the minimum distance from q to the obstacle and ρ_0 the distance of influence of the object. The repulsive potential function U_{rep} is positive or zero and tends to infinity as q gets closer to the object.

If the object boundary is convex and piecewise differentiable, $\rho(q)$ is differentiable everywhere in the free configuration space. This leads to the repulsive force F_{rep}:

$$= \begin{cases} k_{rep} \left(\frac{1}{\rho(q)} - \frac{1}{\rho_0} \right) \frac{1}{\rho^2(q)} & if \;\; \rho(q) \leq \rho_0 \end{cases}$$

$$F_{rep}(q) = -\nabla U_{rep}(q)$$

The resulting force $F(q)=F_{rep}(q)+F_{attr}(q)$ acting on a point robot exposed to the attractive and repulsive forces moves the robot away from the obstacles and toward the goal.

5 Trap Recognition and Recovery

The concept of recovery algorithm [2] comes from the fact when the proposed algorithm fails to overcome a situation along the journey of the robot. It means the robot gets stuck into an infinite loop situation without a break point which is same as following the same path over and over again in the midst of its journey without having an option to come out of the situation so that the robot begin its journey towards goal again after being recovered from the situation like a whirlpool or Trap.

The WFM algorithm is implemented in the following manner: First, the robot calculates the sum of all repulsive forces, F_{rep}, in order to determine the direction of F_{rep},θ_{rep}. Next, the algorithm adds an angle α to θ_{rep}, where $90^0 < \alpha < 180^0$ (or subtracts from θ_{rep}, if following a wall to the right of the robot). In our system $\alpha = 145^o$. Then, the algorithm projects a new virtual attractive force in the resulting direction, which temporarily replaces the attractive force from the target location. The new resultant F_{res} will now point (or eventually converge) into a direction parallel to the obstacle boundary. While wall-following could also be implemented through controlling a fixed distance to the wall, our method is preferable, since it is less sensitive to misreadings of the ultrasonic sensors.

6 Speed Control

The intuitive way to control the speed [2, 3] of a mobile robot in the PFF environment is to set it proportional to the magnitude of the sum of all forces, $F_{res}(q)=F_{rep}(q)+F_{attr}(q)$. Thus, if the path was clear, the robot would be subjected only to the target force and would move toward the target, at its maximum speed. Repulsive forces from obstacles, naturally opposed to the direction of F_{attr}, would reduce the magnitude of the resultant F_{res} thereby effectively reducing the robot's speed in the presence of obstacles.

However, we have found that the overall performance can be substantially improved by setting the speed command proportional to $cos\theta$. This function is given by:

$$V = \begin{cases} V_{MAX} & for\left|F_{rep}\right| = 0 \\ V_{MAX}(1 - \left|\cos\theta\right|) & for\left|F_{rep}\right| > 0 \end{cases}$$

With this function, the robot still runs at its maximum speed if no obstacles are present.

However, in the presence of obstacles, speed is reduced only if the robot is heading toward the obstacle (or away from it. If, however, the robot moved alongside an obstacle boundary, its speed is almost not reduced at all and it moves at its maximum speed, thereby greatly reducing the overall travel-time.

7 The Algorithm

The following algorithm, as shown in Fig. 1, first checks if there is any obstacle. In the absence of obstacle the robot moves straight towards goal with the help of only the attractive force. When some obstacle is detected the algorithm checks whether the robot is within a critical distance from the obstacle. Within the critical distance a repulsive force is generated and thereby, a resultant force is calculated and the robot moves in the direction of the resultant force. If the robot gets stuck in a trap or if the forces are collinear the algorithm switches to wall following mode.

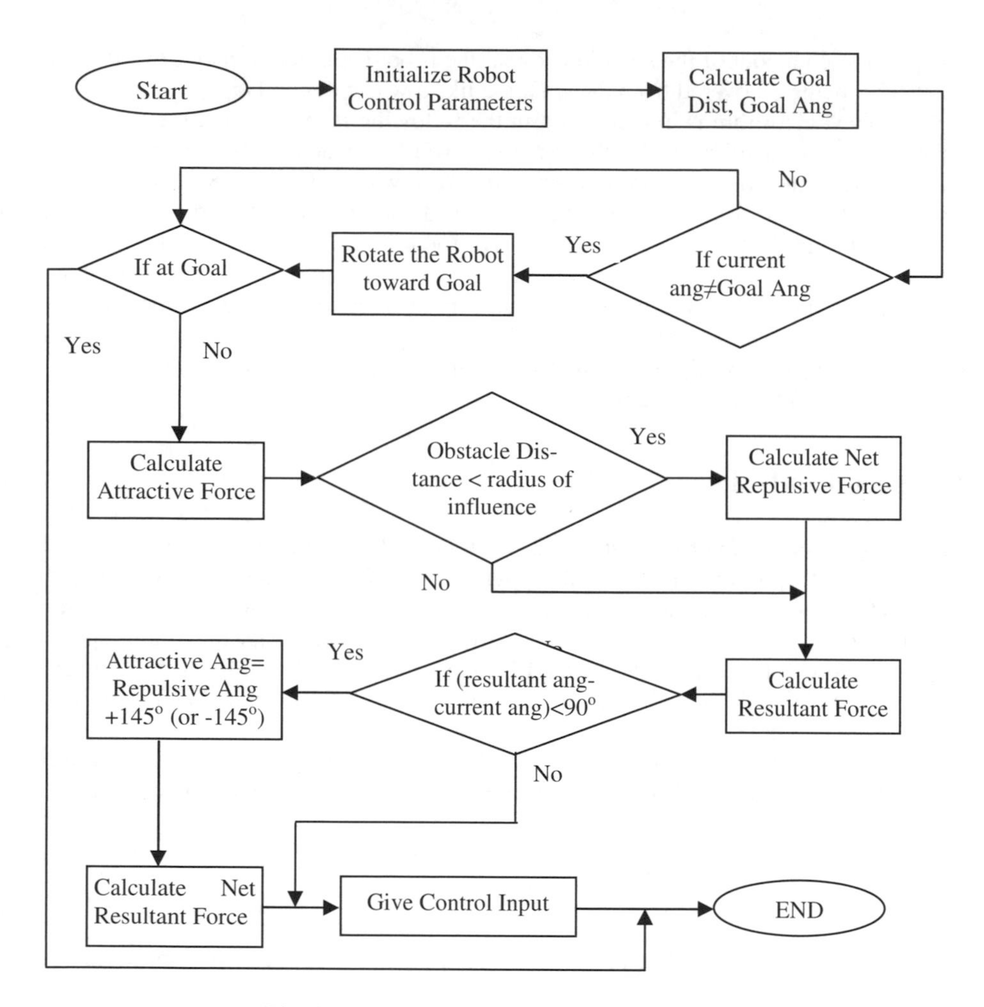

Fig. 1. Schematic Representation of the algorithm

8 Computational Complexity

The computational complexity of the above algorithm depends upon the number of times the set of data from sensor is being used at each iteration. As the goal position is given, the robot calculates the attractive force depending upon its position at each iteration. The presence of obstacle is observed through sensor. Suppose a set of n data is received from sensor. For a set of n data, the algorithm checks n times for a match whether the distance measured through sensor is within the radius of influence. A match refers to the calculation of the repulsive force for that distance and angle. The magnitude and the direction of net repulsive force is finally calculated when all the repulsive forces from the n sensory data for the current iteration is calculated. Hence it can be inferred from the above discussion that the computational complexity of the algorithm is of the order $O(n)$, because the algorithm needs to check from a set of n sensor reading whether the robot is inside the radius of influence, q_0 at each iteration.

9 Simulation

The algorithm is simulated in Matlab taking a working environment of 50 x 50 as shown in the following figures. All the obstacles have been taken as point obstacles. The plot becomes clumsy where the robot moves with less speed. The red points are point obstacles and the black point is the goal point.

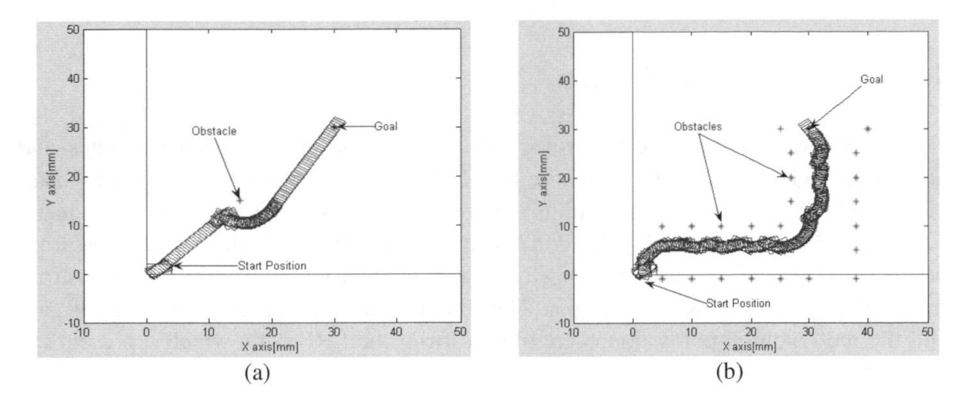

Fig. 2. Schematic Representation of workspace with Robot, Start Point, Goal Point and Obstacle position in 1st Quadrant; Movement through corridor

In Fig. 2.a above, the robot initially faces along the X-axis. Then it rotates towards the goal and then starts moving driven by the attractive force. It faces a point obstacle on the way and the obstacle generates repulsive force upon the robot so that the robot doesn't collide with the obstacle anyway. The robot moves in the direction of the resultant force and thus moves away from the obstacle. Fig. 2.b represents movement of robot through a corridor. From the figure it's noticed that the robot first tries to moves towards goal but it meets a series of obstacles which could be considered as wall, and thus makes oscillatory motion. The oscillation can be avoided if the obstacles are closely spaced.

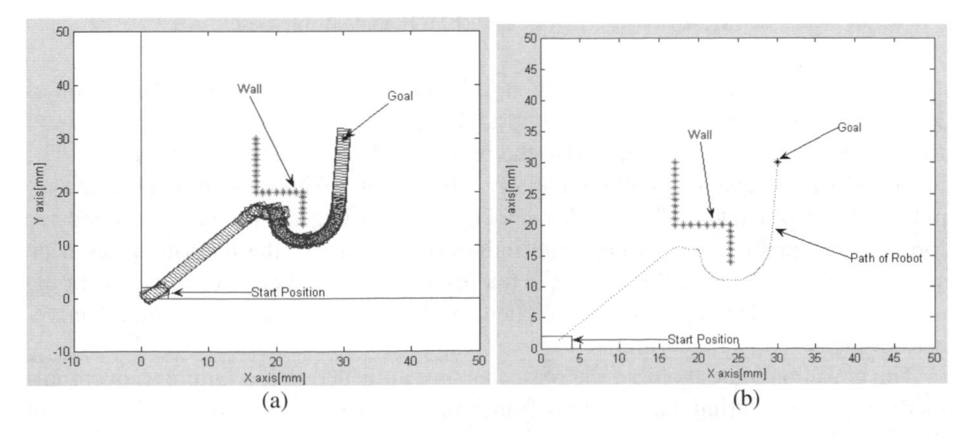

Fig. 3. Recovery from Trap through Wall Follow Through

In Fig. 3. a, b, the robot gets inside a trap and switches to the wall follow through algorithm. Here a virtual force is generated which is same as the magnitude of the original attractive force but the direction of the force is always 145° away from the repusive force. As a result there is always a resultant force which acts in the direction parallal to the wall and the robot moves accordingly until it comes out of the trap. Once, the robot is out of trap, it switches to the normal mode, ie, into the PFF mode.

10 Experimental Validation

The experiments have been performed considering multiple artificial obstacles and natural obstacles, as shown in Fig. 4 (a,b). The robot effectively avoids all the obstacles based on the developed algorithm. The results of testing obstacle avoidance using LRF data are presented in Fig. 4 (c, d). The plot of actual position is achieved from odometry information while moving from start to goal position through an unstructured environment shown in Fig. 4.a. The blue coloured points are actual robot path and the red coloured point clouds are plotted from the LRF data, which represent the positions of the obstacles.

(a) (b)

Fig. 4. (a)The environment; (b) robot avoids obstacle

The experiment data is analyzed through MATLAB for a workspace dimension of 3000 x 8000 sq.mm. In this case, the obstacles are not point obstacles. The value of K_{rep} is kept as high as 10^{12} otherwise the repulsive force generated is very low. The linear velocity is given as 400m/s initially which is inversely proportional to the repulsive force generated. The angular velocity is according the direction of repulsive force. The sampling time is kept 500 miliseconds. Though the algorithm has been simulated for point obstacles but the actual experiment has been performed with 3D obstacles. The videography and the plotted result in Fig. 4 (c,d) shows the effectiveness of the developed algorithm.

The deviation after avoiding the obstacles or wall is due to the failure to overcome inertia of motion during the switching from wall following to potential field section of the algorithm.

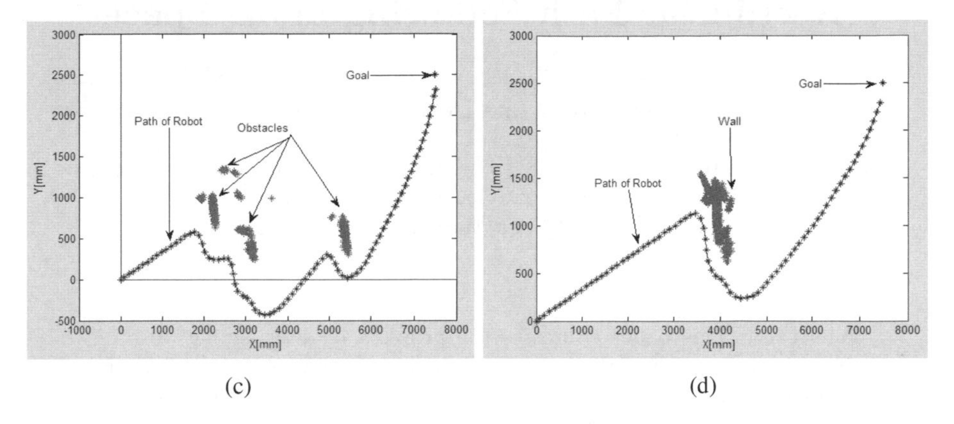

(c) (d)

Fig. 4. Plot of position of robot obtained from Encoder

References

1. Khatib, O.: Real-time Obstacle Avoidance for Manipulator and Mobile Robots. The International Journal of Robotics Research 5(1), 90–98 (1986)
2. Borenstein, J., Koren, Y.: Real-time Obstacle Avoidance for Fast Mobile Robots. IEEE Transactions on Systems, Man, and Cybernetics 19(5), 1179–1187 (1989)
3. Borenstein, J., Koren, Y.: Obstacle Avoidance With Ultrasonic Sensors. IEEE Journal of Robotics and Automation RA-4(2), 213–218 (1988)
4. Siegwart, R., Nourbaksh, I.R.: Introduction to autonomous mobile robots. The MIT Press, Cambridge
5. Krogh, B.H.: A Generalized Potential Field Approach to Obstacle Avoidance Control. In: International Robotics Research Conference, Bethlehem, PA (August 1984)
6. Krogh, B.H., Thorpe, C.E.: Integrated Path Planning and Dynamic Steering Control for Autonomous Vehicles. In: Proceedings of the 1986 IEEE International Conference on Robotics and Automation, San Francisco, California, April 7-10, pp. 1664–1669 (1986)

Rectilinear Path Following in 3D Space

Sikha Hota and Debasish Ghose

Guidance, Control and Decision Systems Lab., Aerospace Engineering Dept.,
Indian Institute of Science, Bangalore, India
{sikhahota,dghose}@aero.iisc.ernet.in

Abstract. This paper addresses the problem of determining an optimal (shortest) path in three dimensional space for a constant speed and turn-rate constrained aerial vehicle, that would enable the vehicle to converge to a rectilinear path, starting from any arbitrary initial position and orientation. Based on 3D geometry, we propose an optimal and also a suboptimal path planning approach. Unlike the existing numerical methods which are computationally intensive, this optimal geometrical method generates an optimal solution in lesser time. The suboptimal solution approach is comparatively more efficient and gives a solution that is very close to the optimal one. Due to its simplicity and low computational requirements this approach can be implemented on an aerial vehicle with constrained turn radius to reach a straight line with a prescribed orientation as required in several applications. But, if the distance between the initial point and the straight line to be followed along the vertical axis is high, then the generated path may not be flyable for an aerial vehicle with limited range of flight path angle and we resort to a numerical method for obtaining the optimal solution. The numerical method used here for simulation is based on multiple shooting and is found to be comparatively more efficient than other methods for solving such two point boundary value problem.

Keywords: optimal path, 3D path planning, rectilinear path following.

1 Introduction

The shortest path calculation between any two given configurations plays a key role in path planning. In the 2D plane, the smooth shortest path for fixed initial and final positions and orientations has been obtained geometrically by Dubins [1]. Reeds and Shepp [2] solved a similar problem using advanced calculus in which a vehicle can move forward as well as backward. Boissonnat et al. [3] proved the same result as Dubins using the powerful Pontryagin's minimum principle. Using Dubins result in 2D plane optimal path planning has been discussed in [4]-[6]. McGee et al. [7] used Dubins path to explore the problem of finding an optimal path for a UAV in constant wind condition. Nelson et al. [8] discuss a 2D non-optimal vector field approach for straight line path following in the presence of wind. Time-optimal trajectory in 3D from any starting point and orientation to any final point and orientation has been discussed in [9]-[11].

P. Vadakkepat et al. (Eds.): FIRA 2010, CCIS 103, pp. 210–217, 2010.

By using the Maximum Principle on manifolds, Sussmann [12] has shown that every minimizer in 3D is either a helicoidal arc or a concatenation of three pieces each of which is a circle or a straight line. He also showed that unlike 2D Dubins path, for sufficiently small distance between initial and final points, there can exist a helicoidal path that is shorter than any CSC (C-circular, S-straight line) path. In [13] and [14], 3D smooth path planning has been discussed using two important properties of 3D curves, curvature and torsion.

In this paper we discuss an efficient algorithm to compute an optimal path (in this paper the optimal path indicates the shortest path) with a prescribed curvature constraint in 3D space for a given initial configurations and the straight line to be followed. It is assumed that the initial point and straight line are situated sufficiently far from each other. For this case, there will always exist a CSC Dubins path from its initial point to any point on the straight line. A CSC path in 3D will be constructed by using the same principle as of Dubins curve in 2D. We will compute the optimal length CSC path from all the existing CSC paths satisfying the initial and final conditions. Later, an efficient suboptimal method will be discussed which yields a path length very close to the optimal one and due to its simplicity and low computational requirements this approach can be implemented on an aerial vehicle with constrained turn radius to reach a straight line with a prescribed orientation as required in several applications. But if the distance between the initial point and the straight line along the vertical axis is high, then the generated path may not be flyable (for an aerial vehicle with limited range of flight path angle) and we need to go for a numerical method for optimal solution. The numerical method used here for simulation is multiple shooting that is comparatively more efficient than other methods for solving two point boundary value problems [15].

2 Problem Definition

The problem is to determine a path of minimum length for a turn rate constrained vehicle between the initial point X_0 (x_0, y_0, z_0) and the straight line to be followed specified by its orientation vector (v_2) and a point X_s (x_s, y_s, z_s) situated on it. The orientation vectors are specified in terms of heading angle (ψ) and flight path angle (γ). The unit orientation vectors at initial point is v_1 $(v_{1x}, v_{1y}, v_{1z}) = [\cos\gamma_0 \cos\psi_0 \quad \cos\gamma_0 \sin\psi_0 \quad \sin\gamma_0]$ and the unit orientation vector for the straight line is v_2 $(v_{2x}, v_{2y}, v_{2z}) = [\cos\gamma_f \cos\psi_f \quad \cos\gamma_f \sin\psi_f \quad \sin\gamma_f]$ (Fig. 1.(a)).

This problem has been formulated as an optimal control problem and is stated below. In cartesian coordinate frame (x, y, z), fixed with respect to the earth, and the path can be represented by the following equations:

$$\frac{dx(s)}{ds} = \cos\psi(s)\cos\gamma(s) \tag{1}$$

$$\frac{dy(s)}{ds} = \sin\psi(s)\cos\gamma(s) \tag{2}$$

$$\frac{dz(s)}{ds} = \sin\gamma(s) \tag{3}$$

$$\frac{d\psi(s)}{ds} = \eta \tag{4}$$

$$\frac{d\gamma(s)}{ds} = \mu \tag{5}$$

where, s represents the curvilinear abscissa on the path and $\psi(s) \in [-180^0, 180^0]$, $\gamma(s) \in [-90^0, 90^0]$. For this problem η and μ will play the role of control inputs. As the vehicle is constrained by its minimum turn radius (r) so the path should have a maximum curvature limit (c_{max}). The curvature profile $c(s)$ can be calculated as given below:

$$c(s) = \sqrt{\eta^2(s)\cos^2\gamma(s) + \mu^2(s)} \tag{6}$$

Let the state vector, $q(s) = (x(s)\ y(s)\ z(s)\ \psi(s)\ \gamma(s))^T$ and the control vector, $u(s) = (\eta(s)\ \mu(s))$. So, the optimal control problem is to minimize the length of path,

$$J(q_0, q_f, u) = \int_0^{s_f} ds \tag{7}$$

where, the state equations are given as,

$$\frac{dq(s)}{ds} = f(q(s), u(s)) \tag{8}$$

and the boundary conditions can be stated as follows: initial condition is q_0 $(x_0, y_0, z_0, \psi_0, \gamma_0)$; final condition, q_f is partially given; (ψ_f, γ_f) are specified and X_f (x_f, y_f, z_f) is situated on the line represented by $X_s + t_{var}v_2$, where t_{var} is the variable parameter. The objective is to minimize the path length with a guarantee of the satisfaction of the constraints

$$-c_{max} \leq c(s) \leq c_{max} \tag{9}$$

3 Solution Approaches

In this section, we discuss a geometric approach to construct a CSC path when initial and final points are situated sufficiently far away. We first compute the minimum length CSC path to reach the straight line. Then, based on the 3D geometry, we propose a suboptimal approach that yields a solution close to the optimal one.

3.1 CSC Path Construction Using 3D Geometry

The CSC path is obtained by using 3D geometry. The intersecting line between the initial and final circular turns is derived analytically. The 3D geometry is shown in the Fig. 1.(b). The initial and final curvilinear paths of minimum

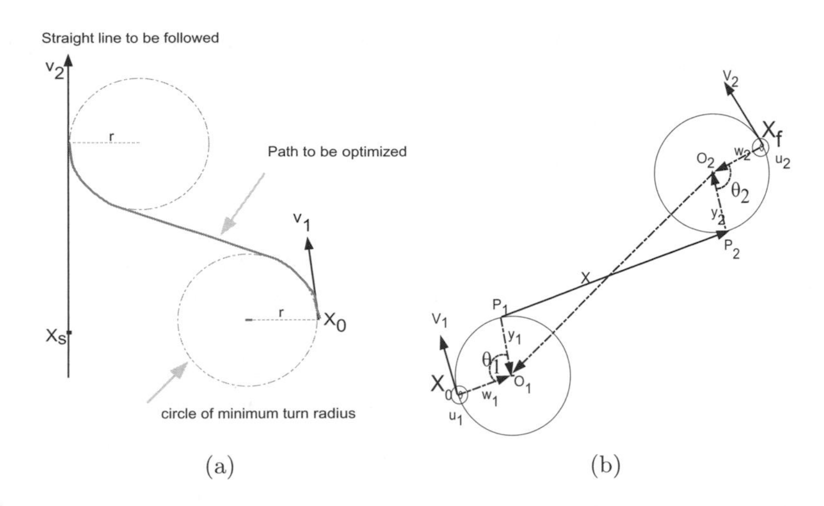

Fig. 1. (a) Problem Definition (b) Initial and final configurations

radius (r) are on different planes and the straight line path segment between these two curves is the intersecting line between these two planes of initial and final maneuvers.

Let the vector that is common for both of these two planes be X (X_x, X_y, X_z). Then, the unit vector x (x_x, x_y, x_z) along this direction is, $x = \frac{X}{||X||}$. The vector perpendicular to the first plane is,

$$U_1 = x \times v_1 \tag{10}$$

The radius vector that is oriented toward the center of the first circle from its initial position is,

$$W_1 = v_1 \times U_1 = v_1 \times (x \times v_1) \tag{11}$$

The center of the first circle is

$$o_1 = X_0 + rw_1 \tag{12}$$

where, $w_1 = \frac{W_1}{||W_1||}$. Then, the radius vector oriented toward the center of the first circle from the point where the tangent line meets the first circle (this is the point from which the vehicle will start to follow a straight line path that is tangential to the initial and final circles) is given by,

$$Y_1 = x \times U_1 = x \times (x \times v_1) \tag{13}$$

The point at which the tangent touches the first circle,

$$P_1 = o_1 - ry_1 = X_0 + rw_1 - ry_1 \tag{14}$$

where, $y_1 = \frac{Y_1}{||Y_1||}$. The vector perpendicular to the second plane is given by,

$$U_2 = x \times v_2 \tag{15}$$

The radius vector that is oriented toward the center of the second circle from its final position is given by,

$$W_2 = -v_2 \times U_2 = -v_2 \times (x \times v_2) \tag{16}$$

The center of the second circle is,

$$o_2 = X_f + r w_2 \tag{17}$$

where, $w_2 = \frac{W_2}{||W_2||}$. Then, the radius vector oriented toward the center of the second circle from the point where the common tangent line meets the second circle is given by,

$$Y_2 = -x \times U_2 = -x \times (x \times v_2) \tag{18}$$

Then, the point at which the tangent touches the second circle is,

$$P_2 = o_2 - r y_2 = X_f + r w_2 - r y_2 \tag{19}$$

where, $y_2 = \frac{Y_2}{||Y_2||}$. We can now solve for X by solving the following equation

$$X = P_2 - P_1 = (X_f + r w_2 - r y_2) - (X_0 + r w_1 - r y_1) \tag{20}$$

Let us introduce two new variables θ_1 and θ_2 which are the first and final turning angles, respectively. This will result in the following equation:

$$\cos\theta_1 = v_1.x \tag{21}$$

$$\cos\theta_2 = v_2.x \tag{22}$$

$$X = (X_f - X_0) - r(x + v_1)\tan\frac{\theta_1}{2} - r(x + v_2)\tan\frac{\theta_2}{2} \tag{23}$$

We can solve these set of equations to get θ_1, θ_2 and X. The total length of the path is $(r\theta_1 + r\theta_2 + ||X||)$.

Note: It is important to note here that when $||X_f - X_0|| > 4r$, CSC path can be obtained for all initial and final orientations.

3.2 An Optimal Path Using Geometrical Method (OGM)

To construct the shortest path from an initial point to straight line as shown in Fig. 2.(a) we need to minimize the length of CSC path that starts from the initial point and meets the straight line.

So, the cost function (C) to be optimized is,

$$C = r\theta_1 + r\theta_2 + ||X|| \tag{24}$$

Where X and X_f are related by the following equation,

$$X = (X_f - X_0) - r(x + v_1)\tan\frac{\theta_1}{2} - r(x + v_2)\tan\frac{\theta_2}{2} \tag{25}$$

And, X_f should lie on the straight line to be followed. So, the constraint that needs to be imposed is as follows.

$$||(X_f - X_s) \times v_2|| = 0 \tag{26}$$

Solving these set of equations (equations (24)-(26)) we can get the shortest path.

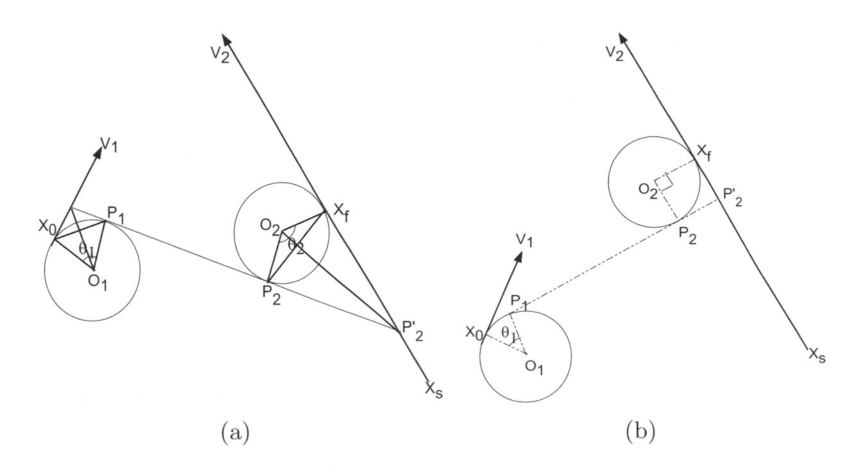

Fig. 2. (a) Geometry of straight line path following (b) Suboptimal Path: straight line segment of the CSC path is perpendicular to the straight line to be followed

3.3 A Suboptimal Path Using Geometrical Method (SGM)

To reduce the computational cost, a suboptimal path can be constructed where the vehicle should take a turn of minimum turn radius from its initial orientation in such a way that it can achieve an orientation direction that is perpendicular to the straight line to be followed. This will reduce the computational time to a larger extent. The meeting point (X_f) that yields the suboptimal path can be calculated in the following way. From Fig. 2.(b) we can write $P_1 = X_0 + r(x + v_1)\tan\frac{\theta_1}{2}$ and,

$$||P_2' - X_s|| = v_2.(P_1 - X_s) \tag{27}$$

$$P_2' = P_1 + x\sqrt{||X_s - P_1||^2 - ||P_2' - X_s||^2} \tag{28}$$

Now, X_f can be obtained as,

$$X_f = P_2' + rv_2 \tag{29}$$

4 Simulation Results and Discussion

For simulation studies we have taken two cases for which the initial and the final conditions are given below. The values of X_0, v_1, v_2 and r will be kept the same for both the cases and different X_s values are considered (see Table 1). We have, $X_0 = [0,0,0]$m, $v_1 = [0.9328, 0.3031, 0.1951]$, $v_2 = [0, .951, .309]$, $r = 5$m. The paths have been generated (Fig 3) using both the methods given in Section 3 and the results have been compared using the numerical method based on multiple shooting (MS) for solving the boundary value problem (equations (1)-(9)). In Table 1, L is the length of the path. The number of intervals taken for multiple shooting is 80. The multiple shooting method, in these cases, for tolerance of 10^{-4}

Table 1. Initial and final conditions for simulation

Case	X_s	L (MS)	L (OGM)	L (SGM)
Case1	[20,0,0] m	22.901 m	22.895 m	22.895 m
Case2	[80,0,0] m	82.906 m	82.896 m	82.896 m

takes about 1.5 min to converge. On the other hand, the geometrical method (OGM) and suboptimal method (SGM) take approximately 25 sec and 1 sec, respectively. The simulations are done in intel pentium 2.60 GHz processor with 1.00GB RAM.

Note: The numerical approach takes more time than the geometrical time-optimal and suboptimal methods as discussed earlier but in case the aerial vehicle is constrained by its maximum and minimum flight path angle ($\gamma_{min} \leq \gamma \leq \gamma_{max}$) and the distance between the initial point and the final straight line along the vertical axis is high, then the computed path using other methods (geometrical optimal and suboptimal methods) may not be flyable and the numerical method is the only option for the shortest path generation for those cases.

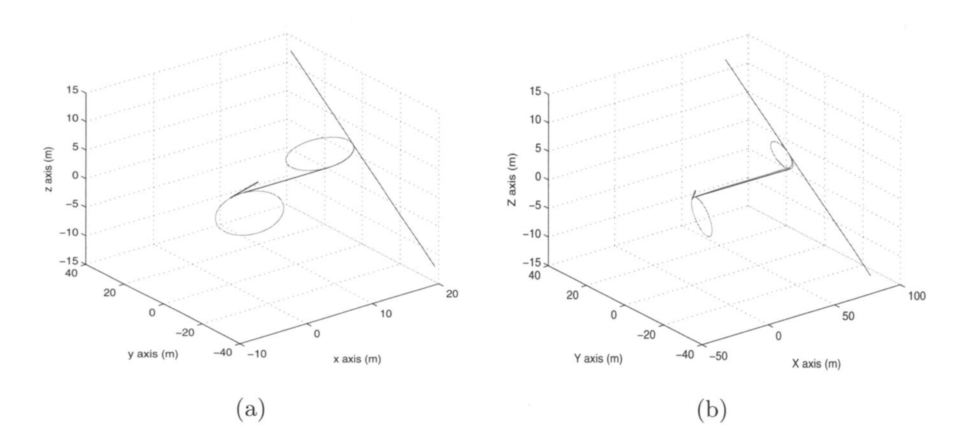

Fig. 3. (a) Case Study 1 (b) Case Study 2 (Red-path using MS, Black-Path using OGM, Magenta-path using SGM)

5 Conclusions

In this paper a straight line path following method in 3D space by an aerial vehicle, has been discussed. Based on 3D geometry, we have derived an optimal and a suboptimal path. This suboptimal solution is efficient from the computational point of view and gives a solution very close to the optimal one. Due to its simplicity and low computational requirements this approach can be implemented on an aerial vehicle with constrained turn radius to reach a straight line with a prescribed orientation as required in several applications.

References

1. Dubins, L.E.: On curves of minimal length with a constraint on average curvature, and with prescribed initial and terminal positions and tangents. American Journal of Mathematics 79, 497–516 (1957)
2. Reeds, J.A., Shepp, L.A.: Optimal paths for a car that goes both forwards and backwards. Pacific Journal of Mathematics 145, 367–393 (1990)
3. Boissonnat, J.D., Cerezo, A., Leblond, J.: Shortest paths of bounded curvature constraints. Journal of Intelligent and Robotic Systems 11(1-2), 5–20 (1994)
4. Hota, S., Ghose, D.: Optimal path planning for a miniature air vehicle using a modified Dubins method. In: UKIERI Workshop on Robust Control of Smart Autonomous Unmanned Air Vehicle, IISc, Bangalore, August 22-23, pp. 64–71 (2008)
5. Hota, S., Ghose, D.: A modified Dubins method for optimal path planning of a miniature air vehicle converging to a straight line path. In: Proc. of the American Control Conference, St. Louis, Missouri, USA, June 10-12, pp. 2397–2402 (2009)
6. Hota, S., Ghose, D.: Optimal rectilinear path convergence of a Miniature Air Vehicle using Pontryagin's Maximum Principle. In: Proc. of the International Conference and Exhibition on Aerospace Engineering, IISc, Bangalore, India, May 19-22 (2009)
7. McGee, T., Spry, S., Hedrick, K.: Optimal path planning in a constant wind with a bounded turning rate. In: Proc. of the AIAA Guidance, Navigation and Control Conference and Exhibit, San Francisco, CA (August 2005)
8. Nelson, D.R., Barber, D.B., Mclain, T.W., Beard, R.W.: Vector field path following for miniature air vehicle. IEEE Trans. on Robotics 23(3) (June 2007)
9. Chitsaz, H., LaValle, S.M.: Time-optimal paths for a Dubins airplane. In: Proc. of the 46th IEEE Conf. on Decision and Control, New Orleans, LA, USA, December 12-14 (2007)
10. Ambrosino, G., Ariola, M., Ciniglio, U., Corraro, F., Lellis, E.D., Pironti, A.: Path generation and tracking in 3-D for UAVs. IEEE Trans. on Control Systems Technology 17(4) (July 2009)
11. Shanmugavel, M., Tsourdos, A., Zbikowski, R., White, B.A.: 3D Dubins sets based coordinated path planning for swarm of UAVs. In: AIAA Guidance, Navigation, and Control Conf. and Exhibit, Keystone, Colorado, August 21 - 24 (2006)
12. Sussmann, H.J.: Shortest 3-dimensional paths with a prescribed curvature bound. In: 34th IEEE Conf. on Decision and Control, New Orleans, pp. 3306–3312 (1995)
13. Bestaoui, Y., Dicheva, S.: 3D flight plan for an autonomous aircraft. In: 48th AIAA Aerospace Sciences Meeting Including the New Horizons Forum and Aerospace Exposition, Orlando, Florida, January 4 - 7 (2010)
14. Shanmugavel, M., Tsourdos, A., Zbikowski, R., White, B.A.: 3D path planning for multiple UAVs using Pythagorean Hodograph curves. In: AIAA Guidance, Navigation and Control Conf. and Exhibit, Hilton Head, South Carolina, August 20 - 23 (2007)
15. Betts, J.T.: Practical methods for optimal control using nonlinear programming. The Society for Industrial and Applied Mathematics, Philadelphia (2001)

Design and Implementation of Autonomous Navigation System for Mobile Harbors

Iksu Shin[1], Yuseok Bang[1], Donghoon Kim[1],
Jongdae Jung[2], and Hyun Myung[1,2,⋆]

[1] Robotics Program, KAIST Yuseong-gu, Daejeon, 305-701, Republic of Korea
[2] Department of Civil and Environmental Engineering, KAIST Yuseong-gu, Daejeon, 305-701, Republic of Korea

Abstract. This paper describes an implementation of an autonomous navigation system for a USV-type Mobile Harbor (MH) prototype and for multiple MHs. The MH is a novel maritime container transport solution that can go out to a ship to load/unload containers at sea and take them to their destination ports. We demonstrate the feasibility of the navigation system for a MH by performing experiments using a prototype in a basin. A devised multiple navigation system enables multiple MHs navigate safely without colliding with each other. In the initial stage of path planning, 3D optimal path planning algorithm in Configuration-Time space is performed. When the MHs navigate using those generated paths, the decentralized and delayed path planning method is performed in real-time by predicting the collision expected regions. To demonstrate the feasibility of this algorithm, simulations were performed.

Keywords: Mobile Harbor, Underwater Surface Vehicle, Navigation System, Localization, Multiple navigation System.

1 Introduction

Recently, the oceanic environment is an exciting and challenging context for the robotics community. Autonomous robots which can explore the deep and wide oceans such as Autonomous Underwater Vehicles (AUVs) and Unmanned Surface Vehicles (USVs) are being developed. However, it is hard to keep them working for a long time autonomously, because of the difficulties in underwater communication and battery capacity. To solve this problem, the USV type robots which can relay the information from underwater vehicle to a base station on the land are being actively researched.

For example, the research about integration of the AUVs and USV to explore in the sea was conducted in Advanced System Integration for Managing the coordinated operation of robotic Ocean Vehicles (ASIMOV) project [1].

The Norwegian University of Science and Technology (NTNU) is also actively studying the field of ship control system, and conducting a lot of experiments

---

⋆ Correspoinding author: Assistant Professor, Ph.D., Department of Civil and Env. Engg. and Robotics Program.

P. Vadakkepat et al. (Eds.): FIRA 2010, CCIS 103, pp. 218–225, 2010.

by applying various control algorithms to their prototypes and ships [2,3]. Thor. I. Fossen and Tristan Perez [4] developed the Marine System Simulator which is a Matlab/Simulink Toolbox library. It includes many types of vehicle models and can help to simulate the response between model ships and changing environments. The laboratory in NTNU is conducting experiments with prototypes of the ship in the basin using the MSS toolbox. Hausler et al. [5] studied the path planning for the multiple marine vehicles on the oceanic environment. They proposed the direct search optimization method which can find the minimum cost for generating path.

Recent trends show continued increase in the global container shipping volume and the introduction of mega-sized containerships, which necessitates the increase in the container cargo handling capacity. A natural consequence of these trends is the need for enhanced port capacity and capability. However, this is not the best solution as it comes with a number of other problems: it raises environmental or security concerns, prohibitively large scale SOC investment and so on. To overcome these problems, KAIST initiated the Mobile Harbor (MH) project [6]. KAIST proposed an MH system as a next generation maritime container transport solution. It is a novel maritime container transport solution that can go out to a ship anchoring in the deep water to load/unload containers on sea and take them to their destination ports regardless of their water depth (see Fig. 1). To safely dock with the container ship or port, an accurate localization and navigation system is essential for the MH.

This paper describes the implementation of an autonomous navigation system for an MH and the hybrid path planning method for multiple MHs. The navigation system for an MH and for multiple MHs are described in Section 2. The implementations of the navigation system are presented in Section 3. Section 4 finally concludes and discusses the directions for future research.

Fig. 1. Mobile harbor can load/unload containers at sea using its crane

2 Navigation System

Fig. 2 shows the overall architecture of the navigation system. To guide a vehicle to a target position, we need a path which consists of way points which the vehicle is supposed to follow. The path generated by path planning is the shortest path from a start position to a target position while avoiding obstacles. To guide the vehicle to follow the desired path, we control the heading angle of the vehicle to track the desired heading angle using PI controller. The Line-of-Sight (LoS) algorithm calculates the desired heading angle using the desired path and the current position of the vehicle.

2.1 Path Planning

Path planning algorithm generates the way points $P_{wi} = (x_{wi}, y_{wi}), i = 1, 2, ..., n$ from the start position $P_s = (x_s, y_s)$ to the target position $P_t = (x_t, y_t)$. However, it requires an additional way point to dock parallel to a containership avoiding rapid movement near the target position. When the target orientation is ψ, the additional way point should be at $P_a = (x_t - l\cos\psi, y_t - l\sin\psi)$ where l is the distance to a target before reaching the target position. There are several methods for the shortest path planning such as Dijkstra's algorithm, Floyd-Warshalls algorithm, and A* algorithm [8]. In our implementation, it is carried out by the use of gridmap-based A* algorithm.

2.2 Line of Sight (LoS) Algorithm

As shown in Fig. 3, the vehicle has the position error between the desired path obtained from path planning and its current position (x^*, y^*). The LoS algorithm is used to minimize the heading angle error and to follow the path as follows:

$$\psi_{err}(t) = \psi_d(t) - \psi^*(t) \approx 0 \tag{1}$$

where ψ^* and $\psi_d(t)$ are the current heading angle and the desired heading angle of the vehicle, respectively.

The LoS point $P_{los} = (x_{los}, y_{los})$ is a pilot point which the vehicle is supposed to track in Fig. 3. Then we can define the LoS vector as a vector from the

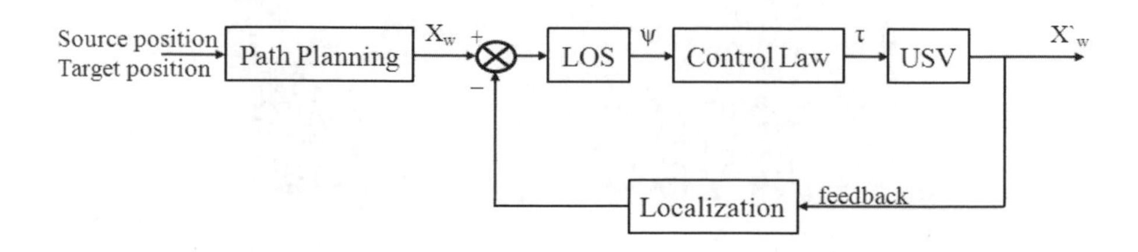

Fig. 2. Navigation system architecture

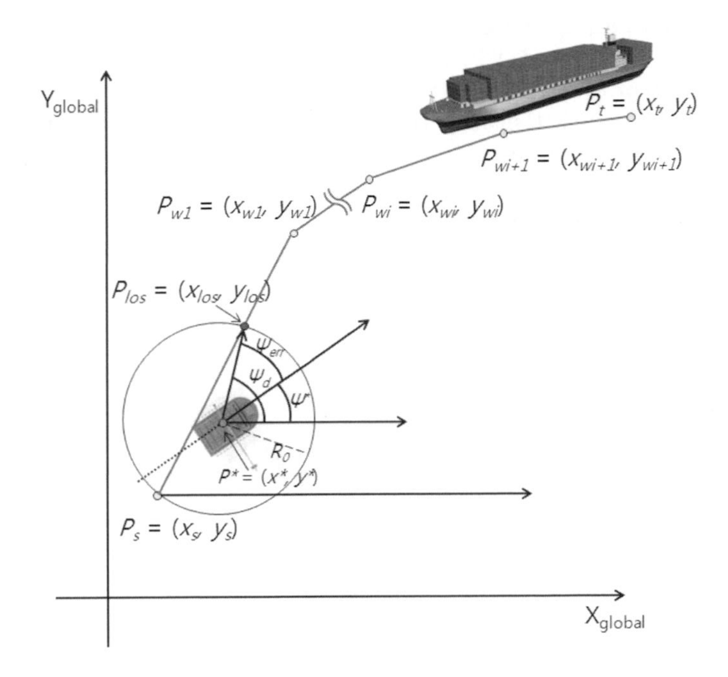

Fig. 3. Coordinate system for the Mobile Harbor

vehicle's current position to the LoS point. The desired heading angle $\psi_d(t)$ can be calculated as:

$$\psi_d = \text{atan2}(y_{los} - y^*, x_{los} - x^*), -\pi \leq \psi_d \leq \pi. \tag{2}$$

A Circle of Acceptance (CoA) is defined as a circle with radius R_0 centered at the current position. The LoS point can be set as an intersection of the CoA and the desired path as illustrated in Fig. 3 [7]. By this approach, the vehicle can return to the desired path rapidly depending on the radius R_0 when the vehicle gets out of the planned path. If R_0 is too short, it may make the movement of the vehicle unstable. Therefore, the value of R_0 should be selected appropriately.

During the tracking of the desired path, the vehicle must switch the way point (x_{wi}, y_{wi}) to the next one (x_{wi+1}, y_{wi+1}) when it comes close to the next way point.

2.3 Navigation System for Multiple Mobile Harbors

To avoid collision between each vehicle, we proposed the hybrid path planning algorithm. The hybrid algorithm consists of 3D A* in Configuration-Time space before starting and the delayed path planning on the move. To generate configuration space, collision-expected region is considered with ship domain region and wake wash. Fig. 4 (a) shows an image of the 3D A* algorithm which can perform MH2's path planning by avoiding the collision-expected region using Configuration-Time space of MH1 and obstacles. As shown in Fig. 4 (b), adding time axis on typical 2D A*, path planning is performed by checking 9 nodes at time $t{+}1$. At first, MHs can perform optimal path planning using 3D A* algorithm. When the vehicles are moving, they perform path planning using the decentralized and delayed path planning method. Fig. 4 (c) shows a graph of the decentralized and delayed path planning algorithm that can avoid expected collision region indicated by circle. If there is an expected collision between MH1 and MH2 at time $t1$, MH2 waits until MH1 passes by at time $t2$.

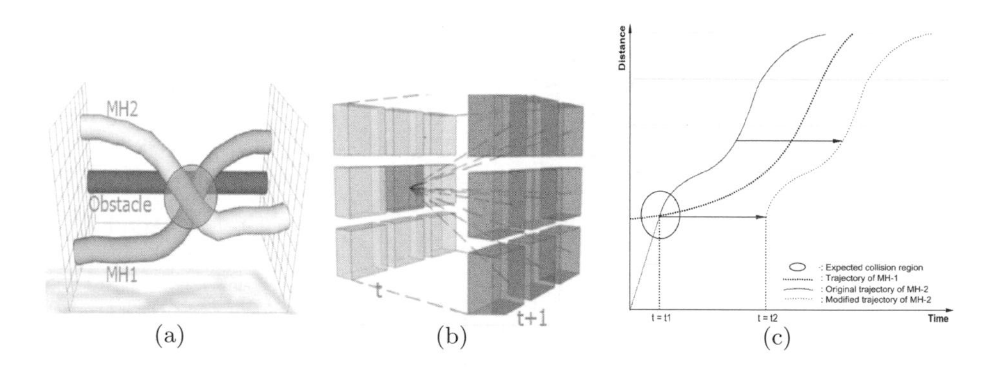

Fig. 4. Multiple Navigation system (a) Perform MH2's path planning by avoiding collision-expected region by obtaining Configuration-Time (C-T) space of MH1 and obstacles. (b) 3D A*Algorithm: Adding time axis on typical 2D A*, perform path planning by checking 9 nodes at time $t{+}1$. (c) Delayed path planning algorithm. $t1$ is the expected collision time and MH2 delayes until time $t2$.

3 Implementation

3.1 Implementation Architecture

In this paper, the autonomous navigation system for an MH and multiple MHs are implemented. A single navigation system is verified by the small scale prototype of an MH which is 1/60 scale. To demonstrate the feasibility of the multiple navigation system, simulations were performed. As shown in Fig. 5, the overall system is composed of a 1/60 scale MH, the indoor localization system, and an external main PC. There is a control software and a visualization software in the main PC. Two tags are installed in the middle of the vehicle to get the position and the heading angle of the MH. The measured range data between the tags and the beacons are transmitted to the control software in the main PC and it is used to calculate the output value of the thrusters. These values are then transmitted wirelessly to the controller board through the Bluetooth module.

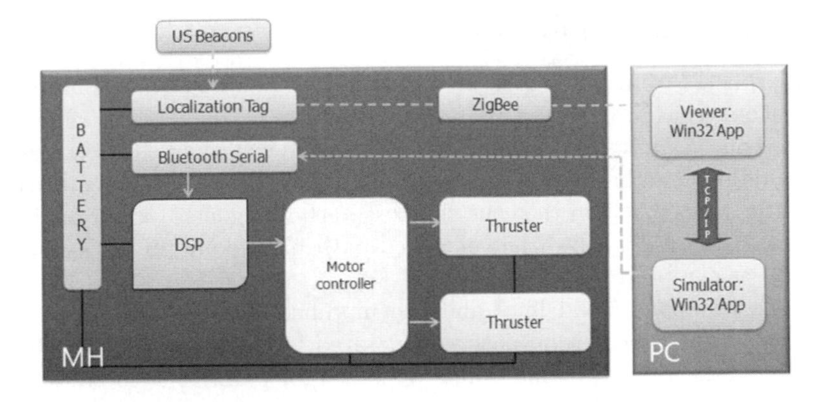

Fig. 5. Implementation architecture of the prototype of an MH

3.2 Beacon Based Localization

The position and the angle data of the MH are acquired from the ultrasonic-based localization module, igs-U, manufactured by Ninety System Co. This system is composed of the beacons (ultrasonic transmitter), a tag (ultrasonic receiver), and a Zigbee (wireless transmitter) module. The distance between a tag and each beacon is measured and the range data is sent wirelessly to the control software in the external PC through a Zigbee module. The sampling time of each beacon is 100 ms, and the beacon can cover a 5m $\times$ 5m $\times$ 2.5m (width $\times$ length $\times$ height) area. We have conducted experiments using 6 beacons and 2 tags in a 10m $\times$ 5m (width $\times$ length) area.

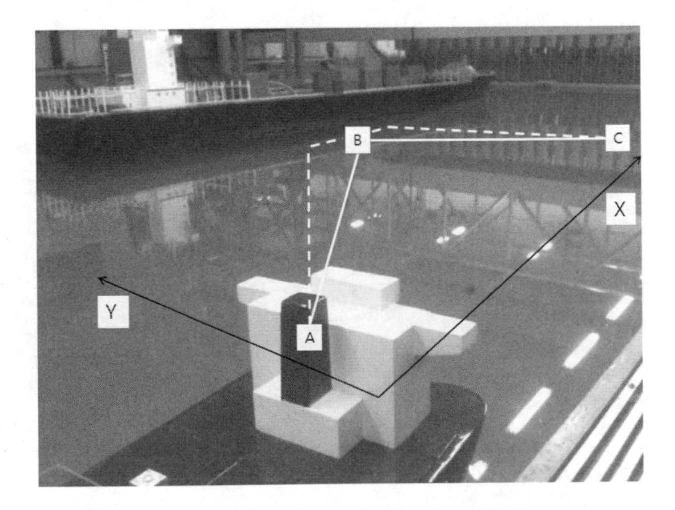

Fig. 6. Solid line: the shortest path from A to C via B, Dotted line: the modifed path for a parallel docking

3.3 Experiments and Results

The experiments were conducted on a 15m × 15m basin at KAIST. We set up the 6 beacons which are located at an interval of 5m. A series of experiments was conducted on the aforementioned basin and a scaled container ship was prepared in the middle of the basin. The USV-type MH was located beside the artificial port on the starboard, and then the shortest path was generated for navigation. After that, the USV tracked the generated path from point $A(X_1, Y_1)$ to point $B(X_2, Y_2)$ as can be seen in Fig. 6.

The shortest path (solid line) and the modified desired path (dotted line) are shown in Fig. 6. The solid path is generated by A* algorithm, which is the shortest path from the starting point to the target point, but the dotted path

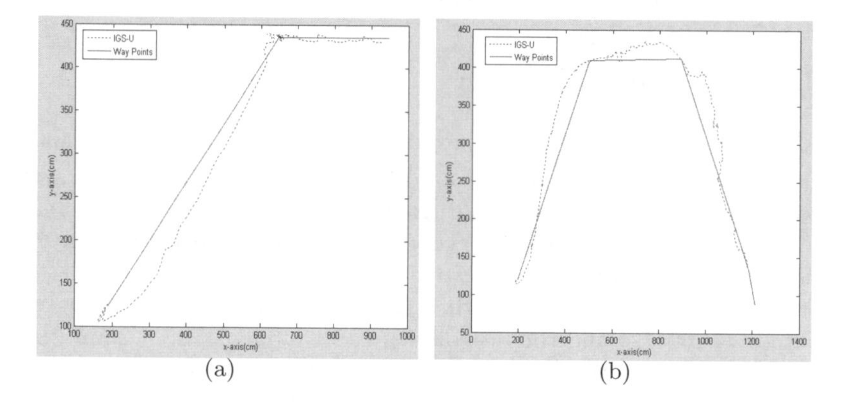

Fig. 7. (a) Solid line: desired path, Dotted line: igs-U data from A(2.5m, 1.0m) to B(9.5m, 4.5m). (b) Solid line: desired path, Dotted line: igs-U data from A(2.5m, 1.0m) to C(12.0m, 1.0m) via B(9.5m, 4.5m).

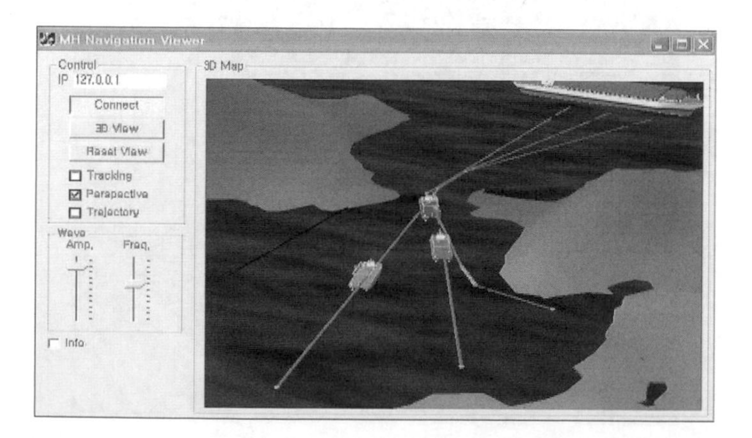

Fig. 8. Snapshot of visualization software: each MH can avoid other vehicles or moving obstacles using the decentralized and delayed path planning algorithm

indicates the regenerated path for a parallel docking. Because it is hard to control the yaw angle at point B by two fixed thrusters, the USV tracks the modified dotted path for docking which was obtained by adding an additional way point P_a as described in Section 2.1.

Fig. 7 (a) shows the actual path (dotted) of the MH obtained from igs-U and the desired path (solid). The USV moved to reduce the error of the yaw angle. The second experiment has been carried out in a similar manner to the first experiment from A(2.5m, 1.0m) to C(12.0m, 1.0m) via B(9.5m, 4.5m). The result is shown in Fig. 7 (b). Fig. 8 shows the simulation snapshot of the hybrid path planning for multiple MHs. In the simulations, each MH can avoid obstacles using the decentralized and delayed path planning method. The algorithm can make one MH wait until the other MH passes by.

4 Conclusion

In this paper, an autonomous navigation system for MHs has been implemented. To verify the proposed algorithms, experiments and simulations have been conducted. A 1/60 scale prototype was implemented and the real-time navigation system was developed. Also, we proposed the hybrid navigation for multiple MHs which can help to avoid other MHs or moving obstacles. As a future work, we are planning to perform experiments with disturbances such as winds, currents, and waves. For this purpose, we should consider the hydrodynamics of the USV and develop an appropriate algorithm for wave filtering.

Acknowledgment. This research was supported by the Technology Innovation Program (Project Number: 10036238), MKE (Ministry of Knowledge Economy).

References

1. Pascoal, A., Oliveira, P., Sivestre, C., et al.: Robotic ocean vehicles for marine science applications: the European ASIMOV project. In: Proc. MTS/IEEE OCEANS 2000, pp. 409–415 (September 2000)
2. Skjetne, R., Smogeli, Ø., Fossen, T.I.: Modeling, Identification, and Adaptive Maneuvering of CyberShip II. In: Proc. IFAC CAMS 2004, Ancona, Italy, pp. 203–208 (2004)
3. Fossen, T.I.: Guidance and Control of Ocean Vehicles, London, U.K. Wiley, Chichester (1994)
4. Marine Systems Simulator, `http://www.marinecontrol.org`
5. Hausler, A.J., Ghabcheloo, R., Kaminer, I., Pascoal, A.M., Aguiar, A.P.: Path planning for multiple marine vehicles. In: Proc. OCEANS (2009)
6. Kwak, B.M.: Mobile Harbor: From New Concept to New Technology. In: UCLA Engineering MAE research seminar series (2010), `http://www.mae.ucla.edu/events/events-archive/2010/mobile-harbor-from-new-concept-to-new-technology-by-byung-man-kwak-kaist`
7. Fossen, T.I.: Marine Control Systems, In: Marine Cybernetics, AS, Trondheim, Norway, pp. 167–169 (2002)
8. Sathyaraj, B.M., Jain, L.C., Finn, A., Drake, S.: Multiple UAVs path planning algorithms: a comparative study. Fuzzy Optimization and Decision Making 3(7), 257–267 (2008)

Path Planning Algorithm Based on the Limit-Cycle Navigation Method Applied to the Edge of Obstacles*

Y.W. Lim[1], Y.H. Kim[1], J.U. An[2], and D.H. Kim[1]

[1] Department of Electronics and Information, Kyunghee University, Seocheon-Dong, Kiheung-Gu, Yongin-Si, 449-701, South Korea
[2] Pragmatic Applied Robot Institute, Daegu Gyeongbuk Institute Science and Technology, TP Venture Plant #1 and 2, 75 2-Gil North Complex, Dalseo-Gu, Daegu, 704-948, South Korea
{leomyth,gemblerz,donghani}@khu.ac.kr, robot@dgist.ac.kr

Abstract. This UGV (Unmanned Ground Vehicle) is not only widely used in various practical applications but is also currently being researched in many disciplines. In particular, obstacle avoidance is considered one of the most important technologies in the navigation of an unmanned vehicle. In this paper, we introduce a simple algorithm for path planning in order to reach a goal while avoiding polygonal-shaped static obstacles. Effectively to avoid such obstacles, a path planned near the obstacle is much shorter than a path planned far from the obstacle. The proposed method can be applied to two situations: when the obstacle and the robot differ in size; and when there are two obstacles. The efficiency of the proposed algorithm was verified through a set of simulations and experiments. Consequently, the proposed limit-cycle method is more effective than the original limit-cycle algorithm.

Keywords: Path planning, Obstacle avoidance, UGV, Limit-cycle, Edge.

1 Introduction

The UGV (Unmanned Ground Vehicle) is not only used in a wide variety of practical applications including industrial and military is but also currently being researched in many disciplines. In the early days of UGVs, the question was about how to move from a starting point to a goal; these days, however, the aim is to improve the energy efficiency. There are two methods for this. One is related to the robot's hardware, such as reducing the robot's weight; the other is related to optimizing the robot's path. In this paper, we emphasize the latter. In particular, obstacle avoidance is considered one of the most important technologies to improve in terms of energy efficiency in the navigation of unmanned vehicles. Robot path planning has been an active area of research, and many methods have been developed to tackle this problem, such as potential field, grid algorithm, neural network, genetic algorithm and immune network methods [1][2][3][4][5]. In this paper, in order for the robot effectively to reach

* This research was carried out under the General R/D Program of the Daegu Gyeongbuk Institute of Science and Technology(DGIST), funded by the Ministry of Education, Science and Technology(MEST) of the Republic of Korea.

P. Vadakkepat et al. (Eds.): FIRA 2010, CCIS 103, pp. 226–233, 2010.

a goal while avoiding rectangular-shaped static obstacles, we introduce a simple navigation algorithm whose essence comes from the original limit-cycle navigation method for path planning.

The article is organized as follows: first, the problem concerning the original limit-cycle navigation method is stated (Section 2). Second, method of mobile robot path planning was proposed in the detail (Section 3). The effectiveness of the proposed path planning method will be demonstrated by simulations in Section 4 and experiments in Section 5. Finally, the conclusion will be drawn out in Section 6.

2 Problem Statement

2.1 Instruction Original Limit-Cycle Method

The limit-cycle navigation method introduces in [6] is one of the well-defined method has remarkable advantages especially for mobile robot. With this method undemanding in terms of computational process time, the robot can be easily navigated and followed in real time. Fig. 1 shows how the limit-cycle functions. The robot moves in the direction of the arrow that is represented by the dotted line. Fig. 1 (a) shows that the robot converged on the limit-cycle clockwise. Fig. 1 (b) shows that the robot converged on the limit-cycle counter-clockwise.

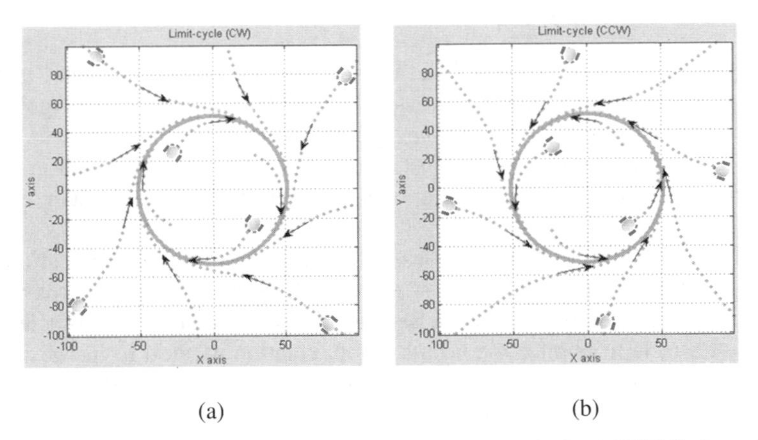

Fig. 1. (a) Clockwise (CW) limit-cycle, (b) Counter-clockwise (CCW) limit-cycle

2.2 Problem Statement of the Original Limit-Cycle

The original limit-cycle navigation method has several problems that stem from false assumptions. One of these problems is the assumption that the obstacle is the same size as the robot, and the other is that the obstacle is the same shape. In this paper, we choose an obstacle of a rectangular shape to differ from the shape of the robot. The dashed dot line shown in Fig. 2 indicates the path generated by the original limit-cycle method, and the dotted line indicates the path that is expected most efficiently and suitably. Accordingly, the original limit-cycle navigation method needs to be improved upon and further developed.

As shown in Fig. 2 (a), if the robot uses the original limit-cycle navigation method in this situation, the robot will not move to goal near the obstacle. Therefore, when an obstacle's size and shape differ from those of the robot, it is inefficient to use the original limit-cycle navigation method.

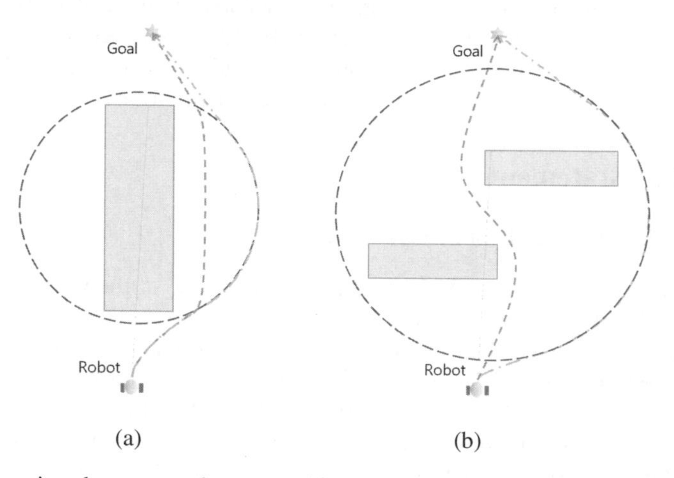

Fig. 2. A comparison between path generated by using the original limit-cycle and anticipating suitable path in the two situations

The other problem stems from falsely assuming that several obstacles are all one obstacle. Accordingly, in this paper, we choose obstacles of a rectangular shape. When using the original limit-cycle navigation method, the robot has to know the obstacles' center of gravity and the most distant point from the center of gravity. The distance from the obstacles' center of gravity to the obstacles' most distant edge is the radius of a limit-cycle.

As shown in Fig. 2 (b), if the robot uses the original limit-cycle navigation method in this situation, the robot will not pass through the gap, although the robot is small enough to pass between the obstacles. Because the robot regards all obstacles as parts of a single obstacle, it applies the limit-cycle navigation method to the obstacles as a whole. Therefore, when obstacles are not sole but several, it is inefficient to use the original limit-cycle navigation method.

3 Path Planning

As Section 2.2 shows, the original limit-cycle navigation method is inefficient in some situations. Accordingly, in this paper, we propose a new path-planning algorithm based on the limit-cycle navigation method applied to the edge of obstacles.

Before using the proposed navigation method in local navigation, we need to define the terms:

- The robot knew the goal's location.
- The robot generates the path of the shortest distance from the starting point to the goal.

- Basically, the robot moves toward the goal and continually checks the goal's location in real time. If the robot does not meet an obstacle on its way to the goal, the robot moves to the goal in a straight line.
- Rotational direction: It decides the turning direction taken to avoid an obstacle. If it moves a right-edge of an obstacle, it turns counter-clockwise (CCW). On the contrary, if it moves left-edge of the obstacle, it turns clockwise (CW).
- When the robot applies the limit-cycle method applied to the edge of obstacles, the robot has limit-cycle which the robot turns safely at the edge of the obstacle. The limit-cycle's radius depends on the size of the robot and the size of the wheels.
- From the robot's point of view, each obstacle has two edges, a right edge and a left edge. The robot selects the edge that is closest and moves in that direction.
- If the robot uses the limit-cycle navigation method to another edge of the same obstacle, it chooses the same turning direction.

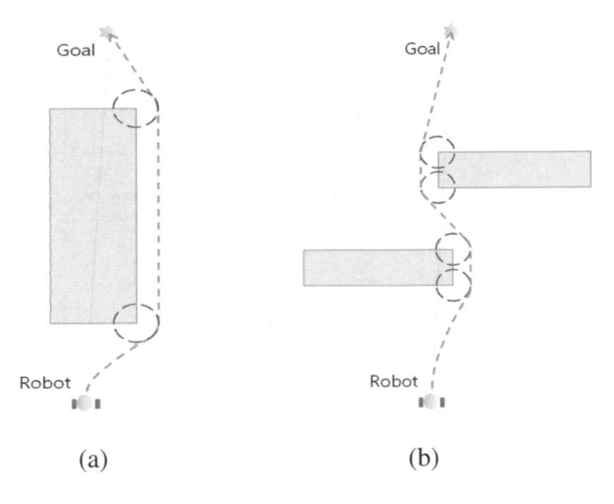

Fig. 3. (a) The path is generated by using the proposed limit-cycle navigation method when the robot meets the rectangular obstacle (b) The path is generated by using the proposed limit-cycle navigation method when the robot meets the two rectangular obstacles

If the robot does not avoid all parts of the obstacle with using the limit-cycle navigation method, it searches for the other edge of the same obstacle or the edge of another obstacle. Then, the robot applies the limit-cycle navigation method to the edge and avoids the obstacle. Under this new algorithm method, when the robot applies the limit-cycle navigation method to the obstacle, the path generated is more efficient than that generated by the original limit-cycle navigation method. The improvement is shown prominently in the following two situations.

As shown in Fig. 3 (a), one of the two situations is when the robot applies the limit-cycle navigation method to a rectangular obstacle. As shown in Fig. 3 (b), the other situation is when there are not one but several obstacles. The flow chart shown in Fig. 4 indicates that the robot is how to move to the goal while avoiding the obstacle.

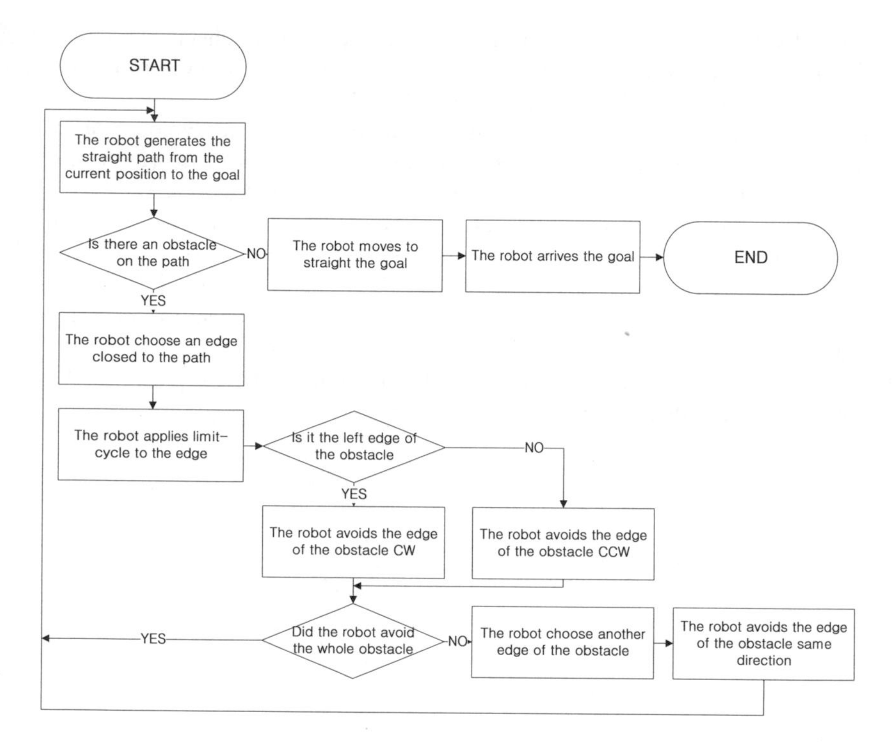

Fig. 4. The flow chart that the robot move to the goal while avoiding an obstacle

4 Simulation Results

In this Section, the proposed path planning algorithm is verified through simulation works with MATLAB. In Section 3, we proposed a path-planning algorithm based on the limit-cycle navigation method applied to the edges of obstacles in two situations: when the obstacle and the robot differ in size; and there are two or more obstacles.

The result of the simulation in Fig. 5 shows that the robot moved from the starting point (1400, 100) to the goal (1600, 2900) while avoiding obstacles. Fig. 5 (a) shows the result of the former simulation. Here, we set up a rectangular obstacle and carried out a simulation to show about how the path is generated by the robot using the proposed limit-cycle navigation method. The result of the latter simulation is shown in Fig. 5 (b). Here, we set up two rectangular obstacles and carried out a simulation to demonstrate how the path is generated by the robot using the proposed limit-cycle navigation method.

Consequently, the paths generated in these simulations were successfully similar to the suitable paths as depicted.

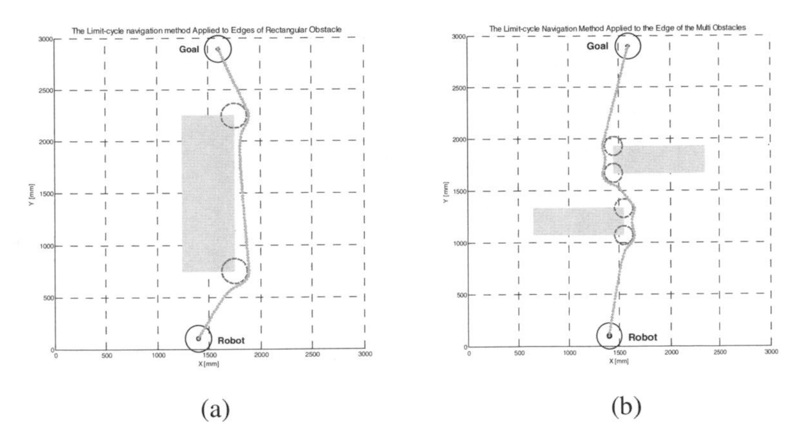

Fig. 5. (a) The path is generated by using the proposed limit-cycle navigation method in the simulation when the robot meets a rectangular obstacle (b) The path is generated by using the proposed limit-cycle navigation method in the simulation when the robot meets two rectangular obstacles

5 Experimental Results

In this Section, the proposed path planning algorithm is verified through experiment with the robot soccer system. The robot can know self position, obstacle's position, shape and size, and the position of obstacle's edges because the robot soccer system worked on the global vision. The experiments shown in Fig. 6 were performed by the robot using the proposed limit-cycle navigation method. Each experiment carried out ten times to get accurate results. Fig. 6 (a) indicates the robot's movement as the first experiment. Fig. 6 (b) indicates the robot's movement as the second experiment.

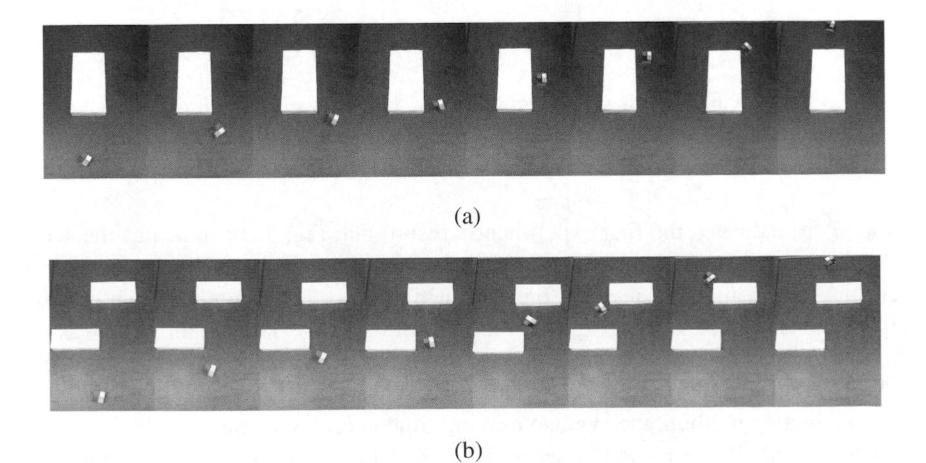

Fig. 6. (a) The robot move to the goal by using the proposed limit-cycle navigation method when it meets the one rectangular obstacle (b) The robot move to the goal by using the proposed limit-cycle navigation method when it meets the two rectangular obstacles

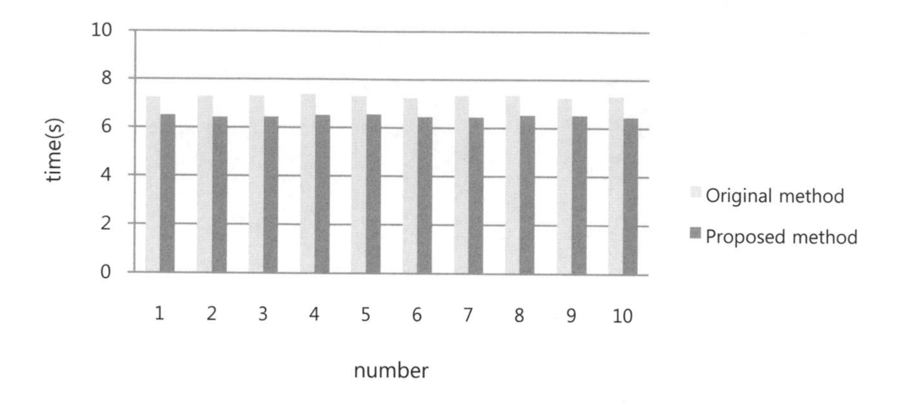

(a)

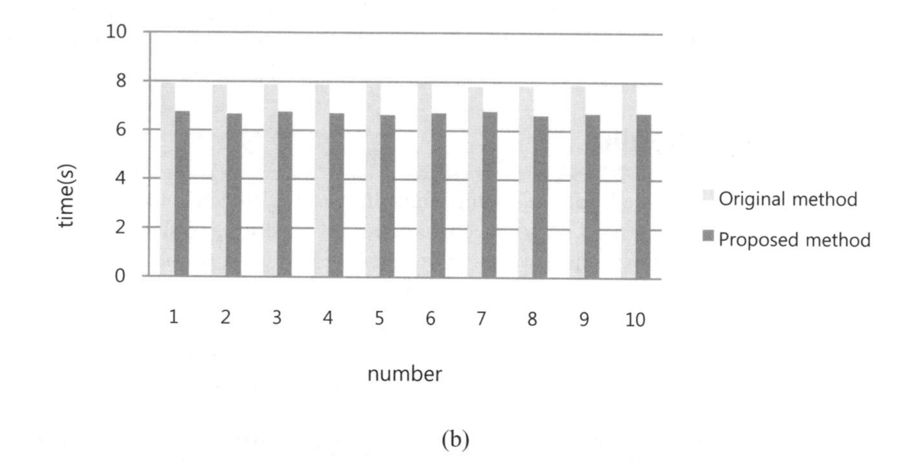

(b)

Fig. 7. (a) The moving time that the robot moves to the goal while avoiding a rectangular obstacle (b) The moving time that the robot moves to the goal while avoiding two rectangular obstacles

Fig. 7 (a) indicates the first experiment's result, and Fig. 7 (b) indicates the second experiment's result. As shown in Fig. 7 (a), when the robot used the original limit-cycle navigation method, average movement time was 7.3 seconds. When the robot used proposed limit-cycle navigation method, the average movement time was 6.48 seconds. As shown in Fig. 7 (b), when the robot used original limit-cycle navigation method, average moving time is 7.89 seconds. When the robot used proposed limit-cycle navigation method, the average moving time is 6.71 seconds.

Consequently, the two experiments show that the proposed limit-cycle method is more effective than the original limit-cycle algorithm.

6 Conclusions and Future Work

When the robot avoids obstacles using the original limit-cycle method, the generated path is based on the obstacles' center of gravity regardless of the location of the starting point, goal and obstacles. Accordingly, using the original limit-cycle method is not efficient in all situations. Above all, it is important to improve the energy efficiency of path planning by finding the most suitable line to connect the starting point to the goal while avoiding obstacles. Therefore, in this paper, we have established a new path-planning algorithm based on the limit-cycle navigation method applied to the edge of obstacles in order to improve upon the drawbacks of the original method. We verified the efficiency of the proposed algorithm through a set of simulations and experiments. The results of the simulations and experiments show that the proposed limit-cycle method is more effective than the original limit-cycle algorithm.

In future work, the proposed limit-cycle navigation method will be examined to cope with various obstacles and irregularly shaped or obstacles as well, and the limit-cycle radius will also be considered in order to improve and expand upon this research. Furthermore, it will be developed to be used in the local vision.

References

1. Ge, S.S., Cui, Y.J.: Dynamic motion planning for mobile robots using potential field method. Autonomous Robots 13, 207–222 (2002)
2. Xiaoxi, H., Leiting, C.: Path Planning Based on Grid-Potential Fields. In: International Conference on Computer Science and Software Engineering (2008)
3. Yang, S.X., Meng, M.Q.H.: Real-time collision-free motion planning of a mobile robot using a Neural Dynamics-based approach. IEEE Transaction on Neural Networks, 1541–1552 (2003)
4. Sedighi, K.H., Ashenayi, K., Tai, H.-M.: Autonomous Local Path-Planning for a Mobile Robot Using a Genetic Algorithm. In: IEEE Congress on (2002)
5. Wang, Y.-N., Hsu, H.-H., Lin, C.-C.: Artificial Immune Algorithm Based Obstacle Avoiding Path Planning of Mobile Robots, pp. 859–862. Springer, Heidelberg (2005)
6. Kim, D.-H., Kim, J.-H.: A real-time limit-cycle navigation method for fast mobile robots and its application to robot soccer. Robotics and Autonomous Systems 42, 17–30 (2003)

An Efficient Random Walk Strategy for Sampling Based Robot Motion Planners

Titas Bera, M. Seetharama Bhat, and Debasish Ghose

Department of Aerospace Engineering, Indian Institute of Science, India
{titasbera,msbdcl,dghose}@aero.iisc.ernet.in

Abstract. Sampling based planners have been successful in path planning of robots with many degrees of freedom, but still remains ineffective when the configuration space has a narrow passage. We present a new technique based on a random walk strategy to generate samples in narrow regions quickly, thus improving efficiency of Probabilistic Roadmap Planners. The algorithm substantially reduces instances of collision checking and thereby decreases computational time. The method is powerful even for cases where the structure of the narrow passage is not known, thus giving significant improvement over other known methods.

Keywords: Randomized Algorithm, Robot Motion Planning, PRM, Random Walk.

1 Introduction

An indispensable concept of modern robot motion planning theory is the concept of configuration space. In these cases, the problem becomes that of finding a path from an initial point to a goal point in the configuration space. However, the explicit mapping from workspace obstacles to configuration space is in general difficult. Early studies showed that the basic version of this problem is PSPACE-complete and the best exact deterministic algorithm known is exponential in the dimension of the configuration space [1],[2]. On the other hand, real world problems generate instances with high dimensional configuration spaces.

Since the mid nineties, in order to break this "curse of dimensionality", random sampling based approaches were introduced [5]. These approaches are based on the generation of random samples to acquire information about the problem instance being solved. The implementation of these algorithms is usually quite simple. The price to pay is algorithmic completeness. Algorithms based on randomly generated samples aims for probabilistic completeness. For details, see [4] and for an excellent survey see [3].

Despite the success of sampling based path planners, motion planning in high dimensional configuration space is still difficult. For PRM like planners one such difficulty arises when configuration space possesses narrow passages. Several sophisticated sampling strategies can remove this difficulty to a large extent, but a satisfactory answer remains elusive.

P. Vadakkepat et al. (Eds.): FIRA 2010, CCIS 103, pp. 234–241, 2010.

In this paper we propose a new method which is suitable to tackle problems of narrow passages in high dimensional configuration space. In Section 2 we formulate the general motion planning problem. In Section 3 we briefly explain the PRM method, problem of narrow passages and related work. Section 4 describes our algorithm for narrow passage sampling and how it is distinguished from other known algorithms. Finally in Section 5 we discuss the implementations and results.

2 Problem Formulation

The configuration of a robot with n degrees of freedom can be represented as a point in an n-dimensional topological space , called the configuration space $\mathcal{C}$. A configuration q is free if the robot placed at q does not collide with the obstacles or with itself. We define the free space $\mathcal{C}_{free}$ to be the set of all free configurations in $\mathcal{C}$. There are a finite number of obstacles in the configuration space, which are closed and bounded sets $\mathcal{O}_i$, $i = 1, 2, \cdots, m$. Let the starting configuration be $x_{start} \in \mathcal{C}_{free}$ and the final configuration be $x_{goal} \in \mathcal{C}_{free}$. The motion planning problem is to find a continuous function $f : [0, 1] \rightarrow \mathcal{C}_{free}$ such that $f(0) = x_{start}$ and $f(1) \in x_{goal}$. The constraints can be nonholonomic and/or differential in nature.

3 PRM and Narrow Passage Problem

Probabilistic road-map methods (PRM) solve motion planning problems that do not involve dynamics of the robot or have negligible dynamics. A classic multi-query PRM planner proceeds in two stages. In the first stage, it randomly chooses samples from $\mathcal{C}_{free}$, called milestones according to a sampling scheme. It then uses these milestones as nodes to construct a graph, called a road-map, by adding an edge between every k pair of milestones that can be connected via a simple collision-free path, typically, a straight-line segment. After the road-map has been constructed, multiple queries can be answered quickly in the second stage. The planner first finds two milestones in the road-map, such that x_{start} and x_{goal} can be connected to these nodes and subsequently search for a path between x_{start} and x_{goal}.

If the configuration space possesses a narrow passage then, to capture the connectivity of $\mathcal{C}_{free}$, it is essential to sample milestones in narrow passages. This, however, is difficult, because of small volumes of narrow passages. Uniform distribution may not work well when the dispersion of the samples is higher than the narrow passage volumes. Intuitively, one should sample more densely near obstacle boundaries because points in narrow passages lie close to obstacles. A method called Gaussian sampler [6] is a simple and efficient algorithm that uses this idea. However, in some cases, many points near the obstacle boundaries lie far away from narrow passages and do not help in improving the connectivity of road-maps. Another approach called OBPRM [9], is an obstacle based sampling where random rays are cast from obstacles and using binary search one looks for collision free points near obstacle boundaries. Other geometric approaches [10], also exist but those are expensive to implement in high-dimensional configuration spaces.

Perhaps the most appealing scheme to answer the problem of narrow passage is the bridge test [7]. Here when a sample lies within an obstacle, one uses this information to build a bridge whose two end points lie in the obstacle while the mid point lies in $\mathcal{C}_{free}$. This method, although it requires a high computational time (because of more number of collision checking), can be very effective. In the following, we give the standard bridge test algorithm (Randomized Bridge Builder).

Algorithm: Randomized **B**ridge **B**uilder (RBB)
For $i \leftarrow 1$ to K
$q_1 \leftarrow$ *Random Configuration*(Uniform Distribution)
  If *Clearance* (q_1) returns False
    then $q_2 \leftarrow$ *Random Configuration*(λ_x)
      If *Clearance* (q_2) returns False
        then $q_3 \leftarrow$ mid point of $\overline{q_1 q_2}$
          If *Clearance*(q_3) returns True
            then Insert q_3 to List G as a new milestone
Return(G)

Here "Clearance" is a collision checker which returns *true* if the sampled configuration is in C_{free} and λ_x is a multidimensional Gaussian distribution with a pre-specified variance σ. The performance of this algorithm depends heavily on the choice of σ. The computation time increases with the increase in number of obstacles, since for each iteration at most 3 times collision detection subroutine need to be called.

4 Random Walk toward Surface

Next we present our algorithm [8], which generates the sample points in the surface of the configuration space obstacle. The idea is, since in high dimension one does not know the suitable value of σ, and there is no a priori knowledge about narrow passage geometry, therefore the most idealistic approach will be to generate points on the surface of the obstacle. The algorithm is based on a simple philosophy, such as, once a sample point is generated within the obstacle one can perform a discrete random walk with a fixed step size a, until the particle comes out of the obstacle.

Algorithm: Random **W**alk Towards **S**urface (RWS)
For $i \leftarrow 1$ to K
$q \leftarrow$ *Random Configuration*
  If $q \in \mathcal{C}_{free}$
    then add to List L

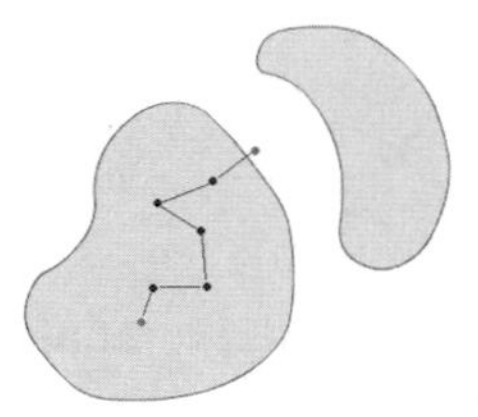

Fig. 1. RWS illustration. Shaded regions are obstacles.

else Flag $\leftarrow 1$
while Flag $= 1$
$\triangleright$ Select Random Direction
$q' \leftarrow$ *Random Walk* (q, a)
If $q' \in \mathcal{C}_{free}$
then add q' to the list L
Flag $= 0$
else $q \leftarrow q'$

Return L

The most important aspect of the algorithm is the reduction in collision checking computations. Both in RBB and Gaussian sampling strategy, to generate a point inside the obstacle, one has to check if the sampled random configuration belongs to either of $m, O_i^{\prime}$s. This increases with increase in number of obstacles. In our case, once a point is generated inside i^{th} obstacle $O_i^{\prime}$, it needs to check if the point belongs to only that obstacle $O_i^{\prime}$ or not. Once it is not within the ith obstacle, it checks for collision with other obstacles. This reduces computation time drastically.

Like Gaussian sampling strategy, depending on the input instances, our algorithm may generate unnecessary points on the surface of the obstacle which may be far away from the critical regions. We propose an optional discrimination method to eliminate unnecessary points. Define a heuristic distance $d(x, y) \propto \frac{1}{N}$ as threshold, where N is the number of points. Once all the points are generated, for every point $p \in N$ a neighbor list is created which contains points that lie within $d(x, y)$. If all the neighbors originate from the same obstacle then p is deleted.

Algorithm: Elimination(L)
$N \leftarrow Size(L)$
For $i \leftarrow 1$ to N
$Q \leftarrow$ *Neighbors* $(q(i), d)$
$I \leftarrow q(i).index$
$M \leftarrow Size(Q)$
For $k \leftarrow 1$ to M
if $I = Q(k).index$
then $p \leftarrow p + 1$

If $p = M$
  then *delete* $q(i)$
else insert $q(i)$ to List S
Return S

We also try to provide an approximate analysis of RWS in terms of mean escape time τ_{es} required to come out of the obstacle. Although our proposed random walk is discrete, but we approximate this with a continuous time diffusion process. This allows us to state a much simpler expression for mean escape time τ_{es}. Since the expression for mean exit time is available in existing literature, therefore we only state the result. For a somewhat elementary proof, see the Appendix. A more general discussion and in depth analysis of diffusion process can be found in [11]. In particular, the expected exit time from a ball of radius R in d dimension is given by

$$\tau = \frac{R^2 - |x|^2}{d}, \quad x \in \mathbb{R}^d \tag{1}$$

where x is the location of the initial position of the particle from the origin. Therefore, maximum of expected exit time is $\tau_{max} = \frac{R^2}{d}$. This implies that as the dimension increases it becomes less and less probable for a random walker to return to the starting point [12]. This means the particles sampled within the obstacles, become more probable to appear onto the obstacle surface more quickly as the dimension goes high ($n > 3$).

There exist marked differences between OBPRM and RWS. In OBPRM one casts rays in random directions from a point inside the obstacles, called a seed. Given a predetermined point B in free space, l distance away from the seed, one searches for a free configuration closest to the obstacle using binary search technique between point B and the seed. This however may lead to configurations that are not in the narrow region. In other words, the successful generation of narrow passage samples on the surface of the ith obstacle depends on the spatial distribution of other obstacles. But in case of RWS algorithm, the successful generation of narrow passage samples on the surface of the ith obstacle depends on itself. Also as said earlier, because of the property of random walks, the RWS is expected to work better as the dimension of the problem goes high.

5 Comparison Results between RWS and RBB

We compare the samples generated in the critical region and time to generate them for both RBB and RWS algorithms in different critical environments. We choose several benchmark problem, tested in a Pentium dual core processor and standard MATLAB runtime environment. Workspace is $[0, 100] \times [0, 100]$. In case of RWS the random walk has a fixed step size of 1 unit. The first benchmark problem involves movement through a narrow corridor for a 4 dof manipulator. For the second problem the degree of freedom is increased to 10. In the third problem, the environment contains narrow passage ways and the fourth problem considers arbitrary shaped obstacle. In each problem, any uniform sampling scheme may cause problems of PRM graph connectivity. The comparison results in Table 1 shows the capability of RWS algorithm compared to RBB while generating configurations in critical regions. We conducted 100 Monte-Carlo simulation for each of the problem and the table shows average results. In Case A RBB fails to generate a configuration in critical slit region after 30 samples, while RWS succeeds in much quicker time. Case B and C show RWS achieve better distribution of configurations near the obstacles and in case D both perform close but RWS generate those samples quickly.

For practical implementation, in case the complex polygonal objects consist a large number of polygons, one needs to have an efficient (linear in time with number of obstacle) collision detection algorithm. For this purpose graphics processing hardware can be used for collision detection.

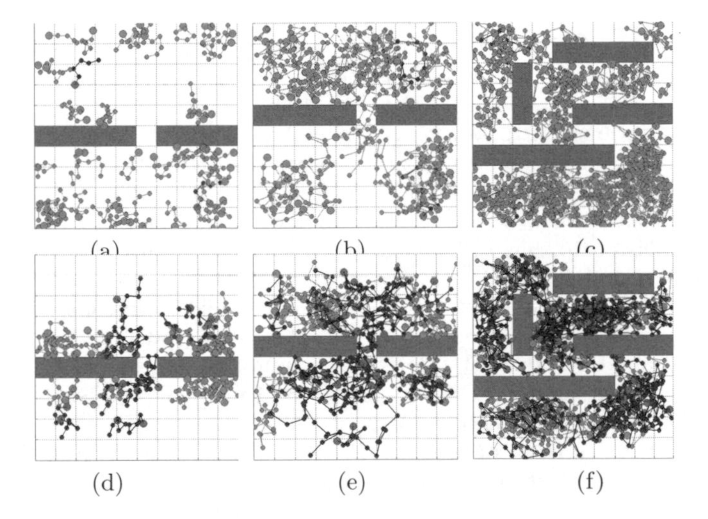

Fig. 2. Case A: (a) RBB, (d) RWS, 30 samples. Case B: (b) RBB, (e) RWS, 50 samples. Case C: (c) RBB, (f) RWS, 100 samples.

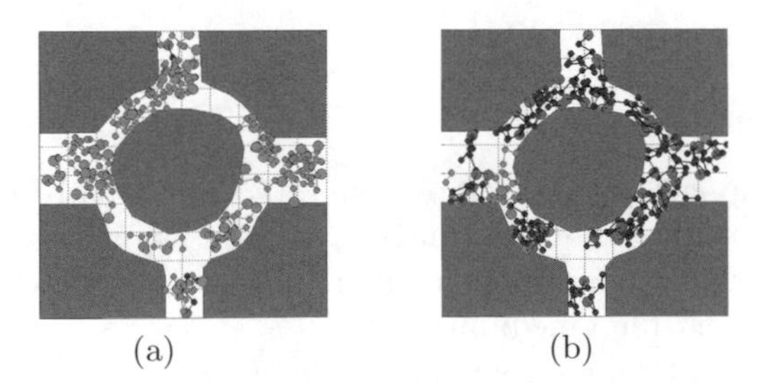

Fig. 3. Case D: (a) RBB. (b), RWS, 30 Samples

Table 1. Average Running Time of RBB and RWS for Different Cases

	RBB	RWS
Case A	30.92s	20.90s
Case B	60.32s	32.50s
Case C	239.32s	72.30s
Case D	52.00s	29.30s

References

1. Lozano Perez, T.: Spatial planning: A configuration space approach. IEEE Transaction on Computers 32(2), 108–120 (1983)
2. Reif, J.H.: Complexity of the mover's problem and generalization. In: Proc. of FOCS, pp. 421–427 (1979)
3. Carpin, S.: Randomized motion planning - a tutorial. International Journal of Robotics and Automation 21(3), 184–196 (2006)
4. Choset, H., et al.: Principles of Robot Motion: Theory algorithms and implementation. MIT Press, Cambridge (2005)
5. Kavraki, L.E., Svestka, P., Latombe, J.C., Overmars, M.H.: Probabilistic roadmap for path planning in high dimensional configuration spaces. IEEE Transaction on Robotics and Automation 12(4), 566–580 (1996)
6. Boor, V., Overmars, M.H., van der Stappen, A.F.: The Gaussian sampling strategy for probabilistic roadmap planners. In: Proceedings of the 1999 IEEE International Conference on Robotics and Automation, pp. 1018–1023 (1999)
7. Sun, Z., Hsu, D., Jiang, T., Kurniawati, H., Reif, J.H.: Narraow Passage Sampling for Probabilistic Roadmap Planning. IEEE Transaction on Robotics 21(5), 1105–1115 (2005)
8. Bera, T., Bhat, M.S., Ghose, D.: Preprocessing configuration space for improved sampling based path planning. In: ICEAE, Bangalore, India (2009)
9. Amato, N.M., Bayazit, O.B., Dale, L.K., Jones, C., Vallejo, D.: OBPRM: An obstacle-based PRM for 3D workspaces. In: Proc. 3rd Workshop Algorithmic Found. Robot., pp. 155–168 (1998)
10. Wilmarth, S.A., Amato, N.M., Stiller, P.F.: Motion planning for a rigid body using random networks on the medial axis of the free space. In: Proc. 15th Annu. ACM Symp. Computational Geometry, pp. 173–180 (1999)
11. Karatazas, I., Shreve, S.E.: Brownian Motion and Stochastic Calculus. Springer, Heidelberg (1988)
12. Hughes, B.D.: Random Walk and Random Environments, vol. 1. Clearson Press, Oxford (1995)

A Appendix

We begin by assuming a random sample generated within a obstacle and calculate the maximum of the mean exit time $\max(\tau_{es})$. This is equivalent to evaluating the maximum mean exit time from a domain Ω, of a particle, which shows brownian motion in a stationary fluid medium. If $b(t,x) \in \mathbb{R}^3$ is the velocity of the fluid at the point x at time t, then a reasonable mathematical model for the position X_t of the particle at time t would be a stochastic differential equation of the form

$$\frac{dX_t}{dt} = b(t, X_t) + \sigma(t, X_t)W_t \tag{2}$$

where, $W_t \in \mathbb{R}^3$ is a white noise. The Itô interpretation of this equation is

$$dX_t = b(t, X_t)dt + \sigma(t, X_t)dB_t \tag{3}$$

where B_t is $m-$dimensional Brownian motion, $X_t \in \mathbb{R}^d$, $b(t, X_t) \in \mathbb{R}^d$, $\sigma(t, X_t) \in \mathbb{R}^{d\times m}$. We call b the drift coefficient and σ the diffusion coefficient.

We can associate a second order partial differential operator A to an Itô diffusion X_t. The basic connection between A and X_t is that A is the generator of the process X_t. Let $\{X_t\}$ be a time-homogeneous Itô diffusion in $\mathbb{R}^d$. The infinitesimal generator A of X_t is defined by

$$Af(x) = \lim_{t\to 0} \frac{E^x[f(X_t)] - f(x)}{t} \tag{4}$$

where, $x \in \mathbb{R}^d$, X_t is the solution of Itô diffusion, $f : \mathbb{R}^d \to \mathbb{R}$ such that limit exists at x. E^x define expectation of x with respect to probability measure.

Consider the stationary case of Itô diffusion when $b = 0$ and $\sigma = I_d$ i.e.,

$$dX_t = dB_t \tag{5}$$

The generator of B_t is

$$Af = \frac{1}{2}\sum \frac{\partial^2 f}{\partial {x_i}^2} \tag{6}$$

where, $f = f(x_1, x_2, ..., x_d)$ twice differentiable, that is, $A = \frac{1}{2}\Delta$ where Δ is the Laplace operator.

Now we state Dynkin's Theorem [11]: Let f be a twice differentiable function. Suppose τ is the stopping time (escape time from a domain Ω, and $E^x[\tau] < \infty$) then

$$E^x[f(X_\tau)] = f(x) + E^x[\int_0^\tau Af(X_s)ds] \tag{7}$$

Now consider a Brownian motion $B = (B_1, B_2, ..., B_d)$ starting at $a = (a_1, a_2, ..., a_d) \in \mathbb{R}^d$ and assume $|a| < R$. We define the first exit time as τ_k of B from the ball

$$K_R = \{x \in \mathbb{R}^d; |x| < R\} \tag{8}$$

According to the Dynkin's theorem we choose $f(x) = |x|^2$ for $|x| \le R$, $X = B$ and $\tau = \sigma_k = \min(k, \tau_K)$, where k is any integer. So,

$$E^a[f(B_{\sigma_k})] = f(a) + E^a[\int_0^{\sigma_k} \frac{1}{2}\Delta f(B_s)ds] \tag{9}$$

$$= |a|^2 + E^a[\int_0^{\sigma_k} d \cdot ds] \tag{10}$$

$$= |a|^2 + d \cdot E^a[\sigma_k] \tag{11}$$

Letting $k \to \infty$, and $\tau_k = \lim \sigma_k < \infty$ a. s. and

$$E^a[\tau_k] = \frac{1}{d}(R^2 - |a|^2) \tag{12}$$

Ordering of Robotic Navigational Tasks in Home Environment

Syed Atif Mehdi and Karsten Berns

Robotics Research Lab, University of Kaiserslautern,
Kaiserslautern, Germany
{mehdi,berns}@informatik.uni-kl.de

Abstract. Autonomous navigation is an essential component of an indoor mobile robot. For an interactive user experience the robot should be able to understand multiple navigational tasks and perform them as optimally as possible. User may issue multiple random destinations to the robot to reach in a scenario like finding the elderly person, but to navigate randomly may result in an inefficient robot manoeuvring and may delay the process of finding the elderly person in the environment. To enhance the navigational efficiency, ordered and organised navigation based on certain cost function is required. In this paper, we are focusing on the need of organisation of navigational tasks keeping in view the unstructured and dynamic nature of the home environment. To achieve speed and accuracy in navigation, a detailed representation of the environment and a better evaluation of the cost function to organise the tasks is required. Navigation is based on grid map generated using laser scanner and sonar sensors mounted on the small sized indoor robot, ARTOS. The *Navigational Cost* is used for organising the tasks and is computed using A* algorithm to determine path to the destination.

Keywords: Assisted Living, Indoor Robot, Autonomous Navigation, Ordering of Navigational Tasks.

1 Introduction

Demographic situation in most industrialised countries reveals a great number of elderly people in the society. These people, despite having mental or physical disabilities, prefer to live in their homes rather than the assisted living facilities. Such people are often unable to perform common household tasks in a comfortable manner and require services of care giving people to perform tasks for them. Sometimes it is not possible to maintain the services of such helping staff for simple tasks such as carrying small objects from one place to another or serving the guests. This issue along with the increasing cost of assisted living facilities suggest use of modern technology for helping elderly people living alone in their homes. Many research groups are working with the aim to lower the nursing costs in the home environment and increase the quality of life of elderly persons. Care-o-bot II [5] and Paro [12] are examples of such efforts.

Autonomous Robot for Transport and Service (ARTOS), Fig. 1, is an initiative to provide services to elderly people, living alone in their homes, considering the factors of

P. Vadakkepat et al. (Eds.): FIRA 2010, CCIS 103, pp. 242–249, 2010.

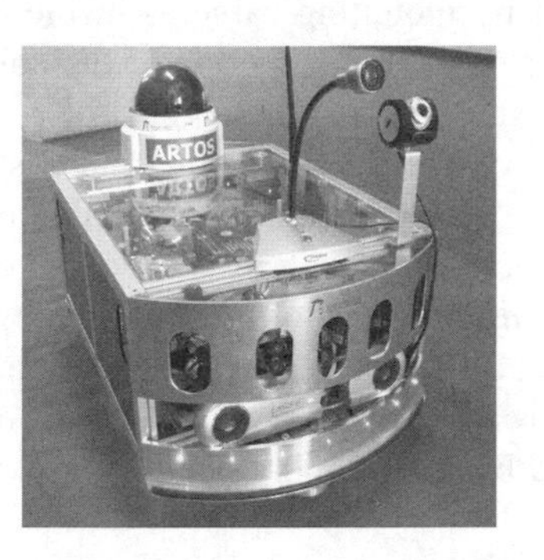

Fig. 1. Autonomous Robot for Transport and Service, equipped with laser scanner and ultrasonic sensors

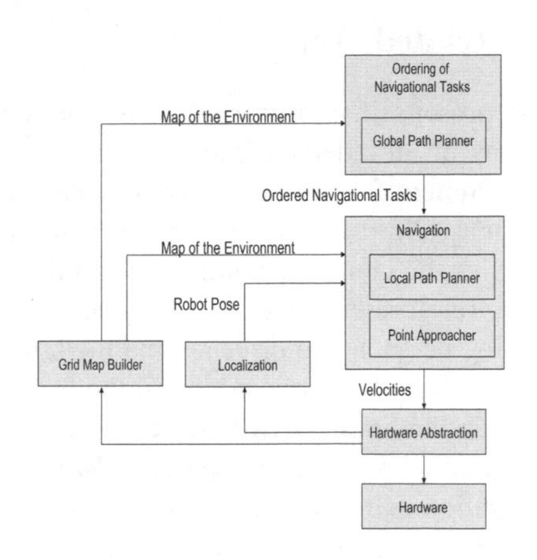

Fig. 2. Overview of the Methodology

cost and quality of service. It is equipped with laser scanner, sonar sensors, RFID reader and a Pan-Tilt-Zoom camera. It is capable of generating map of the environment, navigating autonomously in the environment while avoiding obstacles and carrying objects from one place to the other [1]. The control system is based on MCA-KL[1]. Many components are developed as behaviours of behaviour based control architecture iB2C[2] [9].

Being smaller in size, ARTOS can easily move through door ways and closely placed furniture in the apartment. It is also capable of establishing a telecommunication service between the elderly person and the care-giver using wireless Internet. Moreover, it can be tele-operated by the health-care personnel to detect medical emergencies.

To remotely detect medical emergencies it is necessary that ARTOS should be capable of handling multiple navigational tasks given by remote care-givers, and it should complete all these navigational tasks in shortest possible time. Selecting such tasks randomly can effect the performance of the robot. Unfortunately, the home environment does not provide an easy to navigate place for robots. Therefore, ordering the navigational tasks will make the navigation of the robot faster and easier.

To explain this concept and implementation, the paper is organised in the following way. First of all a short summary of related work is presented. Section 3 presents the methodology of the approach with subsections presenting major modules required for ordered navigation. Experimental results of simulated environment and real environment are being discussed in Sect. 4. Finally at the end, conclusion and future work are presented.

---

[1] MCA-KL: Modular Controller Architecture - Kaiserslautern Branch (http://rrlib.cs.uni-kl.de/).

[2] iB2C: integrated Behaviour-Based Control.

2 Related Work

Tele-presence has been studied by many research groups to monitor the health conditions of an elderly person. It can be achieved by mounting cameras in the house environment for fall or emergency detection. For example [3], is one of the solution that may allow a remote caregiver to analyse the situation. But the number of cameras to cover the complete home area and the expense of altering the environment is quite high. Tele-presence can also be achieved by using a mobile robot controllable by the helping personnel to drive around the home and look for the elderly person (for example see [4]). As discussed in [6] the *Position Point* approach for navigating the robot through the house environment is more comfortable to the users who are less trained for handling robots as compared to *Way Point* approach. This difference between expert navigation and novice navigation can be reduced by letting robot decide the order of destinations to follow.

In a typical office environment the ordering of navigational tasks is simple and straight forward. As mentioned in [2], order all the tasks from the current position to the far most position. In contrast, a home environment poses many restrictions on the robot navigation like some places are easily reachable by the robot and some places are hard to access. Therefore all these destinations cannot be treated equally and navigational tasks cannot be sorted just based on distance closeness. Moreover, the effective usage of battery is an important aspect of robotics [8] and ordering of navigational tasks can also help in reducing energy consumption by effectively moving between the destinations. [11] has used Dijkstra cost matrix to compute the cost of each task and then formulated the problem in Traveller Salesman Problem to obtain the solution. But they have pre-defined tasks and once an execution starts, any change in destination is not accommodated.

In the following, a novel method for ordering the tasks based on the navigational cost calculated using the information from the known map is being presented. The approach takes into account the obstacles in the environment and caters for the dynamic addition and removal of the tasks.

3 Methodology

The approach to address the issue of ordering random tasks, given by the remote person, is based on the usage of information generated by the mapping module [3]. An overview of the methodology is shown in Fig. 2.

3.1 Mapping and Localisation

In order to reach destinations provided by the user, ARTOS need to know the working environment. As discussed in [7], two forms of maps are being generated using laser scanner and the sonar sensors. These maps are merged together in a global map to generate the complete information of the environment. The localisation in the environment is achieved using differential odometry and is improved by reading passive RFID tags installed under the carpet.

[3] Detailed discussion about Mapping and Navigation can be found in [7].

3.2 Navigation and Path Planning

ARTOS can be tele-operated by caregivers to inspect the elderly person at home. There are two ways to control the robot remotely. One way is to use the joystick, provided in the graphical user interface, and the other is to opt for *Position Point* approach and give destination points by clicking on the 2D map of the environment.

ARTOS can navigate autonomously in the environment. The *Local Path Planner* searches the path between the obstacles using A* algorithm to generate the smallest possible path between two points. This information is used by *Point Approacher* to reach the destinations. During navigation, Elastic Bands [10] are being used to keep the robot away from the obstacles. The *Local Path Planner* also updates the path after certain time to accommodate the movement of the robot and newly detected obstacles in the dynamic home environment.

3.3 Ordering of Navigational Tasks

In the scenario of finding the elderly person in the house by the remote caregiver, *Position Point* approach can be used and multiple destinations can be given to search the inhabitant of the house. To accommodate this need, ARTOS is made capable of handling multiple navigational tasks. These tasks can be random destinations in the home environment. While performing these multiple tasks, it is possible that two tasks are close together and can be performed in the same run but the intermediate task between these two tasks is far away and causes the robot to move from the closest tasks to a far away task first and then come back to perform the other task. The overall time taken to complete all the tasks gets better if these tasks are re-ordered and then performed. This approach will save considerable time of navigation and increase the throughput of the tasks. Using Euclidean distance is of limited help in unstructured and dynamic home environment. Moreover, the close destinations (*Distance Closeness*) might not be close (*Navigational Closeness*) for the mobile robot. Consider Fig. 3 where the robot has to move from current position to the destination. The distance between the points is 4 units but for mobile robot *Navigational Distance* or *Navigational Cost* can be 8 units as it has to circumvent the obstacle. Therefore for a home environment, the robot should consider the obstacles in-between to compute the *Navigational Cost* for reaching the destination.

Maintaining costs[4] between points as matrix will not help in home environment since the map is updated continuously. Therefore, it would be beneficial to include the updated obstacles information every time when the cost needs to be calculated.

The idea is to order the tasks based on the *Navigational Cost* determined by the *Global Path Planner*. The *Global Path Planner* examines the grid map and determine the shortest path from start point s to end point d_1 using A*-Algorithm. The algorithm computes the cost by traversing cell by cell in the grid, starting from s, until a path to d_1 has been found. In each processing step, the cell with the lowest cost is chosen as next cell to be processed. The cost $f(x)$ for a cell x is calculated as

$$f(x) = g(x) + h(x) \tag{1}$$

[4] From now onwards, Cost is *Navigational Cost* otherwise stated. *Navigational Cost* and *Navigational Distance* are used interchangeably.

Algorithm 1. Ordering of Navigational Tasks

```
OrderTasks (ListOfTasks) //function takes the list of navigational tasks
//Considering first element as the source, cost is being calculated from source to destination
for i = 2 to length (ListOfTasks) do
   ComputeCost (ListOfTasks[1], ListOfTasks[i])
end for
//After cost calculations, the navigational task list is sorted based on minimal cost
SortBasedOnCost (ListOfTasks)
end OrderTasks

ComputeCost (Source, Destination) //compute the cost of source-destination pair
//Check if Source and Destination are 1. same. 2. near obstacles 3. already occupied
CheckSourceAndDestination(Source, Destination)
//Calculate the cost based on A*-algorithm by determining cost from currentPoint to X
//and predicting cost from X to Destination, that is
//Cost = CalculateActualCost(currentPoint,X) + CalculateEstimatedCost(X,Destination)
//Assign the computed cost to Destination
Destination.Cost = ComputeCostUsingAStarAlgo(currentPoint,Destination)
end ComputeCost
```

where $g(x)$ is the current distance from start point to the current point and $h(x)$ is the estimated distance from current point to the destination.

The computed cost is assigned to the destination. This cost is computed for every destination d_i taking s as the source. The destinations are then sorted based on minimal cost. After reaching destination d_j, the current point is treated as the source and the cost of remaining d_{i-1} destinations are computed and sorted. For closely situated destinations the cost is still low as in case of Euclidean based cost function, but in case of not too close destinations, it will vary depending on the obstacles placed in the environment. The algorithm for ordering the tasks is defined in Algorithm 1. The dynamic addition and removal of tasks has been implemented as follows:

Add a New Task. Whenever a new task is added, it is assigned a maximum cost and is placed at the end of *ListOfTasks*. Afterwards, the cost to reach the new task is calculated from the current task that is being performed. Once the cost is calculated and assigned, the *ListOfTasks* is sorted based on the minimum cost values.

Remove a Task. Removing or cancelling a current task suggests to stop executing the prevailing task and move on to the next task. Since the *ListOfTasks* is sorted based on the *Navigational Closeness* to the current task, the next task to perform will be the one which is closest to the current task.

Next Task to Perform. After completing a task, the next task in *ListOfTasks* is selected and the cost of all the tasks is updated from the new task. This is also required as some changes in the environment might have been recorded in the map and therefore path information needs to be updated.

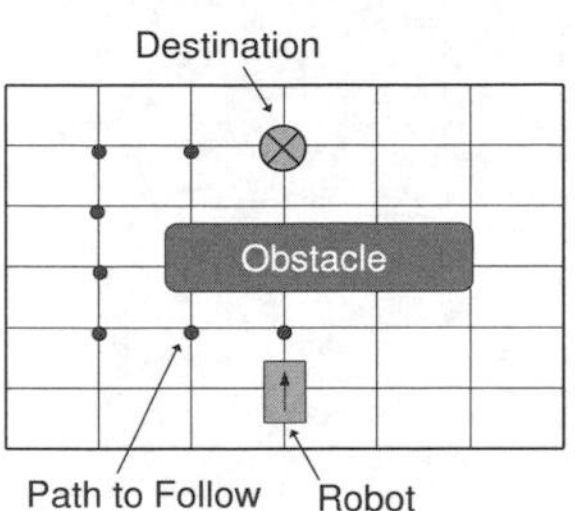

Fig. 3. Path to Follow for the Mobile Robot

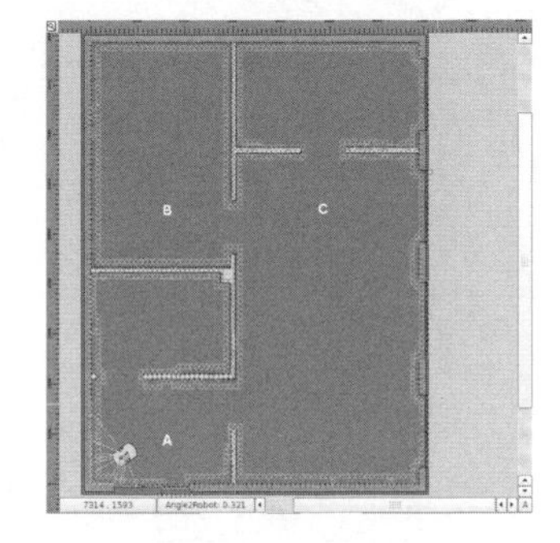

Fig. 4. Simulated IESE Environment

4 Experiments and Results

The algorithm has been tested in simulation and in real environment. The results of both the environments are discussed in the following subsections.

4.1 Simulated Environment

The simulation, of a real apartment developed in Fraunhofer - Institut fuer Experimentelles Software Engineering (IESE), is shown in Fig. 4, where red cells show the obstacles and orange cells represent the neighbourhood of obstacles.

The task of the robot is to reach two destinations namely B, a destination in the Bed Room, and C, a destination in the Living Room, starting from A, a start location near the entrance, as marked in Fig. 4. The coordinates of A, B and C are (1500, 900), (1500, 5500) and (5000, 5500) respectively.

The path followed using Euclidean distance based task organisation results in A to B to C, since A to B is least distance cost. This cost function does not take into account the walls or obstacles between the path. The average time to complete the tasks comes out to be 78330 ms.

The same source and destinations were used for ordering the navigational tasks using the information from the *Global Path Planner*. A higher navigational cost is assigned to reach B as compared to C and, therefore, C is performed earlier resulting in total average time of 55140 ms which is about 29% improvement over previous results.

4.2 Home Environment

The algorithm was also tested in the real home environment at IESE, shown in Fig. 5. The map of the environment as generated by the robot marking obstacles is depicted in Fig. 6. As can be seen, the real environment is cluttered with obstacles giving a limited room for robotic navigation. To ensure consistency, same locations A, B and C are being used as described above.

The path followed using Euclidean distance based cost results in A to B to C with distances as given in Table 1. Here it is clearly visible that no account of obstacles in the

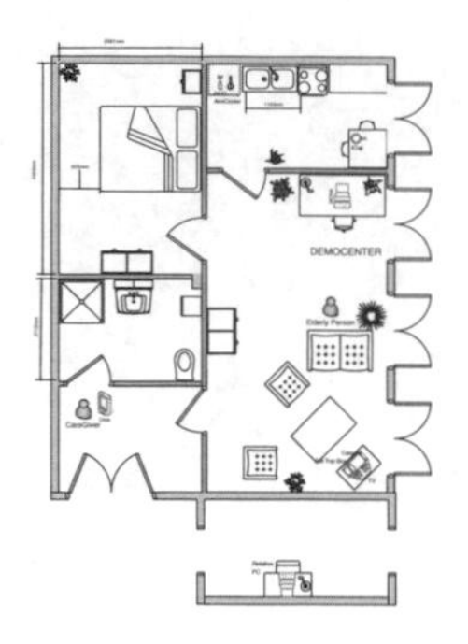

Fig. 5. Testing Facility at IESE, Fraunhofer

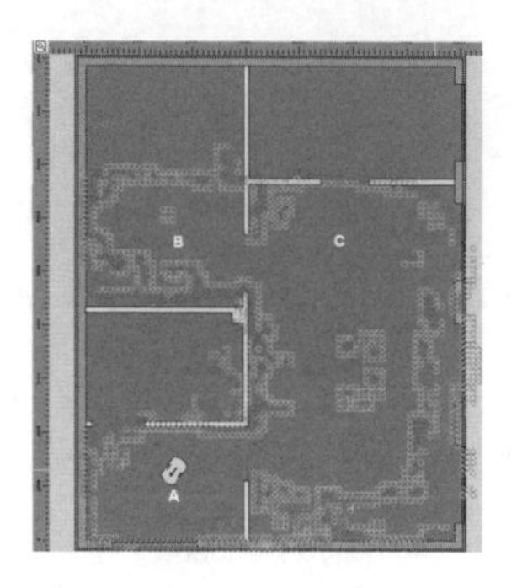

Fig. 6. Generated Map of the Environment

Table 1. Results of Euclidean Distance Based Cost Function in Home Environment

	Cost		Time	
	B	C	B	C
A	4600	5780	77300 ms	-
B	-	3500	-	96200 ms

Table 2. Results of A* Algorithm Based Cost Function in Home Environment

	Cost		Time	
	C	B	C	B
A	6370	7430	56110 ms	-
C	-	3650	-	39330 ms

environment has been taken and the cost remains the same as in simulated environment where there were less obstacles than the real environment. The average total time to reach the destinations is 173500 ms (see Table 1).

The *Navigational Costs* determined by the *Global Path Planner*, on the average, are given in Table 2. The robot follows from A to C and then C to B. The average time required to complete all the tasks is given in Table 2. The total time to complete both the tasks is 95440 ms, a significant (about 44%) improvement over 173500 ms.

In many other experiments conducted in the home environment, the total time to complete navigational tasks by ordering the tasks using *Global Path Planner* was better or equivalent to ordering of tasks using Euclidean based cost.

5 Conclusion

This paper focuses on ordering the navigational tasks for the user to optimally control the robot by considering unstructured obstacles in the environment. The algorithm developed uses A* algorithm to compute the *Navigational Cost* for navigating from source to destination taking the obstacles, in the environment, into account. The tasks are performed depending on the *Navigational Cost*. Experimental results show a significant improvement in time while performing multiple tasks. Moreover, the tasks can be added and removed at anytime without intensive computation or re-planning and also ensuring that the tasks remain ordered and smallest cost tasks are performed earlier.

6 Future Work

Based on the ordering of navigational tasks using map of the environment, the algorithm will be enhanced to order the tasks not only on the basis of *Navigational Cost* but also on the basis of priority. The priorities will make certain tasks more critical and urgent to perform than others. This will enable the robot to perform even more better to the user responses.

References

[1] Armbrust, C., Koch, J., Stocker, U., Berns, K.: Mobile robot navigation support in living environments. In: 20. Fachgespräch Autonome Mobile Systeme (AMS), Kaiserslautern, Germany, pp. 341–346 (October 2007)

[2] Beetz, M., Bennewitz, M.: Planning, scheduling, and plan execution for autonomous robot office couriers. In: Bergmann, R., Kott, A. (eds.) Proceedings of the Workshop Integrating Planning, Scheduling and Execution in Dynamic and Uncertain Environments at the Fourth International Conference on AI in Planning Systems (AIPS), vol. Workshop Notes 98-02. AAAI Press, Menlo Park (1998)

[3] Cucchiara, R., Prati, A., Vezzani, R.: A multi-camera vision system for fall detection and alarm generation. Expert Systems 24(5), 334–345 (2007)

[4] Deegan, P., Grupen, R., Hanson, A., Horrell, E., Ou, S., Riseman, E., Sen, S., Thibodeau, B., Williams, A., Xie, D.: Mobile manipulators for assisted living in residential settings. Autonomous Robots, Special Issue on Socially Assistive Robotics 24(2), 179–192 (2008)

[5] Graf, B., Hans, M., Schraft, R.: Care-o-bot ii - development of a next generation robotic home assistant. Autonomous Robots 16(2), 193–205 (2004)

[6] Labonte, D., Michaud, F., Boissy, P., Corriveau, H., Cloutier, R., Roux, M.: A pilot study on teleoperated mobile robots in home environments. In: IEEE/RSJ International Conference on Intelligent Robots and Systems, October 9-15, pp. 4466–4471 (2006)

[7] Mehdi, S., Armbrust, C., Koch, J., Berns, K.: Methodology for robot mapping and navigation in assisted living environments. In: Proceedings of the Workshop on Robotics and Automation in Assistive Living Systems at the International Conference on Pervasive Technologies Related to Assistive Environments 2009 (PETRA 2009), Corfu, Greece (June 9-13, 2009)

[8] Mei, Y., Lu, Y.H., Hu, Y., Lee, C.: Energy-efficient motion planning for mobile robots. In: IEEE International Conference on Robotics and Automation, Proceedings, ICRA 2004, vol. 5, pp. 4344–4349 (2004)

[9] Proetzsch, M., Luksch, T., Berns, K.: The behaviour-based control architecture iB2C for complex robotic systems. In: Hertzberg, J., Beetz, M., Englert, R. (eds.) KI 2007. LNCS (LNAI), vol. 4667, pp. 494–497. Springer, Heidelberg (2007)

[10] Quinlan, S., Khatib, O.: Elastic bands: Connecting path planning and control. In: Proceedings of IEEE Int. Conference on Robotics and Automation, Atlanta, pp. 802–807 (1993)

[11] Sipahioglu, A., Yazici, A., Parlaktuna, O., Gurel, U.: Real-time tour construction for a mobile robot in a dynamic environment. Robot. Auton. Syst. 56(4), 289–295 (2008)

[12] Wada, K., Shibata, T.: Living with seal robots - its sociopsychological and physiological influences on the elderly at a care house. IEEE Transactions on Robotics 23(5), 972–980 (2007)

Attracting Students to Engineering: Using Intuitive HRIs for Educational Purposes

Pedro Neto[1], Nuno Mendes[1], Nélio Mourato[1], J. Norberto Pires[1], and A. Paulo Moreira[2]

[1] Department of Mechanical Engineering (CEMUC), University of Coimbra, POLO II, 3030-788, Coimbra, Portugal
{pedro.neto,nuno.mendes,n.mourato,jnp}@dem.uc.pt
[2] Institute for Systems and Computer Engineering of Porto (INESC-Porto), University of Porto, Rua Dr. Roberto Frias, 4200-465, Porto, Portugal
amoreira@fe.up.pt

Abstract. Today, industrialized countries are facing a major problem, the lack of skilled engineers. Despite the increasing demand for engineers in the labor market, the number of students going to engineering courses has been declining. This paper reports some initiatives carried out by the industrial robotics laboratory of the University of Coimbra (Portugal) which can contribute to help the youths to obtain an enlarged image of what the engineers can do. These initiatives have focused on the organization of one-day visits to the university campus and participation in events (fairs and exhibitions) to disseminate science and technology. Our participation in such initiatives has been done through the exhibition of a robotic platform to be experienced by the public, usually high school students. In order to allow visitors to intuitively drive the robot, two human-robot interfaces (HRI) has been used. So, to make the process more appealing, the visitors will not only drive the robot but also will play a game with the robot. Besides, it is also important for us receive feedback from the visitors. In such way, a questionnaire about their understanding of the presented robotic platform was proposed. Some points regarding the questionnaires are pointed out for discussion, the students' interest in robotics and the effectiveness of our "hands-on" robotic platform. Results are presented by dividing the answers collected by gender.

Keywords: Robotics, Education, HRI, Student Recruitment.

1 Introduction

Nowadays, industrialized countries are facing a major problem, the lack of skilled engineers. In fact, the demand for engineers in the labor market have increased and the number of students going to engineering courses has been declining, especially in some European countries [1], but also in U.S.A. and Japan [2,3,4]. Apparently, this decline starts when a country reaches first world status and living standards. In these countries, most of the young people are unfamiliar with the "difficult" of the work and do not see the value in undertaking engineering

P. Vadakkepat et al. (Eds.): FIRA 2010, CCIS 103, pp. 250–257, 2010.

courses when apparently they could choose more "appealing" careers [1]. On contrary, in developing countries like China or India, engineering have a high status in the public opinion, and engineers' contributions to society are more visible in these countries [5]. In our opinion, another reason for this trend seems to be that most of the young people have no idea about what engineers can do, and therefore they don't feel motivated to pursue studies in engineering. It is also our mission (engineers and scientists working in robotics field) to contribute to help young people to better understand what engineers do. The industrial robotics laboratory of the University of Coimbra has made efforts to increase the cooperation/collaboration between academia, industry and civil society in general [6]. Over the last few years the laboratory has organized one-day visits to the university campus. These visits are usually attended by high school students and are intended to show the work developed in our laboratory. Moreover, we have organized and participated in events to disseminate science and technology. Our participation in such events has been done through the exhibition of a robotic platform specifically targeted to be experienced by high school students. Generally, visitors are not indifferent to the presence of a robot, perhaps because they have in mind the super intelligent and powerful robots that appear in science fiction films. In fact, robots have been portrayed in a wide range of literary works and films as equivalent to humans, or even as superior human substitutes equipped with sophisticated bodies and intellectual qualities. Nevertheless, most of the visitors know that robots are not so intelligent and powerful, for example, they know that industrial robots are programmed to perform repetitive tasks in car assembly lines. However, most of them don't know that robots have been used for different application purposes such as entertainment, personal use, welfare, education, rehabilitation, etc., [7,8,9].

Robot technology is in constant evolution, but still remains a problem related to the way how humans interact with robots. In this moment, the major part of the robots are controlled and programmed by using your own programming languages, so that only people with technical knowledge (programmers, engineers, etc.) are able to program robots. The goal is to create methodologies to help humans to interact with a robot in an intuitive way. In other words, making a robotic demonstration in terms of high-level behaviors, for example, using speech and/or gestures [10,11], the user should be capable to demonstrate to the robot what it should do. Thus, new and more intuitive ways to robot programming and control are required.

In this paper is presented the robotic platform that we have been used in events to disseminate science. In order to allow visitors to intuitively drive the robot, two human-robot interfaces using a handheld device as a mediator object between a human and a robot (Wii remote and Nunchuck) has been used. So, to make the process more appealing, the visitors will not only drive the robot but also they will play a game with the robot. This kind of events are a good opportunity to be in close contact with visitors, show them that robotics is a multidisciplinary field that evolves knowledge in different areas, but essentially, show that robotics can be "fun". However, it is also important for us receive feedback from the visitors. Questionnaires about their understanding of our explanation were proposed. From the results collected from the questionnaires, we observed the students' interest in robotics and the effectiveness of our "hands-on" robotic platform.

2 Challenges for Robotics

The continuous falling of the birth-rate in developed countries, especially in Europe, is resulting in a reduction in the number of students. Moreover, a significant part of them are going away from engineering fields. This situation may tremendously affect the European industry in the future. The European Union (E.U.) has put forward ambitious plans to face the challenges of globalization, spending significant amounts of money in research and development. The budget provided by the E.U. for different European projects in the robotics field make us strongly believe that robotics will constitute an important key of the European economy in the future decades [1].

In recent years, several events and contests have been realized around robotics, robot parties, competitions, etc., where we have seen the growth of the market for entertainment/educational robots. This type of robots has proved to be a good tool to motivate young people to get involved with robotics [12]. Moreover, many distance-learning platforms and applications have been developed, including e-laboratories platforms in the robotics field [13].

3 Experiments and Results

In our demand for appellative human-robot interfaces (HRI), two human-robot interfaces using a handheld device as a mediator object between a human and a robot (Wii remote and Nunchuck) are presented. The developed interfaces allow users to drive the robot and at the same time play a game with it. The player tilts the playing field by guiding the robot arm, and using the Wii remote or the Nunchuck stick to guide the robot. The game is simple, the user should navigate a spherical object around mazes to reach the end goal (introduce the sphere in a target point), Figure 1. We are using the motion sensor capabilities of the Wii remote to detect the user hand movement (tilt) and thus play the game controlling the robot movements. In another similar approach was used the stick of the Nunchuck to control the robot. In both approaches the robot is moved inside a certain domain of security.

Our robotic platform is composed by an industrial robot Motoman SDA10 equipped with the NX100 controller, a Wii Remote and Nunchuck, a game table attached to the robot wrist, and a computer running the application that manages the system. The above mentioned application receives data from the Wii remote or Nunchuck, interprets the received data and acts in the robot, using for this purpose the MotomanLib, a Data Link Library created in our laboratory to control and manage the robot remotely, Figure 1.

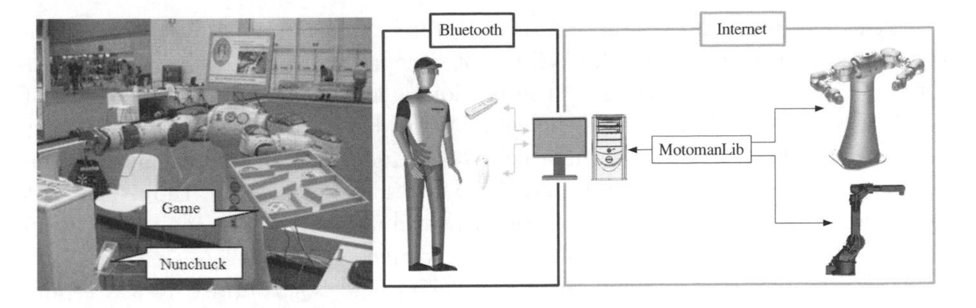

Fig. 1. The robotic platform in an event to disseminate science and technology in Lisbon (left). A schematic representation of the robotic platform, in terms of communication technology. The Wii remote and Nunchuck transmit and receive data without wires (via Bluetooth), giving a greater freedom to the user (right).

3.1 Wii Remote and Nunchuck

The demand for new interaction systems to improve the game experience has led to the development of new devices that allow the user to feel more immersed in the game. In contrast to the traditional gamepads or joysticks, the Wii remote allows users to control/play the game using gestures as well as button presses. It uses a combination of motion sensing and infrared detection to sense its poses (rotations and translations) in 3D space. The Wii remote has a 3-axis accelerometer, an infrared camera with an object tracking system and eleven buttons used as input features. In order to provide feedback to the user, the Wii Remote contains four LEDs, a rumble to make the controller vibrate and a speaker. The Wii Remote communicates with the Wii console or with a computer via Bluetooth wireless link, reporting back data at 100 Hz. The reported data contains information about the controller state (acceleration, buttons, infrared camera, etc.). Several studies have been done using the Wii Remote as an interaction device, particularly in the construction of interactive whiteboards, finger tracking systems and control of robots [14]. In order to extract relevant information from the Wii remote, the motion sensor capabilities of the controller were explored and used to achieve the goals [11].

3.2 Interface Description

In this paper, we present two HRI using a handheld device (Wii remote controller and Nunchuck controller) as a mediator object between a human and a robot, allowing the human to intuitively drive the robot. The developed interfaces should take into account certain required criteria for interactions with robots. Thus, the interaction system should be portable, easy to use, flexible and not requiring a too significant training time.

As mentioned above, the player tilts the playing field by guiding the robot, and using the Wii remote or the Nunchuck stick to guide the robot. To play the game is only necessary to rotate the robot arm around two axes, in this case a rotation around the X axis and other around the Y axis. In order to move the robot the user must tilt the Wii remote according to the desired movement for the table game. Recurring to the motion-sensing and tilting capabilities of the Wii remote, we have that if the controller is held horizontally, it will report acceleration along the Z axis, the acceleration due to gravity g that near to the surface of the earth is approximately $9.8m/s^2$. Thus, even when the user is not accelerating the accelerometer, a static measurement can determine the rotation (tilt) of the human hand that holds the controller.

Analyzing Figure 2-a, when the controller is held horizontally, it will report an acceleration g along the Z axis in the positive direction; $a_z \simeq g$, $a_x \simeq 0$ and $a_y \simeq 0$. But when the controller is rotated around the Y axis (Figure 2-b), $a_x \simeq g$, $a_y \simeq 0$ and $a_z \simeq 0$. On contrary, when the accelerometer is rotated around the Y axis in the reverse direction (Figure 2-c), $a_x \simeq -g$, $a_y \simeq 0$ and $a_z \simeq 0$. A similar approach was done to detect the rotations around the X axis (Figure 2-d and Figure 2-e). Thus, it is easy to detect if the user hand is rotating the Wii remote around the X axis, Y axis, or both at the same time.

According to the accelerations measured from the Wii remote, the application that manages the system sends commands to the robot, moving it. This interface provides really natural interactions because the user only needs to hold the controller and rotate it according the desired robot movement, allowing users to focus on planning high-level tasks instead of low-level actions.

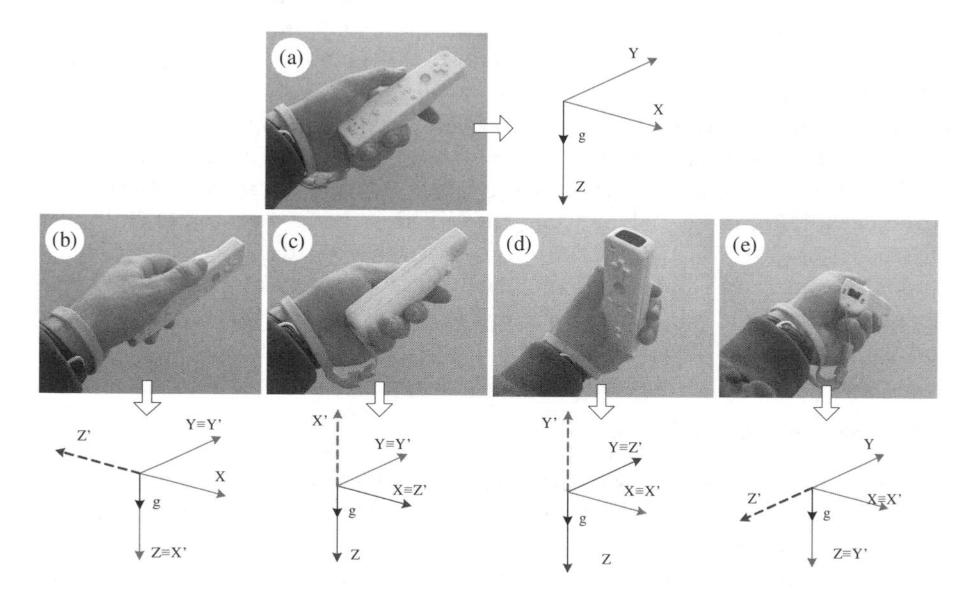

Fig. 2. a – No rotation, b – Rotation around the Y axis in the negative direction (Ry-), c – (Ry+), d – (Rx-), e – (Rx+)

3.3 Questionnaires

In order to receive feedback from students about their interest in robotics and the effectiveness of our "hands-on" robotic platform, a questionnaire was proposed. 80 participants (40 males and 40 females) were recruited for this study in an event to disseminate science in Portugal. The participants were high-school students (14-17 years old). Among the 80 participants, all reported to know the Wii remote. First, it was explained the way the system works in a language understandable to the audience, a brief explanation in 10 minutes. Then, after this explanation the participants responded to the questionnaire. The proposed questionnaire had two questions and three possible answers for each question:

1) Question 1: Do you understand the explanation?

- Answer (a): I completely understand.
- Answer (b): I understand something.
- Answer (c): I don't understand.

2) Question 2: Do you think that robotics is an interesting field?

- Answer (a): Yes.
- Answer (b): No.
- Answer (c): Somewhat.

Each of the 80 participants responded to the questionnaire before interacting with the robot. After all the participants have interacted and played with the robot, they were asked again to respond to the question 2. The results will be presented in next section by dividing the answers collected by gender.

3.4 Results and Discussion

The questionnaire results show that 42.5% of the students evaluated the explanation (question 1) with "I completely understand", 45% with "I understand something" and only 12.5% of the students said "I do not understand", Figure 3 (top left). When asked if robotics is an interesting field (question 2), 52.5% of the students said "yes", 15% said "no" and 32.5% of the students said "somewhat", Figure 3 (top right). This means that in general robotics is an area of knowledge classified as interesting by most of the students. However 15% said "no", robotics is not interesting.

As mentioned above, after the initial explanation and after the participants have interacted and played with the robot, they were asked again to respond to the question 2. Results show that some students change their opinion about their interest in robotics, 60% of the students said "yes" robotics is interesting, 10% said "no" and 30% of the students said "somewhat", Figure 3 (below). This result demonstrates the importance of the experimentation "hands-on" for the students of today. Through the questionnaires it was also possible to conclude that the level of interest in robotics is similar for both males and females (the males report a little bit more interest but in general gender does not influence the decision). However, after playing with the robot the number of males reporting that robotics is interesting increases (from 22 to 28) while the number of females did not significantly change Figure 3 (below). So, we can conclude that practical experimentation is an important factor to motivate students, especially the males.

Every day we see that the Wii remote controller allow visitors to get involved with robot technology, making the process of interaction more appealing than when compared to the traditional keypad controllers.

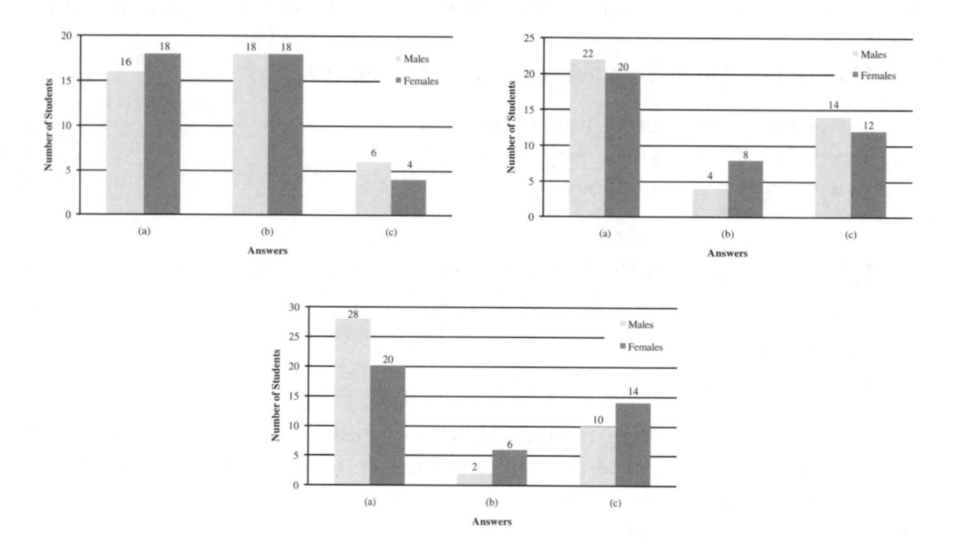

Fig. 3. Results for question 1 (before playing with the robot) (top left). Results for question 2 (before playing with the robot) (top right). Results for question 2 (after playing with the robot) (below).

4 Conclusion

This paper reported some initiatives that can contribute to help the youths to obtain an enlarged image of what the engineers can do. These initiatives have focused on the organization of visits to the university campus and participation in events to disseminate science and technology. Our participation in such initiatives has been done through the exhibition of a robotic platform to be experienced by the public (high school students), allowing users to drive the robot and at the same time play a game with it, using the Wii remote and Nunchuck as HRIs.

The students' interest in robotics and the effectiveness of our "hands-on" robotic platform are pointed out for discussion (questionnaires were proposed). Through the questionnaires it was possible to conclude that the level of interest in robotics is similar for both males and females. However, after playing with the robot the number of males reporting that robotics is interesting increases while the number of females did not significantly change. So, we can conclude that practical experimentation is an important factor to motivate students to get interested in robotics, especially the males. Experience showed us that in many cases, exchanging video streams or sharing educational material is often not adequate to transmit our message to visitors.

References

1. European Federation of National Engineering Associations: Engineering Skills Shortage in Europe. How do we solve it ?(2007)
2. National Science Board: Science & Engineering Indicators 2008. U.S. Government Printing Office, Washington, D.C. (2008)
3. Solis, J., Takanishi, A.: Practical Issues on Robotic Education and Challenges Towards RoboEthics Education. In: 18th IEEE International Symposium on Robot and Human Interactive Communication, pp. 561–565. IEEE Press, Toyama (2009)
4. Solis, J., Nakadate, R., Yamamoto, T., Takanishi, A.: Introduction of Mechatronics to Undergraduate Students Based on Robotic Platforms for Education Purposes. In: 18th IEEE International Symposium on Robot and Human Interactive Communication, pp. 693–698. IEEE Press, Toyama (2009)
5. Lopez-Martin, A.J.: Attracting Prospective Engineering Students in the Emerging European Space for Higher Education. IEEE Trans. on Education 53 (2010)
6. Pires, J.N., Nilsson, K., Petersen, H.G.: Industrial Robotics Applications and Industry-Academia Cooperation in Europe. IEEE Robotics & Automation 12, 5–6 (2005)
7. Geppert, L., Yoshihiro, K.: Dancing with Robots. IEEE Spectrum 41, 34–35 (2004)
8. Sakaki, T.: An Effective Design Method for Welfare Robot and its Application to the Design of Meal-Assistance Robot. In: 17th IEEE Int. Symposium on Robot and Human Interactive Communication, pp. 309–314. IEEE Press, Munich (2008)
9. Siciliano, B., Khatib, O.: Handbook of Robotics. Springer, New York (2008)
10. Yang, H.-D., Park, A.-Y., Lee, S.-W.: Gesture Spotting Recognition for Human-Robot Interaction. IEEE Trans. on Robotics 32, pp. 32, 256–270 (2007)
11. Neto, P., Pires, J.N., Moreira, A.P.: High-Level Programming and Control for Industrial Robotics: using a Hand-Held Accelerometer-Based Input Device for Gesture and Posture Recognition. Ind. Robot, An Int. J. 37, 137–147 (2010)
12. Ali Yousuf, M., Montúfar, R., Cueva, V.: Robotic Projects to Enhance Student Participation, Motivation and Learning. In: 4th Int. Conf. Multimedia, Information and Communication Technologies in Education, pp. 1994–1989. Sevilla (2006)
13. Tzafestas, C.S., Palaiologou, N., Alifragis, M.: Virtual and Remote Robotic Laboratory: Comparative Experimental Evaluation. IEEE Trans. on Education 49, 360–369 (2006)
14. Lee, J.C.: Hacking the Nintendo Wii Remote. Pervasive Computing IEEE 7, 39–45 (2008)

Robots as a Tool for Teaching Differential Equations

Anna Friesel

Copenhagen University College of Engineering,
Electronics and Electrical Power Sections,
Lautrupvang 15, DK-2750 Ballerup, Denmark
afr@ihk.dk
www.ihk.dk

Abstract. This paper describes experiences in teaching mathematics, in particular differential equations and mathematical modeling, as a part of the robot project at the Copenhagen University College of Engineering. Modern engineering students are difficult to motivate in order to learn pure theoretical issues such as solving differential equations. At the same time, the fail rate in mathematics is very high during the first part of our bachelor program. This was the reason to change the structure of our education and incorporate the theory with practical projects. Differential equations, mathematical modeling and control theory are theoretical parts of the robot project. In order to explain the dynamical behavior of the robot, students have to learn more mathematics. The examination results show improved learning potential, when using this method of educating mathematics. The students' evaluations show a very positive effect on their experience with this practical way of learning mathematics.

Keywords: Engineering mathematics, problem-based education, autonomous robot project, teamwork.

1 Introduction

In today's industry and trade, there is an increasing demand for engineers who do not just have excellent competencies in their field but also a good understanding and practical experience in economics, management as well as team-based project work. Engineering education must therefore change its focus from purely scientific and technological, to be a mix of different disciplines, gaining access to scientific and technological innovations. Great emphasis is placed on a combination of formal education, whose role is to impart systematic knowledge of basic disciplines, and informal education which is especially effective in broadening horizons, fostering curiosity, and active learning. Success in today's knowledge-based economy requires that graduates are educated on the basis of critical thinking and problem solving. Several studies carried out in many countries have proven that there still exists a gap between the range of skills that graduates are equipped with and the skills and qualifications that are sought after by employers [1-7]. At the same time the research confirms improvement of the learning outcomes in education when the theory is combined with practical training or projects. Usually the project-based courses are introduced after the students have passed the first 2-3 semesters of the engineering basics, like calculus, physics and computer programming. Students who wish to work professionally with electronics, computers and telecommunication, and who do not wish to continue

P. Vadakkepat et al. (Eds.): FIRA 2010, CCIS 103, pp. 258–265, 2010.

for Master's Degree, are not motivated to use 2-3 semesters for purely theoretical studies. These students usually have high fail rate in mathematics and physics in the first 1-2 semesters and, as a result, many of them stop. Engineering faculties at the universities and colleges in Denmark have experienced decreasing number of applicants over the last 5-6 years, and do not want high drop-out rates of the students in their faculties. Once young people have shown an interest in technology and engineering, it is important to design the education to fit their expectations; the expectation of working on engineering projects and having the influence on the future development of technology. It will be a great benefit to all those involved: students, industry and the development in their countries. Considering the students' desire to work on real engineering projects and the requirements to master the basics of mathematics and physics, the message is clear: the students will be more likely "to stick with derivatives and integrals" if we can offer them some evidence of how necessary these tools are in order to find solutions for engineering problems. If we use this strategy from the first semester the students will see that they are already making progress towards completing engineering work/projects and the positive influence will be shown in an increased pass rate. There are different possibilities to include experimental work together with theory in engineering like:

1. Using simulations programs involving graphics and animations in order to visualize mathematics and physics.
2. Mixing the theory classes with laboratory exercises.
3. Study-tours to industrial companies.
4. Inviting guests from industry for lectures.
5. Problem-based team work.
6. Project-based learning.

Many universities and engineering colleges have the opinion that pedagogical activities must actively involve the students in order to motivate them to learn basics of mathematics and physics [6, 8-11]. Problem-based learning and working with projects has already been implemented in engineering educations in many different universities.

2 Study Structure

Recently, we went through the process to renew the educational study structure in our department and we decided to change the study structure in Electronics at the Copenhagen University College of Engineering (short name in Danish is IHK) towards more applied science, especially mathematics. Similar changes were made for all semesters in our programs, involving projects and teamwork in our basic courses in mathematics and physics. The first four semesters of our program in Electronics and Information Technology (EIT) are shown in Table 1.

Project- based learning requires a high degree of concentration on particular topics, and in order to support this educational method we also changed the weekly time schedule. Students have only two modules/topics during a day: 8:30-12 and 12:30-16:00. Each module includes four lectures of approximately 45 minutes and some necessary breaks in between. One module of tuition is usually related to a course of 5 ECTS credits, and one module of teacher's tuition requires on average 4 hours of self-study for the student.

Table 1. Study program, EIT

<table>
<tr><th>ECTS</th><th>1. semester</th><th>2. semester</th><th>3. semester</th><th>4. semester</th></tr>
<tr><td>2.5</td><td rowspan="2">Object Oriented Prog.1</td><td rowspan="2">Object Oriented Prog.2</td><td rowspan="4">Electro-magnetism</td><td rowspan="4">DSM4</td></tr>
<tr><td>2.5</td></tr>
<tr><td>2.5</td><td rowspan="2">Project 1</td><td rowspan="2">Project 1</td></tr>
<tr><td>2.5</td></tr>
<tr><td>2.5</td><td rowspan="4">DSM1</td><td rowspan="3">DSM2</td><td rowspan="4">DSM3</td><td rowspan="4">REG4E</td></tr>
<tr><td>2.5</td></tr>
<tr><td>2.5</td></tr>
<tr><td>2.5</td><td rowspan="5">Digital Electronics2 Project 2</td></tr>
<tr><td>2.5</td><td rowspan="4">Digital Electronics1 Project2</td><td rowspan="4">Projects</td><td rowspan="4">Robot Project</td></tr>
<tr><td>2.5</td></tr>
<tr><td>2.5</td></tr>
<tr><td>2.5</td></tr>
</table>

Each semester the students carry out one or two projects connected to the theory they learned. During the semester, the students also develop the following non-technical/scientific skills:

- How to work in teams
- How to make a presentation for tutors and other teams on seminars
- Define and describe the fundamental problems and concepts introduced in the course – using proper notation
- Define and describe the fundamental methods for solution introduced in the course – using proper language and notation.
- How to work out written reports in connection to the course's assignments and projects
- How to collect information and acquire new information and knowledge
- How to communicate technical problems in writing and speech
- How to cooperate with other in the team.

3 Theoretical Course Connected to the Robot Project

We implemented the course of differential equations in the curriculum for undergraduates on fourth semester as the theoretical support for the robot project. The differential equations have become a part of the fourth semester course REG4E [12,13]. REG4E includes:

- Differential equations
- Laplace transformation
- Mathematical Modeling
- Basics of Control Theory (PID), continuous and discrete.

The REG4E course is close connected to the PROE4 course, which is the practical project: autonomous robot project. The important goal of REG4E and PROE4 courses is to make students to apply the knowledge of mathematics and control theory in a technical application, in this case in autonomous robot [12,13,15,16]. Each part of REG4E course ends with the mandatory assignment connected very closely to the robot project. Students usually work in teams on the mandatory assignments, the same teams as they formed for the robot project. The documentation of the mathematical modeling of the robot, controller design and simulations are a part of the assignments. These assignments and the reports of the robot project (PROE4), give the basis for the evaluation of these courses. Copenhagen University College of engineering uses CampusNet [14], which is a web-based Virtual Learning Environment (VLE), which can be used to both support and enhance teaching and learning. It provides tools to enable all university members, staff and students, to set up all practical information about studies and provide a way to communicate with the students. The CampusNet platform is similar to WebLearn or Moodle. During the project work the students can communicate very easily and safely through CampusNet, also exchanging knowledge and important materials. Other supporting tools in the learning experience are simulation programs. From the very beginning the students are introduced to MATLAB and P-Spice programs [17,18]. SIMULINK is introduced at the fourth semester, during the robot project. The use of simulation programs enhances learning potential and gives the students the possibility to test different solutions to the project before the practical implementation into the working model. Simulation programs are very important tools to understand the role of nonlinearities in undergraduate programs, where the students are introduced only to linear system's theory, but in their projects have to deal with non-linear problems.

4 The Examination

The examination process in Denmark involves by law an external examiner certified by the Ministry of Education, for all examinations at the university level. For engineering departments the external examiner is very often a company manager. The role of external examiner, among other things, is to keep the engineering curriculum up to date. The external examiner has a great opportunity to discuss the contents of the engineering courses including: pedagogical methods, experimental work, and projects when he or she participates in the examination. The external examiner has to report his/her conclusions about the examination level, and the level of education to the chairman of the external examiners, who is approved by the Ministry of Education. This procedure gives industry an influence on the engineering education in Denmark. The industry can easily have influence to change engineering departments' curriculum according to the needs of the industry. On the other hand the students meet their future managers, and get the knowledge of the requirements in industrial companies at the same time. For project related courses the examination is both group and individual, but the marking is always individual. The evaluation is based on a general impression of the level achieved by the student relative to the objective of both courses. REG4E part of the oral examination is as follows:

1. The individual presentation of a self-chosen topic from the mandatory assignments.
2. The individual examination in the randomly selected topic, one of the 8 topics, including differential equations, mathematical modeling and control theory.

After the student's presentations, internal and external examiners ask supplementary questions to both presentations. After questioning, the student leave the examination room and the examiners discuss the presentations, as a group and as individuals. The marks are given individually, and the students, one by one are invited back into the examination room for the explanation of their marking. After the examination the students can evaluate the course anonymously on the CampusNet.

5 Students' Evaluations

Students make evaluation of theirs courses twice during the semester, in the middle of the semester (to have the possibility for improvements) and after the examination. The final evaluation, after the examination, is made on CampusNet and the results of this evaluation are available to the students participating in the course, the teachers and the head of the department. The evaluations include three parts: course evaluation, teacher's evaluation and general comments. The questionnaire is shown in Table 2.

REG4E and PROE4 ran first time in spring 2009 and second time in fall 2009. The evaluations of REG4E have been answered the first time by 65% of the students and the second time by 69% of the students. The results of the evaluations were as follows:

1. 78% of the students (who answered the questionnaire) were either very satisfied or satisfied.
2. 82% of the students felt that the teaching method supported their learning very well.
3. 89% of the students find that the test/examination method matched the method of teaching.

Table 2. Student's questionnaire

Course evaluation
1. Target fulfillment: **To what degree has your learning measured up to the course description?** **1**(Strongly agree) **2**(Agree) **3**(Neither agree nor disagree) **4**(Disagree) **5**(Strongly disagree)
2. Your own performance How much time do you estimate you have spent on the course compared with the expected time? 1 2 3 4 5
3. The relevance of the course: To what degree do you consider the course relevant for your education? 1 2 3 4 5

Table 2. (*continued*)

4. The academic level of the course:
To what degree has the course challenged your academic abilities/competences?
1 2 3 4 5

5. The study material:
To what degree has the study material supported your learning?
1 2 3 4 5

6. The teaching method:
To what degree has the teaching method supported your learning?
1 2 3 4 5

7. The test/examination method:

7.1 To what degree do you find that the examination method matched the method of teaching?
1 2 3 4 5

7.2 To what degree has your learning measured up to the course description?
1 2 3 4 5

7.3 To what degree do you find that the test/examination method is suitable for testing whether the objective of the course has been fulfilled?
1 2 3 4 5

Teacher's evaluation

1 The teacher's instruction in connection with the preparation of assignments:

1.1 How do you evaluate the quality of the academic instruction you received in connection with the preparation of assignments?
1 2 3 4 5

1.2 To what degree do you find the academic instruction you received, in connection with the preparation of assignments, to be sufficient?
1 2 3 4 5

2 The teacher's instruction regarding teamwork, study methods etc.:

2.1 How do you evaluate the quality of the process instruction you received in connection with the preparation of assignments?
1 2 3 4 5

2.2 To what degree do you find the process instruction you received, in connection with the preparation of assignments, to be sufficient?
1 2 3 4 5

3 The teacher's feedback on student work with assignments etc.:

3.1 How do you evaluate the quality of the feedback you received on questions, work with assignments etc.?
1 2 3 4 5

3.2 To what degree do you find the feedback on questions, work with assignments etc. to be sufficient?
1 2 3 4 5

4 The teacher's presentation of the subject matter and ability to put into perspective:

4.1 To what degree did the teacher's presentation, summarizing and putting the new subject matter into perspective support your learning?
1 2 3 4 5

4.2 To what degree do you find the feedback on questions, work with assignments etc. to be sufficient?
1 2 3 4 5

Table 2. (*continued*)

General comments
1.1 What was good or worked well in the course?
1.2 What was bad or worked badly in the course?
1.3 What are your suggestions for improvements?

6 Conclusion

After two semesters of completed REG4E and PROE4 courses in department of Electronics and Information Technology, we can make the conclusion, that the main objectives have been achieved. The students have got a better understanding of the differential equations and their applications in engineering projects. We have also observed much better understanding of differential equations and mechanics during the fourth semester of their education. The students' own evaluations show an increased motivation to learn mathematics and control theory in this practical approach to the theory, especially for the students with previous practical experience. The examination results are shown in graph on Figure 1.

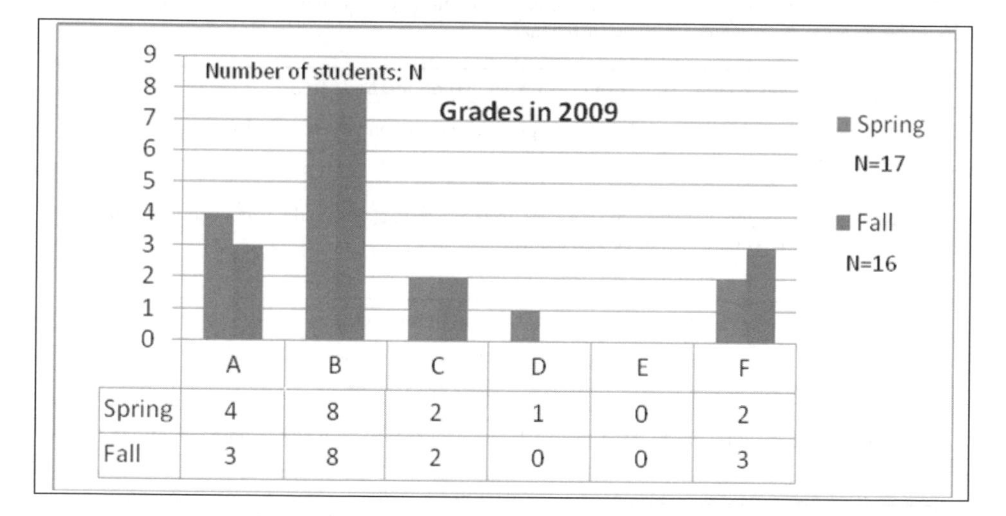

	A	B	C	D	E	F
Spring	4	8	2	1	0	2
Fall	3	8	2	0	0	3

Fig. 1. REG4 examination's results

Acknowledgment

I would like to thank Mogens Pelle at the Copenhagen University College of Engineering, for our good discussions and close cooperation during the process of implementing the new study-structure. Thanks to all the students for valuable discussions on future development of this course, and special thanks to all the students who contribute for the evaluation of the courses. Thanks to I.Stauning, H.Surkau and J.Greve for technical support in connection to the robot project.

References

1. Andersen, A.: Implementation of engineering product design using international student teamwork – to comply with future needs. European Journal of Engineering Education 26(2), 179–186 (2001)
2. Denton, A.A.: The role of technical education, training and the engineering profession in the wealth-creating process. Proceedings of the Institution of Mechanical Engineers, Part B 212, 337–340 (1998)
3. Hedberg, T.: The impact of the Bologna Declaration on European engineering education. European Journal of Engineering Education 28(1), 1–5 (2003)
4. Hillmer, G.: Social & Soft Skills Training Concept in Engineering Education. In: Innovations 2007: World Innovations in Engineering Education and Research, International Network for Engineering Education and Research, pp. 355–366. INEER, Arlington (2007)
5. Skinner, D., Saunders, M.N.K., Beresford, R.: Towards a shared understanding of skill shortages: differing perceptions of training and development needs. Education + Training 46(4), 190 (2004)
6. Waks, S., Sabag, N.: Technology Project Learning Versus Laboratory Experimentation. Journal of Science Education and Technology 13(3), 332–342 (2004)
7. How do you measure success? Designing effective processes for assessing engineering education. ASEE Professional Books (1998)
8. Anderson, L.W., Krathwohl, D.R.: A Taxonomy for Learning, Teaching and Assessing: A Revision of Bloom's Taxonomy of Educational Objectives. Longman, New York (2001)
9. Fink, F.K.: Integration of Work Based Learning in Engineering Education. In: 31st ASEE/IEEE Frontiers in Education Conference (October 2001)
10. Frank, M., Barzilai, A.: Project-Based Technology: Instructional Strategy for Developing Technological Literacy. Journal of Technology Education 18(1), 39–53 (2006)
11. Michau, F., Gentil, S., Barrault, M.: Expected benefits of web-based learning for engineering education: examples in control engineering. European Journal of Engineering Education 26(2), 151–168 (2001)
12. Friesel, A., Guo, M., Husman, L., Vullum, N.: Project in Robotics at the Copenhagen University College of Engineering. In: Proceedings of the 2004 IEEE, International Conference on Robotics & Automation, New Orleans, LA, pp. 1375–1380 (April 2004)
13. Friesel, A.: Learning Robotics By Combining The Theory With Practical Design And Competitio. Undergraduate Engineering Education; AutoSoft Journal, International Journal on Intelligent Automation and Soft Computing; Special Issue on Robotics Education
14. `http://campusnet.ihk.dk`
15. Ahlgren, D.J., Verner, I.M.: Fire-Fighting Robot International Competitions: Education Through Interdisciplinary Design. In: Proceedings of International Conference on Engineering Education, vol. 1, p. 7B3-1 (2001)
16. Ahlgren, D.J., Verner, I.M.: Robot Projects as Education Design Experiments. In: Proceedings of International Conference on Engineering Education, vol. 2, pp. 524–529 (2005)
17. MATLAB, `http://www.mathworks.com/`
18. PSPICE, `http://www.orcad.com/download.orcaddemo.aspx`

A Course Programme in Mobile Robotics with Integrated Hands-on Exercises and Competitions

Ole Ravn and Nils A. Andersen

Automation and Control, DTU Electrical Engineering,
Technical University of Denmark, DTU-Building 326
DK-2800 Kgs. Lyngby
Denmark
{or,naa}@elektro.dtu.dk

Abstract. The paper describes the design of and the considerations for a course programme in mobile robotics at the Technical University of Denmark. An integrated approach was taken designing mobile robot hardware, software and course curricula in an interconnected way. The courses in the programme all feature hands-on elements and competitions to motivate the students and support the learning objectives of the particular course. As the design of such competitions is not trivial some examples and observations are presented.

Keywords: Robotics, education, hardware, software, curriculum, competition.

1 Introduction

Robots capture the imagination whether in movies or in reality and fascinate and motivate people. Working with robotics is a truly multi-disciplinary task including mechanical engineering, electrical engineering and computer science; furthermore robotics lends itself well to demonstrating the engineering work process of 'think, design, simulate, implement, and test' as the models involved are generally quite good. Many basic engineering competences and skill like the power of calibration can be learnt while working with robots.

Today's students in engineering are highly motivated by hands-on elements demonstrating the theoretical elements taught. Due to this combination a large majority of the students benefit by a better understanding of the theory, and furthermore they acknowledge the need for theory and see the connection between theory and solutions to real world problems. This is especially true for highly theoretical topics such as control theory and advanced signal processing.

The addition of a competitive element in the courses further motivates the students to put in more than the officially required number of hours and also trains them in the deadline oriented way that characterize the work situation of many engineers.

At the Technical University of Denmark (DTU) a course programme in mobile robotics has been designed and implemented over the last ten years including those elements. This has lead to an increase in students working in the area of control and

P. Vadakkepat et al. (Eds.): FIRA 2010, CCIS 103, pp. 266–273, 2010.

robotics including student with other background than electrical engineering. This paper describes the ideas behind the course programme.

There are many other examples of single courses in robotics employing some of the same elements [1],[2] such as the MIT Autonomous Robot Design Competition course 6.270. In the DTU mobile robot course programme emphasis is put on algorithms and software for navigation and mission management and it was decided that no hardware whether mechanical or electronic should be build by the students. This reflects an experience that students that try to build a complete mobile robot solution rarely get to the point where the complete system works with advanced software solutions for navigation.

Section 2 describes design consideration, the line of courses and section 3 describes the design of the competitions. Section 4 gives insight into the hardware and software made and finally section 5 gives conclusions of the ten years of development, and references are presented in section 6.

2 Design Considerations and Course Programme

The overall design considerations for the course programme in mobile robotics include:

- There should be significant hands-on elements with real robots in all parts of the programme, due to reasons mentioned earlier.
- Competitions should be a core motivating element but designed especially to support the course elements taught.
- Progression providing new challenges of increasing complexity throughout the programme and differentiated learning so students with different background can supplement each other in teams.
- Each course should have focus on a few specific primary elements.
- Tools like Matlab that support the deeper understanding of the theoretical elements and support the students in a structured way of problem solving not indulging into ad hoc changes.
- Hardware that is simple, robust and flexible to use and supports student in focusing on solving the task at hand.

Yet another benefit from the competition element is that the students will be much more focused on robustness and reliability of the solutions as the performance is evaluated in just two runs on a specific time. Therefore even a simple failure will have consequences like in real life.

The courses in the course programme range from first semester courses to courses at PhD-level, there is even a one day short course for K-12 students visiting DTU with hands-on and focusing on simple geometry and programming.

The course Engineering Practice (31013)[1] on the first semester of the electrical engineering bachelors includes a project introducing calibration as a tool to minimize

[1] Course descriptions on `www.kurser.dtu.dk/~search.aspx?menulanguage=en-GB`.

systematic errors in sensors specifically odometry of mobile robots. The project also introduces programming and team work all in the context of mobile robots thus motivating the students.

The three week intensive course Autonomous Robot Systems (31385) at the third semester on the bachelor program focuses on task decomposition, C programming, script language programming, introduces state-machines, real-time programming and feedback systems. The mobile robot platforms may also be used in the Introductory project (31015) on the fourth semester and in the bachelor project on sixth semester depending on the choice of the students.

At master level there are two technological specialization courses on the study line 'Automation and Robot technology'. Intelligent systems (31380) focuses on using AI and expert systems in mobile robotics and Advanced Mobile Robots (31388) focusing on navigation, sensor fusion and signal and image processing. Finally, many masters projects uses the mobile robot platforms. Recently a Ph.D. course (31389) on advanced navigation, mission management and localization has been introduced.

It is important to stress that the competences acquired by the students in the course program are not limited to mobile robots but can be used in a much broader context e.g. embedded systems.

3 Competitions

Using final competitions as a part of courses is highly motivating for the students Competitions give a very clear picture of how well your team's solution works compared to that of other teams [3]. This often makes students work late nights to achieve a good result. However even if it is a good way of getting and keeping the interest of the students it is of the utmost importance that the competition is carefully designed to support the learning objectives and the course curriculum.

It is important that most of the problems can be solved using the algorithms taught in the course and that the difficulty of implementation matches the level of the students and the nominal time for the course. If failing to do so one runs the risk of supporting the general belief of many students that theory is only usable for solving exam problems while real practical problems are best solved using your own ad hoc methods.

In Automation and Control we have been using final competition on our three courses on mobile robotics. The competition track from the introductory 3-week practical autonomous systems course has a number of obstacles and black and white tapelines on floor to guide the robot from obstacle to obstacle. Points are given for each obstacle the robot successfully copes with. Points are also given for fastest time._It is important to note that the robot should traverse the track totally autonomously, so communication with the vehicle is allowed after start. The schematic plan of the full track is show in figure 1.

1. **Position measurement.** The task is to measure the x component of the position of a box relative to the starting point. Requires a combination of IR sensors and odometry.

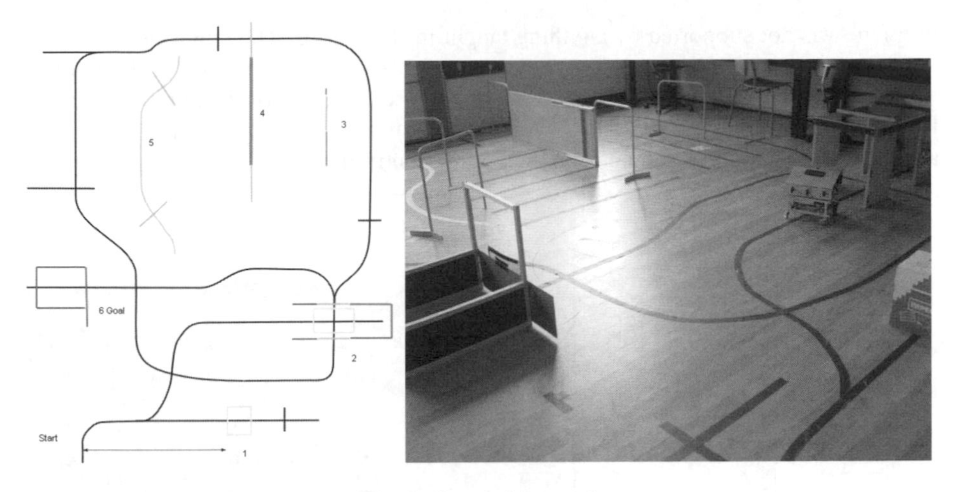

Fig. 1. Competition track

2. **Blocked gate.** The blocked gate is initially blocked by a box that must be pushed away before the gates can be passed. Requires odometry and navigation.
3. **Gate on the loose.** The gate is known to be on a given line but the exact position is unknown. The robot must locate the gate and pass through it.
4. **Wall following.** The wall-obstacle is a variable length wall with a gate at each end. The robot must use the IR-distance sensors to follow the wall and detect the end of the wall.
5. **White line.** As the floor is light it is more difficult to detect a white line than a black line. This obstacle tests the performance of the line detection code.
6. **Goal.** The goal is a garage with a closed gate. The vehicle must push the left side of the gate from the back to open it, before it can enter the garage

Important topics of the course are robot kinematics, odometry, basic motion control, calibration of sensors and odometry, state machines, and reliability.

To solve the competition track it is necessary to be able to follow the lines on the floor. This requires calibration of the reflectance sensor and basic motion control. Correct measurement in 1 requires accurate calibration of both odometry and IR distance sensors. In problem two, accurate navigation relying on a good implementation of the basic motion control is important.

Problem 3 tests the understanding of sample rate and measurement noise. As the IR distance sensors often give false readings especially in sunny weather robots often fail if decisions are based on a single measurement. Furthermore the sample rate of the IR sensor measurement is only 6 Hz which makes it necessary to drive slowly if detection of the gate shall be reliable. As it is seen the competition has been carefully designed to support the curriculum of the course.

In the course ‘Advanced Mobile Robots’ two competition tracks have been tried. In the first the robot should locate cardboard boxes, read a number on the box and push the box to a predestined position. Also balls should be categorized by color and pushed to given positions. This competition was a disaster. The main problem was that precision pushing of boxes and balls turned out to be extremely difficult and this problem was not supported by anything taught in the course. The teams used much of their time inventing clever pushing devices made from plastic gadgets and tape. Furthermore all the teams had a very low score at the competition leading to a general feeling of failure. After two tries with poor results this competition problem was abandoned. The present competition track is shown in fig 2.

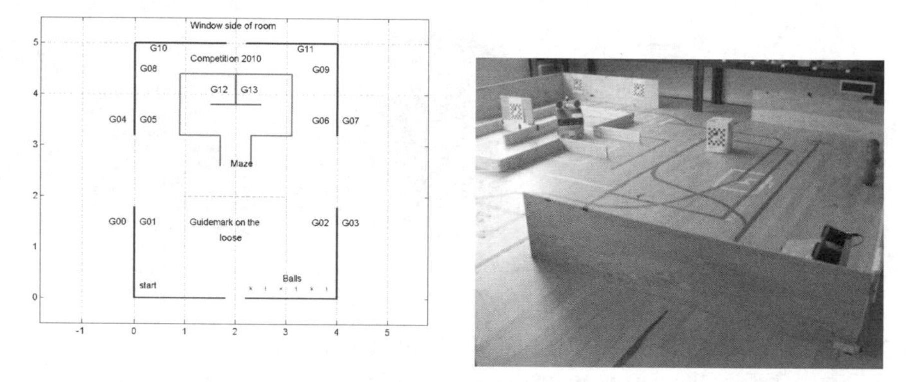

Fig. 2. Competition track

The competition track consists of a box of 40 cm high wooden walls with passages in the ends and the sides. 14 poses (position and direction) of guide marks readable by the robots are given. The robot must start within the start square in the lower left corner. The number of the first guide mark pose is given 10 minutes before the competition. The robot must go to that position and read the number of the next guide mark position. The robot must go through the route determined by the guide marks put in the box just before the competition. One of the guide marks on the route will be in the maze which gives higher score. At some point during the run the robot must find the extra guide mark (position unknown), read its number and calculate the position in the box coordinate system. Up to six balls (red or blue) are placed in the lower left corner of the box and points are given for telling the correct number of blue balls and the correct number of red balls.

The course curriculum contains kinematics, image processing, localization, planning, obstacle avoidance and programming. Each of the topics is covered by compulsory exercises before solving the final project leading to the competition and a 20-30 minutes presentation of the final project. To solve the competition described above on-line planning is necessary as the route is unknown before the competition starts. As the route will be long localization with respect to the box coordinate system is a must and to get full points for the guide mark on the loose the accuracy must be better than 10 cm. Image processing is used for deciding ball colors and the task is not trivial due to varying lighting conditions. Programming consists of writing C++ plug-ins for the laser- and camera server and coordination and motion control in SMR-CL [4]. As seen most of the tasks of the competition track are directly supported by topics from the curriculum but the students are left free to use any solution of their own choice (of cause limited be the given robot and available programming languages). The new competition track proved a great success as most of the time was used to implement solutions based on the taught methods combined with the ingenuity of the student. The performance of the teams in the competition was satisfactory especially given that the best score was only one point less than the highest possible score.

4 Equipment

When designing a course program with emphasis on hands-on exercises it is important to provide the correct hardware and software for the exercises. As we choose to put emphasis on teaching algorithms for mobile robots a hardware platform must be provided along with suitable software. Due to the high prices for commercial robot platforms in 2000 we decided to build our own platform.

An important design parameter was that the robot should be so small that it can easily be handled by one person without any safety problems and on the other hand it should be big enough to carry equipment such as cameras and small manipulators. It was also important that the design was kept simple so that it could be produced on a small workshop with a reasonable effort.

The chassis is made from standard aluminum profiles that are assembled by bolts. The size was chosen to 28 by 28 cm due to the size of a μATX motherboard and the maximum size requirement of the Eurobot contest. The platform is shown on fig 3.

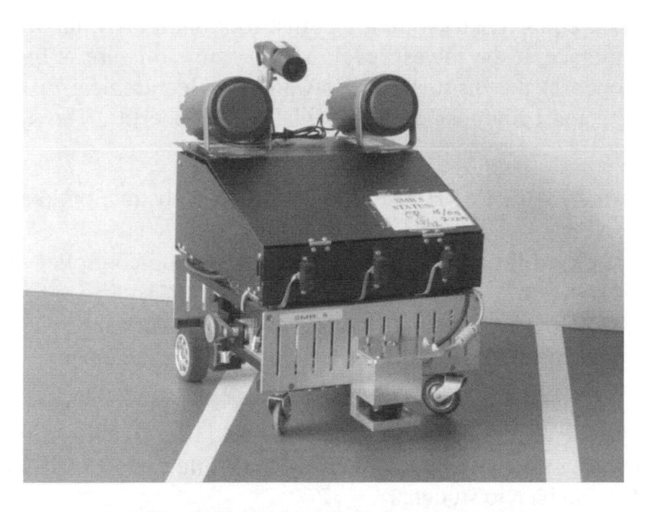

Fig. 3. Small Mobile Robot platform

As programming and test of algorithms were the main targets of the courses it was important that the computer was powerful and had a substantial amount of memory for data logging. It was also considered essential that up- and download of code and data should be as easy as possible to avoid waste of teaching time. Therefore the final choice was a standard μATX PC-motherboard with a PC-Card wireless LAN card connecting it to the Internet.

The computer and all the electronics except the motor controllers are placed in the upper part of the robot. As the platform is intended to support both the standard courses and bachelor and master projects it was decided to go for a modular design so that it is easy to add new components and sensors to the robot.

Due to its simplicity a master slave serial RS485 bus is used for communication between the modules and the PC. This makes it possible to use a standard RS232 driver on the PC as the masters write line may be enabled all the time.

The power supply is ATX compliant and enables the SMR to run on battery or external power while recharging. The SMR should be able to run for approximately 2 hours when fully charged. The battery is a 12 V sealed lead-acid battery and has a capacity of 7 Ah. The following modules are currently connected to the RS485-bus:

- Motor power module. Includes velocity control based on encoder measurements.
- Reflectance sensor, measures surface reflectance. Used for line following.
- 6 IR distance sensors. Used for obstacle avoidance and sensing the environment.
- Wheel Encoders. 2000 tics per wheel revolutions i.e. app 0.1 mm resolution.
- Rate gyro for measuring the turning speed of the robot

Furthermore the robot is equipped with

- Gubby Firewire camera for vision tasks
- Hokuyo URG laser scanner for localisation and obstacle detection.

The software is based on a Slackware Linux distribution. RTAI-linux is used to assure real-time performance at the lowest level. The control software is based on Mobotware [5] developed at the institute. Mobotware is a hierarchical distributed system based of plug-ins and communication through TCP-IP sockets. The system has three core modules:

- Robot Hardware Deamon (RHD) Flexible hardware abstraction layer for real-time critical sensors
- Mobile Robot Controller (MRC) Real-time Closed–loop controller of robot motion and mission execution
- Automation Robot Servers (AURS) Advanced framework for processing of complex sensors and soft real-time mission planning and management.

The hierarchical structure makes it possible for the students to program the system at all levels and the use of plug-ins makes it easy to add new functionality without compromising the basic structure of the system (which often happens if source code for the whole system is given to students)

To make it possible to use the robots with very limited programming knowledge an interpreted robot control language is included. The SMR-CL language here has proved very simple to learn even for non-technical users and is targeted specifically for robot applications giving an efficient program being able to run on even small computers. SMR-CL is inspired by Colbert [6] but where Colbert relies on C syntax SMR-CL requires very little programming skills. Specific robot issues like specifying velocity and acceleration can be done as options to commands giving a compact and easily readable code and a simple basic structure gives transparency for non-programmers. SMR-CL provides built-in features like multiple stop conditions and handles real-time issues at the appropriate level.

5 Conclusion

This paper describes the design of a comprehensive course programme in mobile robotics with emphasis on inclusion of substantial parts of hand-on exercises and competitions. The programme is very popular and has been attended by more than 800 students over the years. As reported by several authors' inclusion of competitions in standard engineering curricula can be very motivating for the students. However ten years of experience at DTU Automation and Control shows that the provided hardware and software must be carefully matched to the students skills otherwise it might easily lead to frustration over using too much time on the tools and to little on the subject of the course. Likewise the utmost care must be taken to design the competition contents such that it is supported by the topics taught in the courses. Failing to do so might easily support the opinion of many students that theory is best suited for exam problems while real practical problems are best solved using your own ad hoc methods. But if well designed competitions are unsurpassed in increasing student interest and effort.

References

[1] Friesel, A.: Teamwork and Robot Competitions in the Undergraduate Program at the Copenhagen University College of Engineering, In: FIRA, CCIS, vol. 44, pp.279-286. Springer, Hiedelberg (2009)
[2] Valavanis, K.P.: Special Issues on Robotics in Education. IEEE Robotics and Automation Magazine 10(2,3) (2003)
[3] Avanzato, R.: Assessment and outcomes of robot competitions at Penn State Abington. Computers in Education Journal 19(3) (2009)
[4] Andersen, N.A., Ravn, O.: SMR-CL A Real-time Control Language for Mobile Robots. In: CIGR 2004 , Beijing, PR China (2004)
[5] Beck, A.B., Andersen, N.A., Andersen, J.C., Ravn, O.: MobotWare – A Plug-in Based Framework for Mobile Robots, In: IAV 2010, Lecce, Italy (2010)
[6] Konolige, K.: COLBERT: a language for reactive control. In: Brewka, G., Habel, C., Nebel, B. (eds.) KI 1997. LNCS, vol. 1303, Springer, Heidelberg (1997)

RoboWaiter Competition: Linking Robotics Education to Social Responsibility

David J. Ahlgren[1] and Igor M. Verner[2]

[1] Department of Engineering, Trinity College, Hartford, CT, U.S.A.
David.Ahlgren@trincoll.edu
[2] Department of Education in Technology & Science,
Technion—Israel Institute of Technology
ttrigor@technion.ac.il

Abstract. This paper reports on RoboWaiter, the first contest in assistive robotics organized and managed with active participation of people with disabilities. RoboWaiter was offered in conjunction with the Trinity College Firefighting Home Robot Contest in 2009 and 2010. Members of the sponsoring organization, the Connecticut Council on Developmental Disabilities, worked closely with Trinity to formulate the contest theme and rules. RoboWaiter offers participants a unique challenge of multidisciplinary design projects that culminate in the competition held in Connecticut. RoboWaiter's main goals are to promote awareness of the needs of people with disabilities and to provide an engineering design challenge for all levels from middle school to Ph.D. students and engineering professionals. The contest also sought to encourage students to consider social dimensions of the engineering profession. In this paper we describe the RoboWaiter rules and the resulting engineering challenges, provide an assessment of progress so far, and consider future directions.

Keywords: Assistive robotics, robot competition, educational robotics, socially responsible education.

1 Introduction

According to the Bureau of Industry and Security, U.S. Department of Commerce, more than 17% of Americans have a disability, and half of that cohort has a severe disability. The number of persons with severe disabilities is increasing and will continue to grow as the population ages [1]. Many of these persons with disabilities use an assistive technology device, which is an "item, piece of equipment, product or system, whether acquired commercially off the shelf, modified, or customized, that is used to increase, maintain, or improve the functional capabilities of persons with disabilities."[2]. The U.S. Bureau of Commerce report points out that there is a rapidly growing assistive technology industry that employs more than 20,000 workers [1]. Still, the number of people currently using assistive technology is only a fraction of those who could benefit from it [2]. Therefore we believe that it is essential to increase public awareness about the need for assistive technology and especially to

P. Vadakkepat et al. (Eds.): FIRA 2010, CCIS 103, pp. 274–281, 2010.

focus efforts to motivate engineering students who will form the scientific and technical core for future progress.

There is a strong movement toward developing robotic wheelchairs and manipulators, rehabilitative and instructional robots, and other assistive robotic agents that aid people through physical and social interactions. For example, researchers at MIT have developed rehabilitation robots that train muscles compromised by a stroke [3]. Another example of an assistive agent is a robot dog, developed at Georgia Tech and intended to replace service dogs. This robot fetches objects and opens drawers and is controlled by verbal commands and a laser pointer [4].

The promising landscape in assistive robotics motivated the development of an assistive robot event to be offered in conjunction with the Trinity College Fire-Fighting Home Robot Contest (TCFFHRC), an annual international event that attracts 120 teams each year from around the world. The TCFFHRC is a non-profit event that is open persons of all ages and levels of expertise [5, 6]. Three members of the Connecticut Council on Developmental Disabilities and the lead author began the RoboWaiter planning effort in the summer and fall of 2008. The group chose the contest theme and generated the rules, which may be found at the contest website [7].

The chosen theme reflects real concern to Council members: the need for a person with disabilities to obtain food from a refrigerator during an emergency, when a personal assistant cannot be present. This planning process was the genesis of a unique contest, RoboWaiter, whose goal is to develop autonomous robots that serve as food servers (waiters). This unique event was the only assistive robot competition in the world in 2009-2010 [8].

2 RoboWaiter Contest

As stated in the official contest rules, "The RoboWaiter Contest challenges teams to create a robot that can retrieve a plate of food and transport it to a table in a reliable and efficient manner. The arena simulates a home kitchen with the usual fixtures and a pair of dolls simulating the humans served by the robot." [1]. In 2010 there were two levels of competition, the Standard Division and the Advanced Division. Both levels take place in a scale model kitchen that measures 2.5 m x 2.5 meters. The arena includes common scale-model kitchen items including a table, a sink, a refrigerator, a chair, and a doll representing a second, elderly, person.

2.1 RoboWaiter Standard Division

In this division objects are placed in known positions except for the chair, which is constrained in one dimension only. In the standard division a single shelf, placed 21 – 23 cm above the floor represents the refrigerator, and the plate is a plastic pet food can top approximately 10 cm in diameter. Starting at a known home position marked by a 30 cm white circle, RoboWaiter robots navigate to the shelf, pick up the plate, and deliver it to the table where the doll, representing the person with disabilities, sits in a wheelchair. Robots are designed to avoid collisions with the sink, the chair, and the doll. The robot is guided by a beacon, located at the center of the front edge of the shelf, consisting of three bright red LEDs spaced two centimeters apart. The center of the plate is aligned with the center LED. To aid navigation toward the able, there is one bright red LED on each of the three exposed sides of the table.

In the competition each robot has three trials (maximum of four minutes each), and order of finish is based on reliability, as measured by the number of successful trials, and then by run time within groups having equal number of successful trials. Run operating successfully in three optional modes may reduce times: Return Trip Mode (time multiplier 0.8), Food Mode (food is placed in the plate, multiplier 0.8), and Arbitrary Start location mode (staring position randomly chosen by the judge, 0.85). For a full description of the RoboWaiter rules, the reader is referred to the contest website [7].

2.2 RoboWaiter Advanced Division

The new 2010 Advanced Division represented a first step toward creating a "smart home" in which autonomous robots interact with intelligent appliances. The Advanced Division uses the same arena as the Standard Division, but the Advanced Division includes a scale-model refrigerator, whose door may be opened and closed by the robot. The refrigerator has two shelves; each equipped with three centered light-emitting diodes (LEDs) facing outward, and it is equipped with a modulated 300 mW IR beacon located at the center of the refrigerator door. The robot opens the door when it "trips" a proximity sensor embedded in the arena floor. This sensor is located centrally in the beacon's radiation pattern at a distance of 65 cm from the door. In addition to the proximity sensor, the floor sensor has three bright white LEDs that the robot uses to verify its position and to anticipate the door's closing or opening.

Advanced division robots begin their task when the judge issues an audio start signal, a pure tone at 3.5 KHz or at 7.5 kHz. When the robot detects the 3.5 kHz signal, its goal is to pick up the plate from the bottom shelf, and if 7.5 kHz, from the top shelf. After decoding the start signal, the robot opens the door by finding and actuating the floor sensor. Once the door is open, the robot must pick up the plate from the designated shelf and navigate back to the sensor to close the door. To finish the autonomous service task, the robot travels to the table, avoiding obstacles along the way, places the plate on the table, and returns to the home position.

2.3 Design Challenges

Realizing a RoboWaiter robot is a challenging multidisciplinary task that requires successful design of sensing, interfacing, mechanics, and software sub-systems. Each sub-system presents a significant problem that requires research, detailed design processes, and careful testing, and to achieve a high level of performance each sub-system must be fully tested. The goal is to develop a reliable, fault-tolerant robot that succeeds on all three trials. A brief discussion of each sub-system follows.

2.4 Robot Base

The robot base must provide reliable and programmable robot motion throughout the RoboWaiter arena. Quickness and the ability to perform dead reckoning are desirable traits. In the 2009 and 2010 RoboWaiter events, teams used a variety of commercial and non-commercial robot bases; all base designs, commercial or designed by participants, were acceptable under the contest rules. Examples of commercial/kit robots used in the contest include the iRobot Create and the Lego NXT. A reliable base is the starting point for a successful RoboWaiter machine, and great care must be exercised to understand and fully test the base.

The robot base must provide reliable and programmable robot motion throughout the RoboWaiter arena. Quickness and the ability to perform dead reckoning are desirable traits. In the 2009 and 2010 RoboWaiter events, teams used a variety of commercial and non-commercial robot bases; all base designs, commercial or designed by participants, were acceptable under the contest rules. Examples of commercial/kit robots used in the contest include the iRobot Create and the Lego NXT. A reliable base is the starting point for a successful RoboWaiter machine, and great care must be exercised to understand and fully test the base.

2.5 Sensors

The RoboWaiter rules impose many requirements on the robot's sensing system. The sensing system enables navigation, obstacle detection, and plate detection. Sensors must detect physical features of the arena, find obstacles, detect audio starting signals, detect both IR and visible light beacons, and decode start signals. RoboWaiter robots have used IR ranging sensors (Sharp GP2D12, for example) and laser rangefinders for navigation and phototransistors and cameras for beacon detection. Decoding of start signals has been performed via microphones and preamplifiers, tone detecting hardware, and software FFT and FPGA-based decoders. The RoboWaiter challenge provides fertile ground for sensor choice, deployment, and evaluation.

2.6 Arm and Gripper

Development of a mechanism to grip, hold, and carry the plate is perhaps the most challenging task for the RoboWaiter designers. In 2009 and 2010 teams developed a variety of gripping devices. A device developed by a successful team in 2009 employed a slot that mated with the edge of the RoboWaiter plate. Another second successful design in 2009 used a clamping device that was lowered onto the table from above, enclosing the plate. The gripper then closed onto the plate and held it securely. Several teams in the 2010 RoboWaiter contest used Lego-based grippers; an example is shown in Figure 1. This arm was actuated by a servo that drove two meshed gears to achieve opening and closing action. A second servo allowed the arm assembly to be rotated by 180° as a whole, allowing the assembly to fold back over the robot during navigation maneuvers to reduce overall robot size (Figure 1). When retrieved, the plate rested on thick rubber bands that provided softness at the plate-arm interface but were strong enough to hold the plate. This mechanism grabbed the plate successfully in the 2010 contest.

Design of an arm and gripper for the RoboWaiter Standard Division contest is an open problem requiring research and application of mechanical engineering practices including 3D CAD modeling and animation, analysis of forces and moments, and choice of materials. Gripper design for the Advanced Division aims at controlled motion in three dimensions, and successful grippers have yet to be demonstrated.

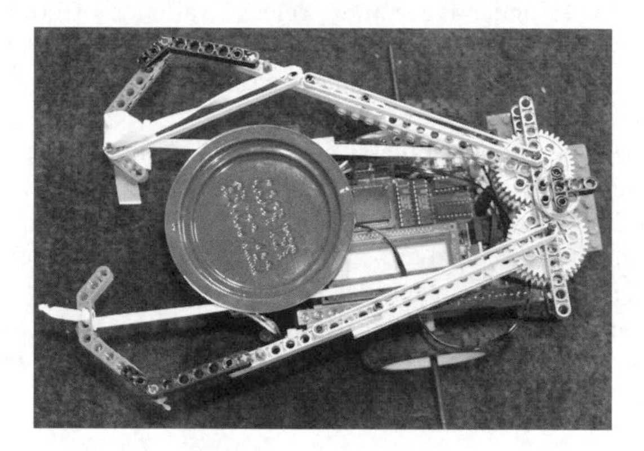

Fig. 1. Lego-Based Arm and Gripper holding RoboWaiter plate

2.7 Software

The successful RoboWaiter software system provides the means to verify, integrate, and tune all sub-systems, and it allows a full solution to the contest tasks. Programs must be written to quickly evaluate and debug motor drives, sensors, and the arm/gripper device. In addition, the software sub-system must enable testing and evaluation of required robot behaviors; these include navigation, plate detection, obstacle avoidance, play conveyance, and return to home functions. Participants in 2009 and 2001 used a variety of programming platforms, which included Interactive C, Lego NXT tools, and LabView. A wide choice of tools is available, and none has a clear advantage.

3 Results and Assessment

Nine teams entered the 2009 RoboWaiter contest. Three robots transferred the plate to the table at least once. The winning entries in were "James the Bot", built by a team of engineering students from Grand Valley State University, Michigan; "Pinchy", designed by a professional software developer from Florida, and "Tsunami", developed by a high school student from Ort Givat Ram in Jerusalem. In 2010 there were fourteen robots, eleven in the Standard Division and three in the new Advanced Division. None of the 2010 robots completed the task, but five teams achieved "honorable mention" status by performing the retrieval task in part. The difficulties in 2010 focused on navigation and inability to avoid the chair. A robot that used a video camera to sense the red LEDs forgot to mask the camera's field of view, causing red objects outside the maze to be recognized as plate beacons. The failure of robots to execute the task completely in 2010 led to this comment from one participant:

> *"RoboWaiter is a much more difficult and realistic task than fire-fighting…it is both an engineering challenge and a challenge that has application to help humanity."*

In order to solicit wider feedback regarding RoboWaiter, we conducted a follow-up study in 2009 and 2010. Our special interest was to get evidence on to what extent the contestants are motivated by the engineering, humane, and social challenges of the assistive robotics competition. Here, "humane challenge" represents the educational challenge of teaching robotics by involving students in assistive robotics projects; thus we encourage sensitivity to the genuine needs of people with disabilities, and to foster feelings of compassion and sympathy toward them. The term "social challenge" concerns strengthening students awareness about social problems of people with disabilities and thinking about how assistive robotics can help in solving their problems.

3.1 Educational Survey

In 2009 and 2010 we collected information through our observations and through a post-contest educational survey. The survey sought perceptions and attitudes of persons who developed RoboWaiter robots as well as supporters, who are persons with disabilities or their families who attended the contest, served as volunteers, or participated in developing the contest. The survey presented each participant with a list of reasons for competing and asked each person to rate the importance of each reason using the scale from 1 (not important) to 5 (very important). A similar survey section was given to contestants and the supporters.

The answers from the survey groups are shown in Tables 1 and 2. As found, all of the possible reasons were important, with Engineering challenge and curiosity at the top and with social challenge at the bottom.

Table 1. Possible reasons for participation in RoboWaiter 2009 & 2010 (N=18)

Possible Reason	Avg.	Std. Dev.
1. Engineering challenge	4.28	1.23
2. Humane challenge	3.83	0.99
3. Social challenge	3.28	1.02
4. Interest in assistive robotics	3.83	1.04
5. Curiosity	4.18	0.81
6. Interest to major in the subject	3.76	1.15
7. Job, scholarship or advanced studies opportunities	3.75	1.39

Evaluations given by the supporters were high for all the possible reasons, but the highest scores were given to interest in robotics, attracting attention to needs of users of assistive technology and introducing students to the assistive technology subject.

Table 2. Possible reasons for supporting RoboWaiter 2009 & 2010 (N = 5)

Possible Reason	Avg.	Std. Dev.
1. Interest in robotics	4.20	1.30
2. To foster development of assistive technology (AT)	3.60	1.34
3. To attract attention to needs of users of AT	4.20	1.10
4. To increase awareness among potential users of AT	4.00	1.00
5. To inspire dialogue between users and developers of AT	4.00	0.71
6. To introduce students to the AT subject	4.20	0.84

In 2010 the authors also carried out an educational survey of all contest participants, whether entered in the firefighting contest or the RoboWaiter contest. The survey asked individuals to rate contest-related development of disciplines, abilities, and skills in 22 different areas. The survey suggests that RoboWaiter and firefighting participants experience similar learning outcomes but that RoboWaiter excels in five areas: learning in mechanics, robot programming, robot navigation, robot manipulation, robot interaction with its environment, and robot interaction with humans. These areas are related to the assistive task and to the need for developing a precise, controllable arm and gripper.

3.2 Written Interviews

After the competition we conducted a series of written interviews with a number of RoboWaiter contestants and supporters. A few responses are presented below.

Professional engineer from Florida, participant:

I had not given assistive robotics much thought before the competition. As an engineer, I tend to get focused on a task and forget about people. The project gave me lots of new ideas. I think it was a very effective way to get people thinking about what robotic technology can do for people with disabilities.

College student from Connecticut, participant:

In RoboWaiter learning robotics and programming, as well as social, moral, and humane issues take place. These experiences help students realize what their education is leading up to, and how they can end up changing the lives of so many people…It was great to have participation of people with disabilities. There is no downside to them being there and seeing how robots could potentially help. Their participation can also help in coming up with new ideas, as they converse with the builders.

Member of CTCDD, person with disabilities:

I would like to learn more about how robots can assist people with disabilities in various places such as in homes, cars, employment, etc. I feel RoboWaiter is more of education and an awakening to the participants that robots could be essential to people with disabilities of any age…more teams should have the opportunity to come to the competitions because RoboWaiter is a wonderful experience.

4 Conclusion

In this paper we described RoboWaiter, the world's first assistive robot competition, discussed the associated engineering challenges, and presented the results of contest surveys. The first year of RoboWaiter motivated the development of the Advanced Division, which marries autonomous robots with "smart home" environments. RoboWaiter offers the opportunity for many fruitful and realistic innovations to be developed and implemented by contest participants. The mission is to increase robot intelligence while providing an increasingly realistic environment in which robots can operate. Surveys and participant feedback indicate that RoboWaiter presents a difficult engineering challenge. The contest engaged persons with disabilities in developing the contest theme and the contest rules, and these persons served as competition judges and volunteers. Feedback from these individuals indicated that robot competitions will draw attention to the need for new technologies, promoting technology development, and motivating engineering education.

Acknowledgement

The generous support of the Connecticut Council on Developmental Disabilities is gratefully acknowledged.

References

1. U.S. Department of Commerce Technology Assessment of the U.S. Assistive Technology Industry, http://www.bis.doc.gov/defenseindustrialbaseprograms/osies/defmarketresearchrpts/assisttechrept/4intro.htm
2. Assistive Technology Outcomes and Benefits (2004), http://www.atia.org/i4a/pages/indexcfm?pageID=3681
3. Hogan, N., Krebs, H.I., Rohrer, B., Palazzolo, J.J., Dipietro, L., Fasoli, S.E., Stein, J., Frontera, W.R., Volpe, B.T.: Motions or Muscles? Some Behavioral Factors Underlying Robotic Assistance of Motor Recovery. VA Journal of Rehabilitation Research and Development 43(5), 605–618 (2006)
4. Nguyen, H., Kemp, C.C.: Bio-inspired Assistive Robotics: Service Dogs as a Model for Human-Robot Interaction and Mobile Manipulation. In: Proceedings of the IEEE RAS/EMBS International Conference on Biomedical Robotics and Biomechatronics (BioRob 2008), pp. 542–549 (2008)
5. Verner, I., Ahlgren, D.: Robot Contest as a Laboratory for Experiential Engineering Education. ACM J. on Educational Resources in Computing. Special Issue on Robotics in Undergraduate Education, Part 1 4(2), 2–28 (2005)
6. Verner, I.M., Ahlgren, D.J.: Robot Projects and Competitions as Education Design Experiments. Int. J. Intelligent Automation and Soft Computing, Special Issue on Educational Robotics 13(1), 57–68 (2007)
7. Trinity College Fire-Fighting Home Robot Contest, http://www.trincoll.edu/events/robot
8. Robot Competitions in 2009-2010 Academic/School Year, http://www.robots.net

A Comparison between Growing and Variably Dense Self Organizing Maps for Incremental Learning in Hubel Weisel Models of Concept Representation

Neo Choon kiat Daniel[1], Kiruthika Ramanathan[1], Shi Luping[1], and Prahlad Vadakkepat[2]

[1] Data Storage Institute, (A*STAR) Agency for Science, Technology and Research, DSI Building, 5 Engineering Drive 1, Singapore 117608
[2] Department of Electrical and Computer Engineering
National University of Singapore

Abstract. Hubel Weisel models of pattern recognition are thought to be anatomically and physiologically faithful models of information representation in the cortex. They describe sensory information as being encoded in a hierarchy of increasingly sophisticated representations across the layers of the cortex. They have been shown in previous studies as robust models of object recognition. In a previous work, we have also shown Hubel Weisel models as being capable of representing a hierarchy of concepts. In this paper, we explore incremental learning with respect to Hubel Weisel models of concept representation. The challenges of incremental learning in the Hubel Weisel model are discussed. We then compare the use of variably dense self organizing maps to perform incremental learning against the original implementation using growing self organizing maps. The use of variable density self organizing maps shows better results in terms of the percentage of documents correctly clustered. The percentage improvement in clustering accuracy is in some cases up to 50% over the original GSOM implementation for the incrementally learnt module. However, we also highlight the issues that make the evaluation of such a model a challenging one.

Keywords: Concept representation, Cognitive models, Hubel Weisel Architectures, Self Organizing maps.

1 Introduction

There exist several bottom-up approaches to hierarchical models of object recognition that are based on the anatomy of the visual cortex. They make use of Mountcastle's (1978) theory of uniformity and hierarchy in the cortical column and the model of simple to complex cells of Hubel and Weisel (1965), modeling how simple cells from neighboring receptive fields feed into the same complex cell, meaning that the complex cell has phase invariant response.

Mountcastle (1978) showed that parts of the cortical system are organized in a hierarchy and that some regions are hierarchically above others. In general, neurons in

P. Vadakkepat et al. (Eds.): FIRA 2010, CCIS 103, pp. 282–289, 2010.

the higher levels of the visual cortex represent more complex features with neurons in the IT representing objects or object parts (Hubel and Weisel, 1965). Hubel Weisel models have been developed for object recognition (Cadieu et al., 2007; Fukushima, 2003) proposing a hierarchy of feature extracting simple (S) and complex (C) cells that allow for positional invariance. The combination of S-cells and C-cells, whose signals propagate up the hierarchy allows for scale and position invariant object recognition.

Our work revolves around the following question: If the structure of the cortical column is uniform and hierarchical in nature and if the model of simple to complex cells can be used to model the visual cortex as discussed in prior works, then can such a model also be used to represent other modalities of information such as the concepts derived from text? We are therefore aiming to design a bottom up hierarchical memory for the representation of concepts, much the same way as it is designed for the representation of images. In our prior work and in this paper, we define a concept as being a keyword in a document.

Prior work (Ramanathan et al., 2010) describes how a bottom up hierarchy of Growing Self Organizing maps (GSOM) (Alahakhoon et al., 2000) was created to explore bottom up learning in Hubel Weisel models specifically for representing a hierarchy of concepts. The paper showed that the Hubel Weisel model of concept memory captured certain cognitive properties of semantic cognition (Rogers and McClelland, 2003) such as hierarchical representation, multiple similarities (Sloutsky, 2003) and chain retrieval (Widrow, 2009).

It should be noted that the hierarchical self organizing maps described in this paper is different from that described in Rauber et al (2002). While Rauber et al., describe a top down hierarchical differentiation of documents based on clusters, our work aims to capture multiple similarities within concepts (Sloutsky, 2003, Sloutsky et al., 2007) as discussed in psychological studies of conceptual acquisition. The hierarchy that is studied here is therefore bottom up.

In this paper, we explore the incremental nature of conceptual memory and the challenges in implementing incremental learning in Hubel Weisel architectures. More specifically, we compare the addition of a variably dense self organizing map (VDSOM) (Shimada et al., 2008) implementation and explore its effects when compared to the original GSOM implementation.

The rest of the paper is organized as follows. In section 2, we describe related literature, and in section 3, the algorithm for Hubel Weisel concept memory. We also motivate the need for variably dense neuron growth during incremental learning and describe the implementation of VDSOM that was implemented. In section 4, we describe the simulations, evaluation methods and the results obtained. Section 5 concludes and discusses further work

2 Related Work

Hubel Weisel models of concept representation builds upon research in two areas – Hubel Weisel architectures and concept representation. On the spectrum of cognitively inspired architectures for object recognition are the Hubel Weisel models. Beginning from the Neocognitron (Fukushima, 2003) and HMAX (Cadieu et al.,

2007), various bio inspired hierarchical models achieve object recognition and categorization. The primary idea of these models is a hierarchy of simple (S) and complex (C) cells. S cells work as feature extracting cells, whose input connections are variable and modified in learning. Each S cell then comes to respond selectively to particular features in the receptive field. The S cell is therefore a feature extractor which, at the lower levels, extracts local features and, at the higher layers, extracts global features. C cells allow for positional errors in the features. A C-cell is therefore more invariant to shift in position of the input pattern. The combination of S-cells and C-cells, whose signals propagate up the hierarchy allows for scale and position invariant object recognition.

From concept learning and representation literature, Quillian (1965) proposed conceptual representation as a top down hierarchy of related concepts. However, recent literature (Rogers and McClelland, 2008, Sloutsky 2003) have proposed conceptual learning as a bottom up process, especially in early development. Sloustky (2007) discussed that the cognitive process of conceptual representation is grounded in powerful learning processes such as statistical and attentional learning, and not in innate knowledge. Sloutsky (2003) discusses how children group concepts based on, not just one, but multiple similarities, which tap the fact that those basic level categories have correlated structures (or features). This correlation of features is also discussed in McClelland and Rogers (2003) who argue that information should be stored at the individual concept level rather than at the super ordinate category level allowing properties to be shared by many items. The Hubel Weisel model of conceptual memory (Ramanathan et al., 2010) is the first model that implements the bottom up hierarchical processing of Hubel and Weisel out of the domain of object recognition has attempted to model hierarchical representation of keywords to form concepts.

3 System Architecture

3.1 The Hubel Weisel Model of Conceptual Memory

In this section we describe briefly the Hubel Weisel model of conceptual memory (Ramanathan et al., 2010) which is organized in a bottom up hierarchy. This means that the component features are represented lower in the hierarchy before the representation of concept objects. Each level in the hierarchy has several modules. These modules model cortical regions of concept memory. The modules are arranged in a tree structure, having several children and one parent. In our paper, we call the bottom most level of the hierarchy level 1, and the level increases as we move up the hierarchy. The keywords from a document form the inputs to the system. We feed these directly to level 1. Level 1 modules resemble simple cells of the cortex, in that they receive their inputs from a small patch of the input space. Several level 1 modules tile the input space. A module at level 2 represents the union of the input space of all its children level 1 modules. However, a level 2 module obtains its inputs only through its level 1 children. This pattern is repeated in the hierarchy. Thus, the module at the tree root (the top most level) covers the entire input space, but it does so by pooling the inputs from its child modules. In the visual cortex, the level 1 can be considered analogous to the area V1 of the cortex, level 2 to area V2 and so on.

To understand how the model learns, let us consider the inputs and outputs of a single module $m_{k,i}$ in level k of the system as shown in Figure 1a. Let **x**, representing connections $\{x_j\}$ be the input pattern to the module $m_{k,i}$. **x** is the output of the child modules of $m_{k,i}$ from the level k-1, and **a** represents the weights of the competitive network. The vector **a** is used to represent the connections $\{a_j\}$ between **x** and the cells in the module $m_{k,i}$.

During learning, each neuron in $m_{k,i}$ competes with other neurons in the vicinity. The competitive learning algorithm in our model is implemented using Growing Hierarchical Self Organizing Maps (GSOM) (Alahakhoon et al., 2000).

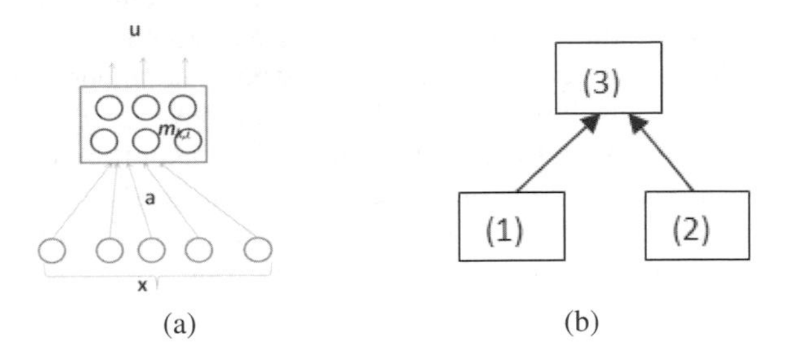

(a) (b)

Fig. 1. a. Inputs and outputs to a single module $m_{k,i}$. b. The concatenation of information from the child modules of the hierarchy to generate inputs for the parent module.

When all the modules at level k finish training, the second stage of learning occurs. This comprises the process by which the parent modules learn from the outputs of the child modules. Here, consider the case shown in Figure 1b where the module 3 is the parent of modules 1 and 2. Let **x(1)** be the output vector of module 1 and **x(2)** be the output vector of module 2. **x(i)** represents the Euclidean distance from the input pattern to the each output neuron i of the child modules. The input to module 3, $\mathbf{I(3) = x(1)||x(2)}$, is the concatenation of the outputs of modules 1 and 2. A particular concatenation represents a simultaneous occurrence of a combination of concepts in the child module. Depending on the statistics of the input data, some combinations will occur more frequently, while others will not. During the second stage of learning, the parent module learns the most frequent combinations of concepts in the levels below it. A GSOM is again used in the clustering of such combinations. The learning process thus defined can be repeated in a hierarchical manner.

3.2 Incremental Learning

In real life, the information presented to memory is dynamic in nature, and incremental and online learning are desired property of Hubel Weisel systems as well. However, Incremental learning poses a challenge in Hubel Weisel based computational models due to two reasons. (1) Damage to the knowledge represented by old neurons which is fundamental in competitive learning. (2) Propagation of information in the hierarchical architecture. The number of output neurons of each child node increases with the introduction of the incremental batch. The input dimensions of the parent node are therefore changed and incompatible with the dimensions of the previous batch. Farahmand et al. (2009) and Fukushima (2004) discuss the problems associated with incremental and online learning in hierarchical architectures.

In this and the subsequent sections of the paper, we will use the terms *batch 0* to represent the first batch of information. *Batch 1* refers to the subsequent set of information. Once the system learns the information from documents in *batch i*, only the hierarchical structure and the neuron architecture are retained. All other information regarding the information presented is discarded.

Ramanathan et al (2010) implemented incremental learning in the Hubel Weisel architecture. The methodology and algorithms are described in their paper. While the algorithm was effective, loss of old memories still occurred, despite the introduction of pseudo vectors. The behavior of the model during incremental learning is further discussed in Ramanathan et al (2010). One aspect of memory representation in the model, however, was found curious and is discussed as below.

Figure 2a illustrates the expected results when the model learns, in batch 0, the concepts of mammals, birds and reptiles. The top layer module (module 4) groups the concepts by macro similarities in the data, while the bottom layers group the concepts by their micro similarities. In this case, the child modules (module 1 to module 3) may reflect groupings of micro similarities such as habitat or diet, wherein both some animals and some birds (say dog and sparrow) may share the same habitat (city).

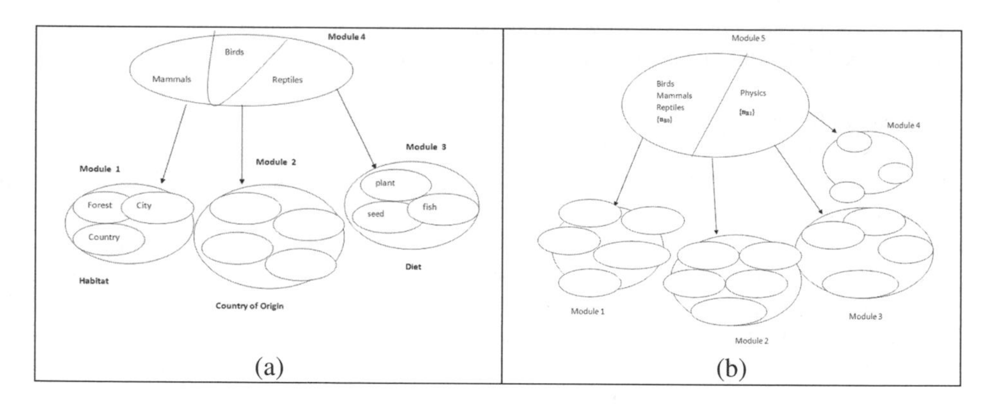

Fig. 2. (a) Description of micro and macro similarities in the Hubel Weisel concept memory (b) Conceptual clusters after incremental learning

Now, a set of documents containing concepts very different from the concepts represented in Figure 2a is introduced into the system. In the example shown in figure 2b, the new concepts belong to the domain of Physics. It was empirically observed in this case that the concepts of batch 0 tended to group together after the introduction of batch 1.

However, it was also observed that the mean quantization error (QE) of the neurons representing the concepts in batch 0 ($\mathbf{n}_{B0}$) is much greater than the mean quantization error of the neurons representing concepts in batch 1 ($\mathbf{n}_{B1}$). This is illustrated by the equation below. Where i is the index of neuron vector **n,** and **x** is an input vector is represented by the neuron $\mathbf{n}_i$.

$$QE_{\mathbf{n}_i} = \frac{1}{n_{\mathbf{x} \in \mathbf{n}_i}} \sum_{\mathbf{x} \in \mathbf{n}_i} \|\mathbf{n}_i - \mathbf{x}\| \tag{1}$$

When $QE_{n_{B0}} >> QE_{n_{B1}}$ we can justify the need for more neurons to represent the information in the neuron with a larger QE, hence the motivation for a variably dense implementation of competitive learning. To implement variably dense competitive learning in our GSOM modules, we adapted the algorithm proposed by Shimida et al (2008).

4 Experimental Results

In this paper, we evaluate the model in terms of its information representation properties. In the domain of data clustering, a model is generally evaluated evaluated in terms of clustering accuracy and quantization error. We will adopt these metrics in the evaluation of the GSOM and VDSOM based models for incremental learning.

Table 1. Summary of document sets used in training

Dataset name		**Research paper corpus**	**Time 60**
Number of data patterns		165	420
Reduced concept vector size		1500	1000-5000
Input dimension to level 1 modules		500	100
Variable density based learning parameters	τ	0.8	0.9
	η	0.1	0.1
GSOM training parameters	**Learning rate**	0.1	0.1
	Smooth factor	0.05	0.05
	Spread factors	{0.2 , 0.2}	{0.8,0.8}

Two datasets were used in measuring the effectiveness of the models proposed. The research paper corpus consisted of 165 documents taken from conference papers in various domains including History, artificial intelligence, Electrical systems and photonics. The TIME60 corpus consisted of 420 news articles and is considered a benchmark dataset in text mining applications. It has previously been used in the WEBSOM application. Table 1 summarizes the properties of the document data used to train the system.

4.1 Results on the Research Paper Corpus

Three experiments were carried out on the research paper corpus. In each experiment random samples of data were selected from the corpus. Each experiment's inputs consisted of data from four different categories of papers and the increments were designed to be roughly equal. Each experiment was repeated 20 times, using a randomized order of samples for batch 0 and batch 1 respectively. The mean and standard deviations of results obtained are as given in Table 2.

The results in Table 2 show the percentage of documents misclustered in the top most incrementally learnt node in the hierarchy, and compares the performance with that of the GSOM implementation. The general observation showed a significantly higher clustering accuracy for the variably dense implementation when compared to the GSOM implementation, with an average of 50% improvement in clustering accuracy at the top layer in the hierarchy. However, the average number of neurons generated in the variably dense self organizing maps is significantly higher when compared to the GSOM implementation.

Table 2. Experimental results on the research paper corpus

	GSOM results			Variably dense self organizing map implementation results		
	Clustering accuracy (%)		Avg # of neurons generated	Clustering accuracy (%)		Avg # of neurons generated
	mean	stdev		mean	stdev	
Expt 1	21.6	0.03	9	83.9	0.06	15
Expt 2	45.3	0.05	8	75.2	0.1	17
Expt 3	21	0.04	6	77	0.08	19

4.2 Results on the Time 60 Corpus

Table 3 illustrates the results of incremental learning on the TIME60 corpus. As the actual documents are not categorized, we evaluate the performance of the categorization purely based on the average quantization errors. In both experiments, the incremental learning was tested using 300 input patterns in batch 0 and 120 input patterns in batch 1. The experiments were repeated 20 times with the input batches randomized. The average results are reported in Table 3.

From Table 3, we can observe that the average quantization error of the inputs in the top layer is the least in batch learning. The Variably dense neuron implementation performs better than the GSOM implementation for incremental learning. However, this may also be due to the large number of neurons generated.

Table 3. Results of incremental learning for the TIME60 corpus

	# of neurons		Average Quantization Error	
	Mean	stdev	mean	stdev
Batch version	19	2.2	1.93	0.184
GSOM	16	3.4	2.43	0.292
VDSOM implementation	25	5.3	2.12	0.195

5 Discussions and Conclusions

Implementing variably dense neurons in the Hubel Weisel model of concept memory improves the general performance of GSOM based modules in incremental learning performance. While the algorithm works well on small corpuses, of under 500 input vectors, for large number of documents, it is desirable that a more robust adaptation function be developed to adapt the newly formed neurons in order for them to more effectively represent the dis-similar data. This would prevent the generation of too much useless neurons as dis-similar input vectors fail to find their best matching unit and hence calling the growth function to generate additional neurons to represent them.

One issue with the evaluation of misclustering in the Hubel Weisel memory model is the evaluation of clustering accuracy in the bottom layers. Even in a labelled corpus, labels are only available for macro similarities. Therefore the micro similarity based clustering that is performed in the bottom layers cannot be directly quantified or measured. Further work entails developing solutions to this aspect of the problem.

References

Cadieu, C., Kouh, M., Pasupathy, A., Conner, C., Riesenhuber, M., Poggio, T.A.: A Model of V4 Shape Selectivity and Invariance. Journal of Neurophysiology 98, 1733–1750 (2007)

George, D., Hawkins, J.: A hierarchical Bayesian Model of Invariant Pattern Recognition in the Visual Cortex. In: International Joint Conference on Neural Networks, vol. 3, pp. 1812–1817 (2005)

Farahmand, N., Dezfoulian, M., GhiasiRad, H., Mokhtari, A., Nouri, A.: Online Temporal Pattern Learning. In: IJCNN 2009 (2009)

Fukushima, K.: Neocognitron capable of incremental learning. Neural networks 17, 37–46 (2004)

Mountcastle, V.: An Organizing Principle for Cerebral Function: The Unit Model and the Distributed System. In: Edelman, G.M., Mountcastle, V.B. (eds.) The Mindful Brain, MIT Press, Cambridge (1978)

Hubel, D., Weisel, T.: Receptive fields and functional architecture in two non striate visual areas (18 and 19) of a cat. Journal of NeuroPhysiology 28, 229–289 (1965)

Fukushima, K.: Neocognitron for handwritten digit recognition 1. Neurocomputing 51C, 161–180 (2003)

Alahakhoon, D., Halgamuge, S.K., Srinivasan, B.: Dynamic Self Organizing maps with controlled growth for Knowledge discovery. IEEE Transactions on Neural Networks 11(3), 601–614 (2000)

Lagus, K.: Text retrieval using self-organized document maps. Neural Processing Letters 15(1), 21–29 (2002)

Fellbaum, C.: Wordnet: An electronic lexical database. MIT Press, Cambridge (1998)

Mc Clelland, J.L., Rogers, T.T.: The parallel distributed processsing approach to semantic cognition. Nature Reviews Neuroscience 4(4), 310–322 (2003)

Sloutsky, V.M.: The role of similarity in the development of categorization. Trends in Cognitive Sciences 7, 246–251 (2003)

Sloutsky, V.M., Kloos, H., Fisher, A.: When looks are everything – appearance similarity versus kind information in early induction. Psychological Science 18(2), 179–185 (2007)

Rauber, A., Merkl, D., Dittenbach, M.: The growing hierarchical self organizing map, exploratory analysis of high dimensional data. IEEE Transactions on Neural Networks 13(6), 1331–1341 (2002)

Rogers, T.T., McClelland, J.L.: Precis of Semantic Cognition, a Parallel distributed Processing approach. Brain and Behavioral Sciences 31, 689–749 (2008)

Ramanathan, K., Shi, L., Chong, T.C.: A Hubel Weisel model for hierarchical representation of concepts in textual documents. In: The Annual Congress of the Cognitive Science Society (2010) (accepted)

Quillian, M.R.: Semantic Information Processing, pp. 227–270. MIT Press, Cambridge (1965)

Shimada, A., Kanouchi, M., Arita, D., Tanuguchi, R.: Robust estimation of human posture using incremental learning self organizing Map. In: IJCNN 2008 (2008)

Widrow, B., Etemadi, M.: Cognitive memory: human and machine. In: IJCNN 2009, pp. 3516–523 (2009)

Hand Posture Recognition Using Neuro-Biologically Inspired Features

Pramod Kumar P, Stephanie Quek Shu Hui,
Prahlad Vadakkepat, and Loh Ai Poh

Department of Electrical and Computer Engineering,
National University of Singapore,
4 Engineering Drive 3, Singapore - 117576
{pramod,stephanieq,prahlad,elelohap}@nus.edu.sg
http://www.ece.nus.edu.sg

Abstract. A novel algorithm using biologically inspired features is proposed for the recognition of hand postures. The C_2 Standard Model Features of the hand images are extracted using the computational model of the ventral stream of visual cortex. The features are extracted in such a way that it provides maximum discrimination between different classes. The C_1 image patches specific to various classes, which are good in the interclass discrimination, are identified during the training phase. The features of the test images are extracted utilizing these patches. The classification is done by a comparison of the C_2 features extracted. The algorithm needs only one image per class for training and the overall algorithm is computationally efficient due to the simple classification phase. The real-time implementation of the algorithm is done for the interaction between the human and a virtual character *Handy*. Ten classes of hand postures are used for the interaction. The character *Handy* responds to the user by symbolically expressing the identified posture and by pronouncing the class number. The programming is done in the windows platform using c# language. The image is captured using a webcam and the presence of the hand is detected by skin color segmentation in the YCbCr color space. A graphical user interface is created to train the algorithm, to display the intermediate and final results, and to present the virtual character. The algorithm provided a recognition accuracy of 97.5% when tested using 20 hand postures from each class. The hand postures are performed by different persons, with large variation in size and shape of the hand posture, and with different lighting conditions. The recognition results show that the algorithm has robustness against these variations.

1 Introduction

Hand posture recognition is an important area of research in visual pattern recognition, having wide applications in human-robot interaction (HRI), human-computer interaction (HCI) and in virtual reality (VR). Visual interaction is an easy and effective way of interaction, which does not require any physical contact and does not get affected by noisy environments.

P. Vadakkepat et al. (Eds.): FIRA 2010, CCIS 103, pp. 290–297, 2010.

Hand gestures are generally classified as *static* and *dynamic* gestures. Static hand gestures (also known as a hand posture) are those in which the hand position does not change during the gesturing period. Static gestures rely on the information about the flexure angles of the fingers. In dynamic hand gestures (also known as a hand gesture), the hand position is temporal and it changes continuously with respect to time. Dynamic gestures rely not only on the finger's flex angles, but also on the hand trajectories and orientations. Dynamic gestures can be viewed as actions composed of a sequence of static gestures that are connected by continuous motion. A dynamic hand gesture can be expressed as a hierarchical combination of static gestures. In this paper the recognition of static hand gestures is considered.

Over the past three decades, there are lots of success stories in vision based pattern analysis. However, the mainstream computer vision has always been challenged by human vision, and the mechanism of human visual system is yet to be understood well, which is a challenge for both neuroscience and computer vision. The human visual system rapidly and effortlessly recognizes a large number of diverse objects in cluttered, natural scenes and identifies the specific patterns, which inspired the development of computational models of biological vision systems. Recent developments in the use of neurobiological models in computer vision tries to bridge the gap between neuroscience, computer vision and pattern recognition. These developments include the modeling of primate visual cortex [1,2], and its use in image feature extraction, object recognition, face recognition, and natural scene understanding.

The visual object recognition is mediated by the ventral visual object processing stream in the visual cortex [1,4]. Serre *et al.* [2,3] proposed a computational model of the ventral stream, based on the standard model of visual object recognition [1]. This model provides a hierarchical system which comprises four layers that imitates the feed-forward path of object recognition in the ventral stream of primate visual cortex.

The computational model proposed by Serre *et al.* extracts the scale and position tolerant C_2 Standard Model Features (SMFs). A major limitation of this model in real-world applications is its processing speed. Also the algorithm needs a separate classifier for the classification of the pattern. The present paper proposes a modification of this algorithm which does not need a separate classifier stage, and which is computationally efficient. The prototype C_1 patches used for learning are selected from specific classes. The patches which have good interclass differences are identified and the discriminative C_2 features are extracted. The classification is done by a comparison of the extracted C_2 features.

Section 2 of this paper briefly explains the C_2 feature extraction system. The proposed algorithm is explained in Section 3. The real-time implementation of the recognition algorithm and the experimental results are discussed in Section 4. Finally Section 5 concludes the paper.

2 The Feature Extraction System

Hubel and Wiesel discovered the organization of receptive fields, and the properties of *simple* and *complex* fields in cat primary visual cortex [5]. The receptive fields of neurons in the primary visual cortex of mammals is imitated using Gabor filters [6]. Gabor filter based features have good discriminative power between different textures and shapes in the image.

Riesenhuber and Poggio proposed a hierarchical model of ventral visual object-processing stream in the visual cortex [1]. Serre *et al.* implemented a computational model of the system and used it for robust object recognition [2,3]. Later these features, namely C_2 standard model features (SMFs), were used for hand writing recognition [8] and face recognition [9]. The proposed algorithm utilizes the C_2 features for the multi-class recognition of hand postures.

Table 1. Different layers in the C_2 feature extraction system

Layer	Process	Represents
S_1	Gabor filtering	simple cells in V1
C_1	Local pooling	complex cells in V1
S_2	Radial basis functions	V4 & posterior inferotemporal cortex
C_2	Global pooling	inferotemporal cortex

The computational model proposed by Serre *et al.* consists of four layers (Table 1). Layer 1 (S_1) consists of a battery of Gabor filters with different orientations (4) and sizes (16 sizes divided into 8 bands). This imitates the simple cells in the primary visual cortex V1 which filters the image for the detection of edges and bars. Layer 2 (C_1) models the complex cells in V1, by applying a MAX operator locally (over different scales and positions) to the first layer results. This operation provides tolerance to different object projection size and it's position in the 2-D plane of the visual field. In layer 3 (S_2), radial basis functions (RBFs) are used to imitate the V4 and posterior inferotemporal (PIT) cortex. This aids shape recognition by comparing the complex features at the output of C_1 stage (which corresponds to the retinal image) with patches of previously seen visual image and shape features (in human these patterns are stored in the synaptic weights of the neural cells). Finally, the fourth layer (C_2) applies a MAX operator (globally, over all scales and positions) to the output of layer S_2, resulting in a representation that expresses the best comparison with previously seen images. The output of layer 4 are the C_2 SMFs, which are used for the classification of the image.

Simple cells in the third layer implement an RBF, which combines bars and edges in the image to more complex shapes. RBFs are a major class of neural network model, comparing the distance between input and a prototype [7]. Each S_2 unit response depends in a Gaussian-like way on the Euclidean distance between a new input and a stored prototype. The prototype patches of different sizes (center of the RBF units) are drawn randomly (random image and position) from the training images at the level of the second layer (C_1). Each patch contains all the four orientations. The third layer compares these patches by calculating the summed Euclidean distance between the patch and every possible crop (combining all orientation) from the image of similar size. This comparison is done separately with each scale-band representation in the second layer.

The final set of shift and scale invariant C_2 responses is computed by taking a global maximum over all scales and positions for each S_2 type, i.e., the value of the best match between a stored prototype and the input image is kept and the rest is discarded.

3 The Proposed Hand Posture Recognition Algorithm

Seree *et al.* suggested that it is possible to perform robust object recognition with C_2 SMFs learned from a separate set of randomly selected natural images. However it was agreed that *object specific* features (the features learned from images which contain the target object) perform better compared to the features learned from random images.

The proposed algorithm extracts the features in an object class specific way. The modification is done at the third layer of the C_2 feature extraction system. Instead of selecting the patches from random positions and random images, they are selected from geometrically significant positions of specific hand posture classes. In [2] support vector machines (SVM) and boosting based classifiers are used for the classification, whereas the proposed algorithm does the classification by a simple comparison of the extracted features. Also the patches which have more inter-class differences are identified and selected which increased the classification accuracy and reduced the computational cost (by reducing the number of features). Fig. 1 shows a block diagram of the proposed algorithm. The different blocks of the diagram are explained in the following subsections.

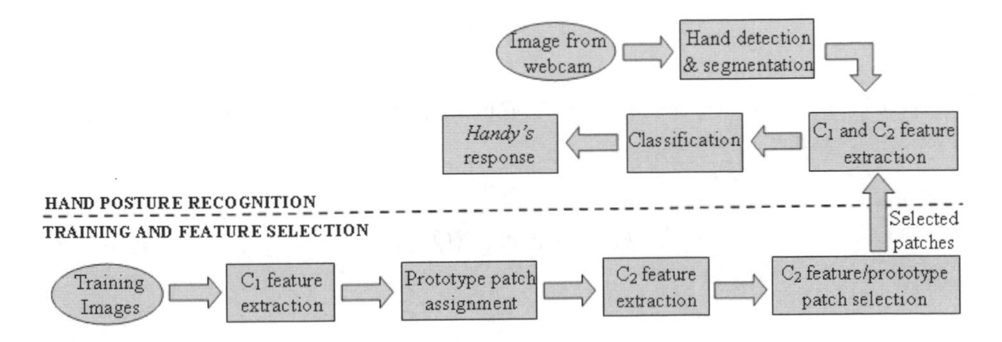

Fig. 1. Block diagram of the proposed algorithm

3.1 Assigning Prototype Patches Specific to Posture Classes

The C_1 features of the training images (one image/class) are extracted. The prototype patches (center of the RBF units) of different sizes are assigned to the C_1 images of specific classes. $N(=10)$ patches, each with $Q(=4)$ different patch sizes are extracted from $M(=10)$ classes. The patches are located at geometrically significant positions in the hand image (Fig. 2). Each patch contains C_1 responses with $O(=4)$ different orientations. The C_2 features are extracted from the C_1 images using the prototype patches. Each of the extracted C_2 feature corresponds to a particular patch with a specific size and totally there are QxMxN C_2 features.

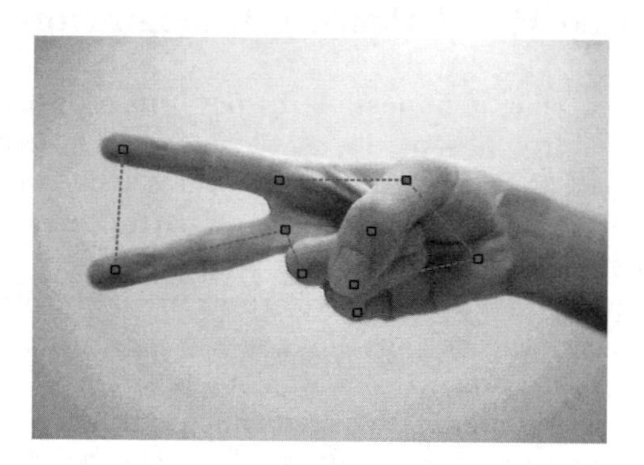

Fig. 2. Positions of prototype patches in a sample hand image

3.2 Classification

Once the C_2 features are extracted the classification is done as follows. Let

$$P_{ij} = \sum_{k=1}^{Q} C_2(ijk), \tag{1}$$

where $C_2(ijk)$ is the C_2 feature corresponding to the jth patch from the ith class with k^{th} patch size, then,

$$C = \mathrm{argmax}_{C_i}\{Q_{C_i}\}, \tag{2}$$

where

$$Q_{C_i} = \sum_{j=1}^{N} P_{ij}, \tag{3}$$

and C is the class to which the hand posture belongs to.

3.3 Selection of the Discriminative C_2 Features

The C_2 features are selected based on the interclass differences in the C_2 features of the training images. Each C_2 feature corresponds to a particular prototype patch. The prototype patches which helps to distinguish different classes are identified and selected to make the extracted C_2 features discriminative. Let

$$P'_{ij} = \max_{C_l \neq C} \{P_{ij}(C_l)\}, \tag{4}$$

where $P_{ij}(C_l)$ is the response P_{ij} (1) of an image from class C_l, then $P_{ij} - P'_{ij}$ represents a *margin of classification(MC)*. The prototype patches from each class are sorted according to the descending order of the corresponding *MCs* and the first *N'*(= 5 in this paper) patches are selected.

The features of the test image are extracted using these selected discriminative patches to get the discriminative C_2 features. The hand posture classification is done using (2) and (3) (*N*=*N'* in 3).

4 Real-Time Implementation and Experimental Results

The proposed algorithm is implemented in real-time in windows XP platform, for the interaction between human and a virtual character *Handy*. The programming is done using C# language. The virtual character is created using Adobe flash. Twelve different animations (10 to represent different predefined classes, 1 to represent non-class, and 1 to represent idle mode) are developed. A text to speech converter is used to generate voice response. *Handy* welcomes the user and gives the command to connect the camera, if it is not connected already. Once the camera is connected and the posture is shown, *Handy* identifies it and responds to the human by symbolically showing the posture. Also it pronounces the class number. If the posture does not belong to any of the trained classes *Handy* shows exclamation and asks *what is that?*.

The RGB image is captured using a webcam. The image is converted to YCbCr color space and the hand is detected using skin color segmentation (HSV color space also tried, but YCbCr color space provided better results). The segmented image is converted to gray scale and used for further processing. The hand image is captured only if the movement of the hand is less than a threshold value, which is adjustable. Low threshold value provides high motion sensitivity (more motion sensitivity means the hand posture will be detected only if the hand motion is lesser). The hand motion is detected by subtracting the subsequent frames in the video. A graphical user interface (Fig. 3) is developed which has provisions for displaying the skin colored area, left / right hand selection, varying the motion sensitivity of the hand, and, the training and testing of the recognition algorithm. The user interface displays the input video (on left). The posture class number, the corresponding hand posture image from the library, and the response of *Handy* are also displayed (on right) in the user interface.

The algorithm is trained for 10 classes of hand postures (Fig. 4). Only one image per class is used for the training and the testing is done by showing each

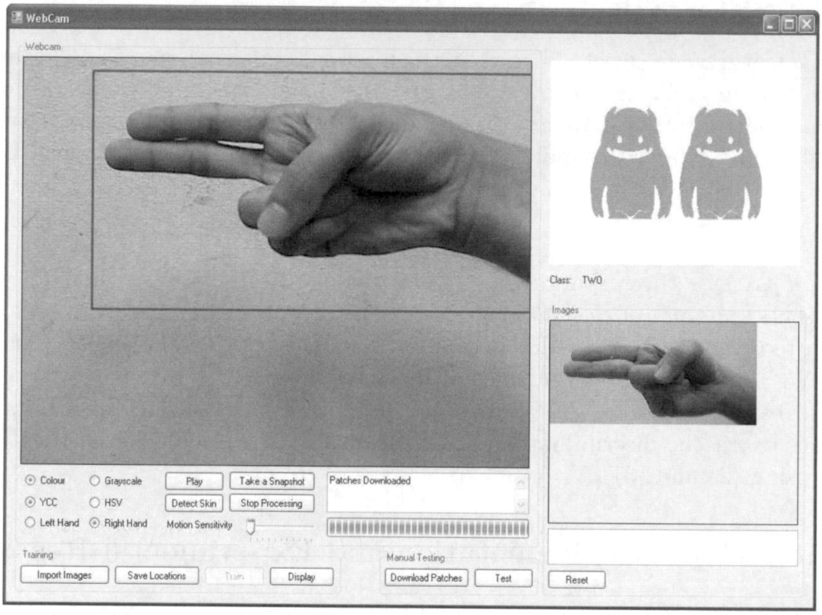

Fig. 3. The user interface

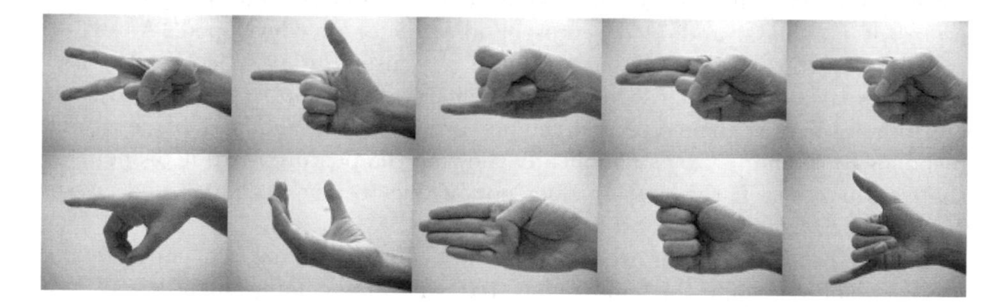

Fig. 4. Hand posture classes used in the experiments

posture 20 times. The hand postures are performed by different persons, with large variation in size and shape of the hand posture, and with different lighting conditions. The algorithm recognized the postures with an accuracy of 97.5%. The average recognition time taken is few milliseconds.

5 Conclusion

A computationally efficient and reliable hand posture recognition algorithm is proposed for the interaction between human and computers. The proposed algorithm is motivated by the computational model of the ventral stream of visual cortex. The C_2 SMFs of the hand images are extracted in such a way that it provide good discrimination between different classes, and the classification is done by a comparison. The algorithm is implemented in real-time for the interaction between human and a virtual character *Handy*. The algorithm recognized the hand postures at a faster speed, with reliable accuracy. The recognition accuracy of the proposed classifier shows its good discriminative power. Also the proposed system has person independent performance.

The proposed algorithm easily recognizes hand postures with light and dark backgrounds. However the performance of the algorithm in complex backgrounds is to be improved. This work is suggested as an extension of the present paper.

References

1. Riesenhuber, M., Poggio, T.: Hierarchical models of object recognition in cortex. Nature Neuroscience 2(11), 1019–1025 (1999)
2. Serre, T., Wolf, L., Bileschi, S., Riesenhuber, M., Poggio, T.: Robust object recognition with cortex-like mechanisms. IEEE Transactions on Pattern Analysis and Machine Intelligence 29(3), 411–426 (2007)
3. Serre, T., Wolf, L., Poggio, T.: Object recognition with features inspired by visual cortex. In: Conference on Computer Vision and Pattern Recognition, San Diego, CA, pp. 994–1000 (2005)
4. Goodale, M., Milner, A.: Separate Visual Pathways for Perception and Action. Trends in Neuroscience 15(11), 20–25 (1992)
5. Hubel, D.H., Wiesel, T.N.: Receptive Fields, Binocular Interaction and Functional Architecture in the Cat s Visual Cortex. J. Physiology 160, 106–154 (1962)
6. Jones, J.P., Palmer, L.A.: An evaluation of the twodimensional gabor filter model of simple receptive fields in cat striate cortex. Journal of Neurophysiology 58(6), 1233–1258 (1987)
7. Bishop, C.: Neural Networks for Pattern Recognition. Oxford Univ. Press, Oxford (1995)
8. Van der Zant, T., Schomaker, L., Haak, K.: Handwritten-word spotting using biologically inspired features. IEEE Transactions on Pattern Analysis and Machine Intelligence 30(11), 1945–1957 (2008)
9. Lai, J., Wang, W.X.: Face Recognition Using Cortex Mechanism and SVM. In: 1st International Conference Intelligent Robotics and Applications, Wuhan, China, pp. 625–632 (2008)

Wavelet Based Medical Image Fusion Using Filter Masks

Susmitha Vekkot

Department of Computing, London Metropolitan University,
North Campus,Holloway Road, London, UK
s.vekkot@gmail.com

Abstract. This paper deals with convolution based image fusion using filter masks and reviews the performance of each with respect to qualitative and quantitative strategies. Fusion is performed using discrete wavelet transformation at two levels. The low and high frequency coefficients obtained are subjected to separate fusion rules. The low frequency approximation coefficients are selected based on a pixel selection rule while high frequency details are selected by convolution using averaging, gaussian, unsharp, prewitt and sobel filter masks of varying sizes. The performance evaluation in each case is conducted using objective strategies like RMSE and PSNR and results are graphically interpreted. Thus a comprehensive analysis is conducted to ensure the best fit mask for medical diagnosis and treatment applications.

Keywords: discrete wavelet transforms, filter mask, averaging, gaussian, prewitt, sobel.

1 Introduction

Seeing is believing. Visual stimulus is the best way to identify the textures and global features of any environment. In most scenarios, more than one image is necessary to precisely represent the spatial and spectral information. Fusion of images obtained from various sources enriches information content, thereby enhancing image quality for human perception, machine learning and computer vision. Robotics make use of images from wide variety of sensors deployed in the area of interest, which are fused to get a perceptible interpretation of scenarios. Image fusion has immense robotic applications in areas like micro-surgery, remote sensing, military and surveillance. Robotic sensors can be deployed in hazardous terrain where human intervention is often dangerous or impossible [1].

An image can be represented either by its original spatial representation or in frequency domain. But a simultaneous depiction of images in both spatial and frequency domains is not possible by virtue of Heisenbergs uncertainty [2]. Therefore, time-frequency based wavelet transform analysis has gained increased popularity in image processing domains focusing on feature extraction and combination. Wavelet transform captures localized data across multiple resolutions, thereby providing reliable and non-redundant directional representation for feature extraction [3]. Fusion can be performed at pixel, feature or decision level.

P. Vadakkepat et al. (Eds.): FIRA 2010, CCIS 103, pp. 298–305, 2010.

The decision map for fusion of salient features can be based on criteria like energy [4] or variance [5]. This paper is a sequel to [2] and focuses on fusion of medical test images using Discrete Wavelet Transform (DWT) involving mask based filtering for detail coefficients of decomposition. Next section gives the background information on transformation used and demonstrates the idea of image decomposition using two dimensional DWT. Section 3 details the filter masks used, flowchart for fusion and experimental results. Section 4 presents the quantitative analysis and discusses the performance of filter masks on medical test images. The paper is concluded in section 5.

2 Background

Transformations are essential to extract information from signals which are not available by representation in spatial domain. The concept of wavelet based signal representation had its origin in early 1930s which saw the introduction of basis functions with varying scales for signal analysis. The term wavelet was first coined by Haar in his thesis [6]. Mallat [7] gave a thorough mathematical treatment to the concept of wavelet orthonormal bases and introduced a pyramid algorithm to wavelet transform decomposition.

2.1 Discrete Wavelet Transforms

Wavelets are mathematical functions used for signal analysis at different resolutions or scales. The irregularity of wavelets makes them better basis for analysis of discontinuous real-world signals. The most attractive features of wavelets are their good localisation properties in both frequency and spatial domains. A prototype mother wavelet undergoes translation and scaling operations to give self similar wavelet families. Temporal analysis is performed using a high frequency (compressed) version of the mother wavelet and frequency analysis by a low frequency (dilated) version of the same [6]. Discretisation of translation and scale parameters can be performed by taking,

$$a = a_0^j \quad and \quad b = m a_0^j b_0, \qquad j, m \varepsilon Z \tag{1}$$

Wavelet family can be defined as in (2),

$$\psi_{j,m}(t) = a_0^{-j/2} \psi(a_0^{-j} t - m b_0) \tag{2}$$

DWT of a function f(t) can be given by (3),

$$DWT_f(j,m) = \int_{-\infty}^{+\infty} f(t) \psi_{j,m}^*(t) dt \tag{3}$$

where $\{\psi_{j,m}\}$ forms an orthonormal basis for $L^2(\Re)$.

DWT uses a subband coding algorithm using dyadic coefficients and provides sufficient analysis information [7].

2.2 Two-Dimensional DWT Decomposition

Fig. 1 shows the CT test image being decomposed into its approximations (low-low) at level L (LL_L) and horizontal (low-high) (LH_l), vertical (high-low) (HL_l) and diagonal (high-high) (HH_l) details at each level l where $l = 1, 2...L$. Here, we can see that only the approximation coefficients at each level are further decomposed and hence we get details at levels $l = 1, 2, ...L$ and approximation at last level L after the wavelet decomposition.
We have used 'sym8' wavelet for decomposition which is a member of symlet wavelet family. Symlets are near symmetric wavelet families consisting of compactly supported orthogonal wavelets with highest number of vanishing moments [8] and the symmetry property makes them ideal bases for image processing applications. In this paper, we have used 'sym8' wavelets in the family with 8 vanishing moments.

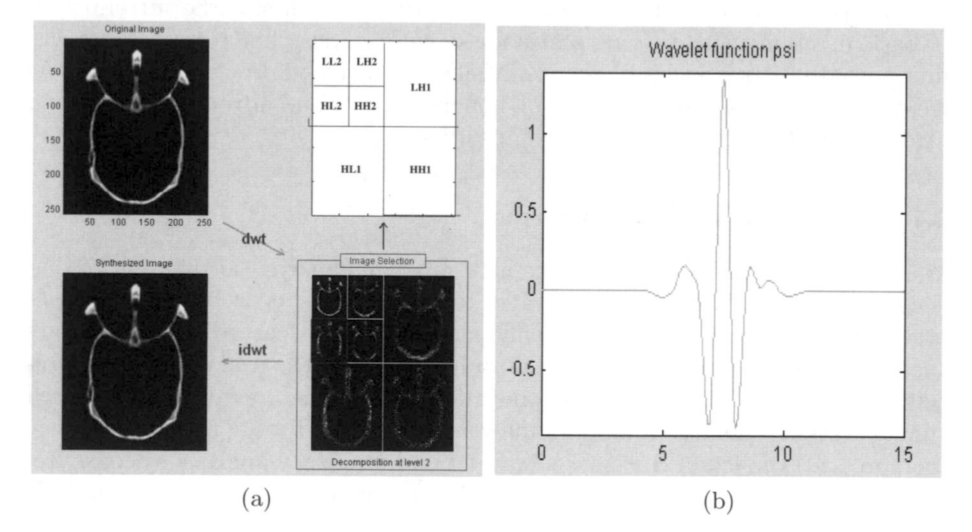

(a) (b)

Fig. 1. (a) is a demonstration of 2 dimensional DWT decomposition of CT scan image at level $L = 2$ using sym8 wavelet. After total decomposition, we get approximations (LL) at level 2 and horizontal, vertical and diagonal details (LH,HL and HH) at levels 1 and 2 while (b) gives the structure of 'sym8' wavelet used for decomposition.

3 Feature Enhancing Mask Based Image Fusion

This paper explores the possibility of an optimum choice of filter mask to perform a hybrid fusion method involving integration of pixel and mask based rules in a single fused image [2]. Filter masks of different types and sizes are used for convolution which are listed below:

- 7X7 averaging low pass filter mask which is used for getting a smoother image by applying a blur.
- 7X7 gaussian low pass filter mask with a standard deviation 0.5.
- 3X3 unsharp filter mask which gives a good contrast image with sharp boundaries. It is a contrast enhancement filter.
- 3X3 prewitt filter which is used to emphasise the horizontal edges in the image. It can also be used to emphasise vertical edges by transposing the filter.
- 3X3 sobel filter emphasises the horizontal edges but has a smoothing effect on the image.

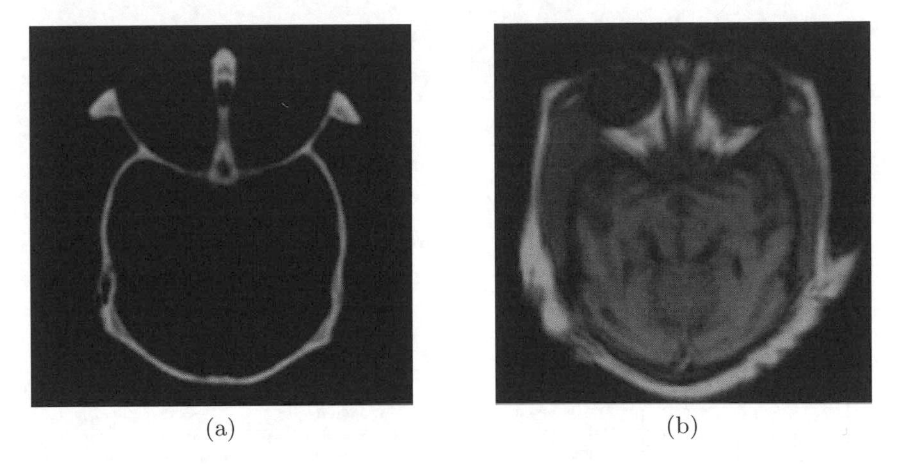

Fig. 2. (a) represents the CT scan image of size 256X256 and (b) represents MRI scan image of size 256X256 used for test. (a) gives the structural details and (b) gives the anatomical details of the body part under scan.

The Computed Tomography (CT) and Magnetic Resonance Imaging (MRI) medical source images for testing the filter masks are given by Fig. 2 and the flowchart for image fusion used in this paper is given by Fig. 3. Fig. 4 gives the results obtained by fusing CT and MRI scan images using each of the filter masks.

4 Results and Discussion

From Fig. 4, Fig. 5, and Table 1, the averaging and gaussian filter masks give relatively clearer images with good features. In Fig 4(c) with unsharp filter, though the contrast and brightness of image are enhanced, internal features are not very clear as in Fig. 4(a) and (b). Prewitt and sobel filter masks enhance horizontal edges in the image. Compared to unsharp filtered image in Fig. 4(c), Fig. 4(d) and (e) give clearer anatomical information.

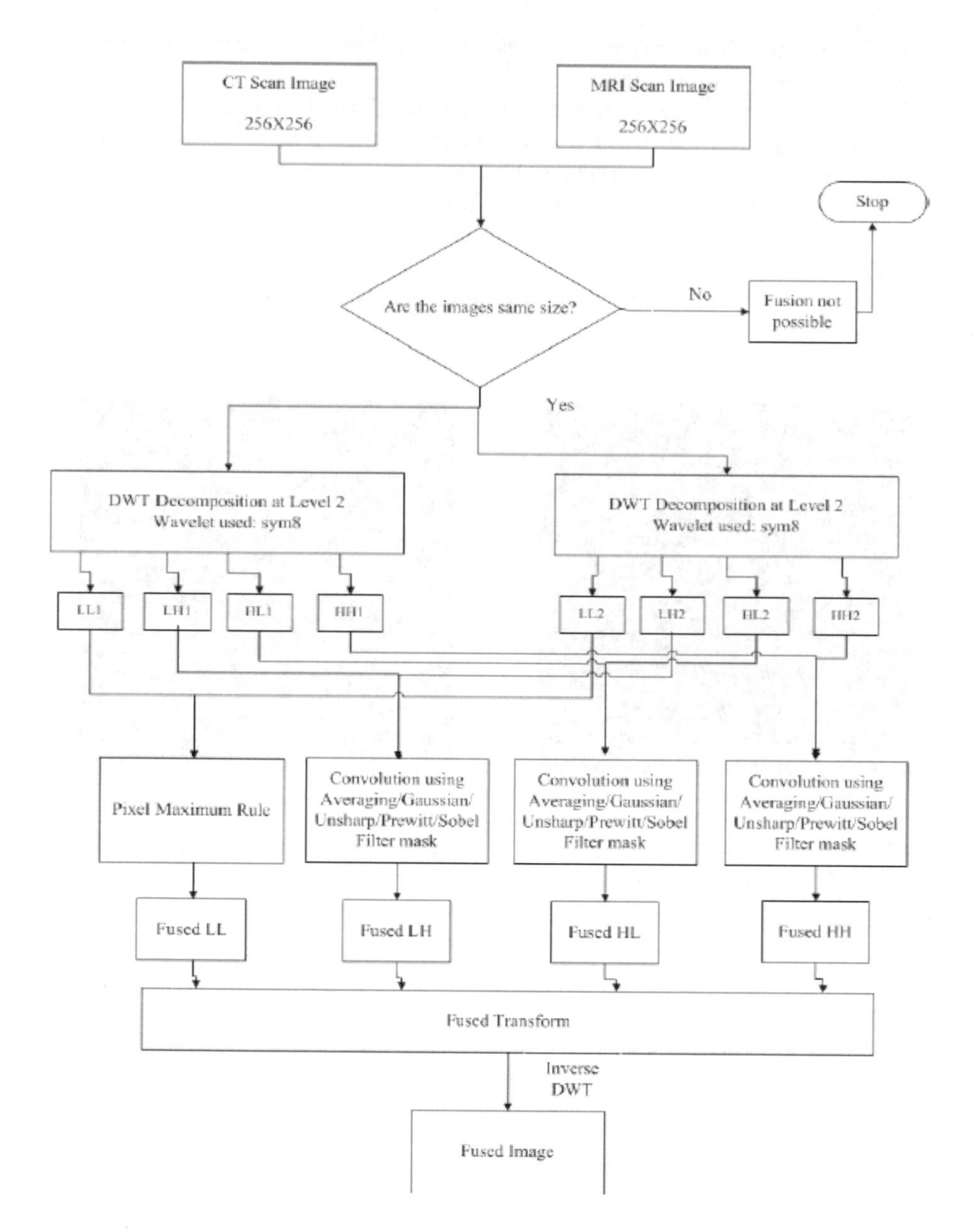

Fig. 3. Wavelet based image fusion: The source images are decomposed to low-low(LL), low-high(LH), high-low(HL) and high-high(HH) bands before fusion rules are applied. Approximations(LL) bands from both images are subjected to pixel based maximum selection rule while horizontal, vertical and diagonal details(LH, HL and HH) are subjected to convolution using each of the filter masks 7X7 averaging, 7X7 gaussian, 3X3 unsharp, prewitt and sobel. The resulting fused transform is reconstructed into fused image by inverse DWT.

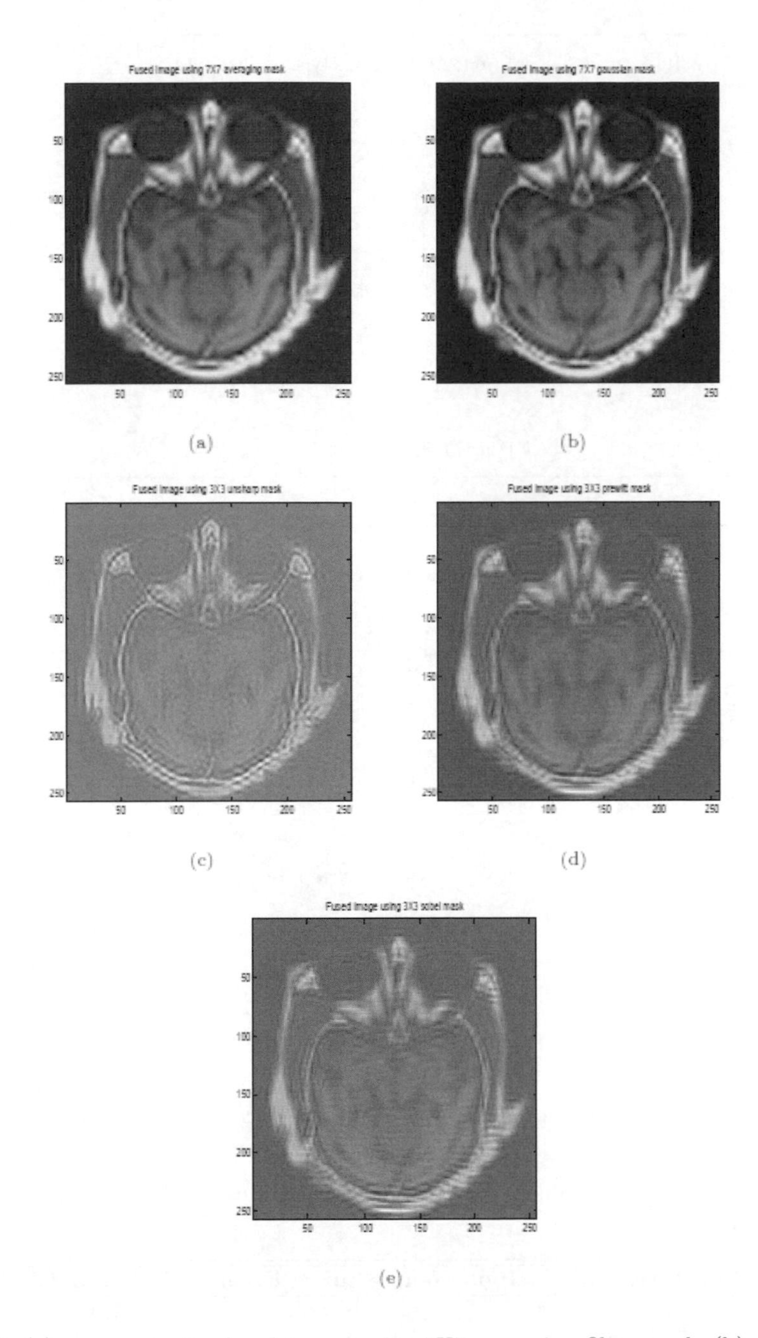

Fig. 4. (a) represents the fused image using 7X7 averaging filter mask, (b) represents fused image using 7X7 gaussian filter mask, (c) the fused image using 3X3 unsharp filter mask and (d) and (e) give fused images using prewitt and sobel filter masks for horizontal edge enhancement respectively. sym8 wavelet is used for decomposition and reconstruction as it is most suitable for image processing.

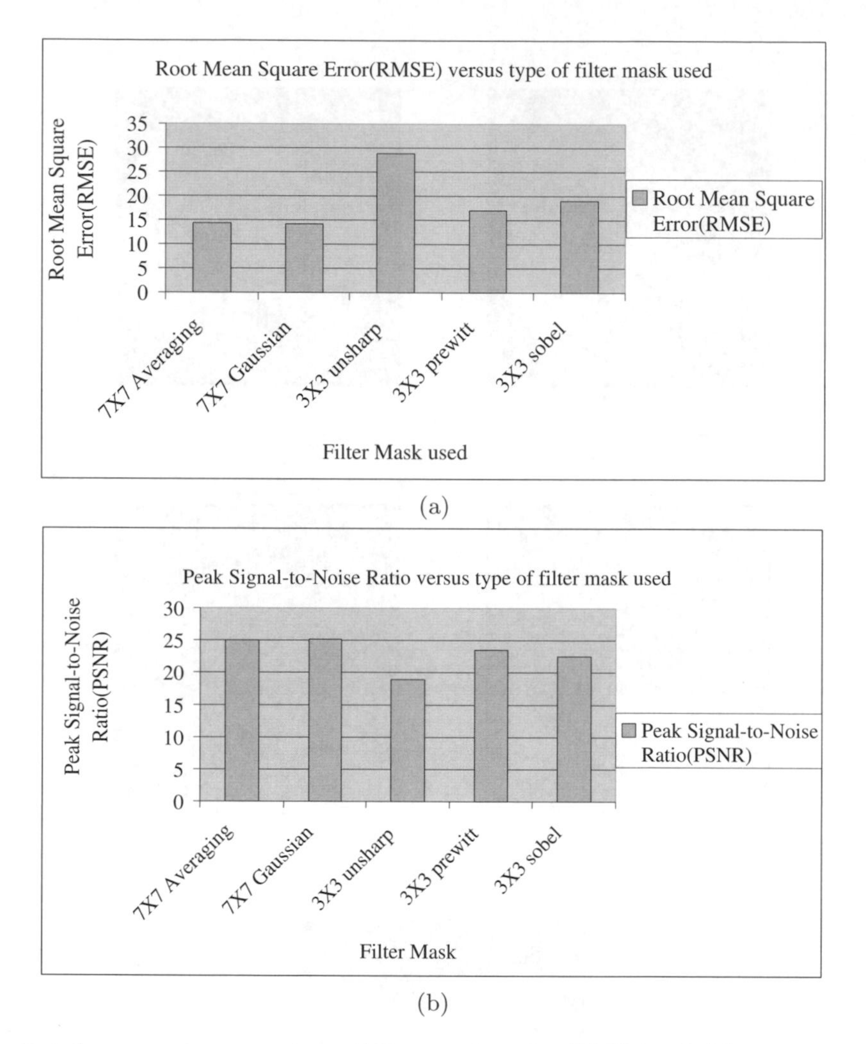

Fig. 5. (a) represents comparison of filter masks using RMSE and (b) represents the comparison using PSNR. The highest PSNR and least RMSE are obtained for 7X7 gaussian filter mask and the least PSNR and most RMSE for 3X3 unsharp filter.

Table 1. Experimental results for medical image fusion using filter masks of varying features and sizes

Serial No.	Fiter Mask	Root Mean Square Error	PSNR in decibels
1	7X7 Averaging	14.28	25.04
2	7X7 Gaussian	14.16	25.11
3	3X3 Unsharp	28.78	18.95
4	3X3 Prewitt	16.94	23.55
5	3X3 Sobel	18.93	22.59

5 Conclusion

The paper demonstrates the necessity of an optimum choice of filter mask depending on the applications in medical diagnosis and treatment. This paper expands the idea in [2] and takes it a step further by allocating different filter masks and reviewing the efficiency of each, in relation to both qualitative and quantitative measurement criteria. Based on the results, an informed decision regarding the choice of filter masks can be made depending on the requirements for a particular application.

Acknowledgments. The CT and MRI medical test images used for image fusion in this research are courtesy of Oliver Rockinger [9].

References

1. Luo, R.C., Kay, M.G.: Data fusion and sensor integration: state of the art 1990s. In: Abidi, M.A., Gonzalez, R.C. (eds.) Data Fusion in Robotics and Machine Intelligence, pp. 7–135. Academic Press, San Diego (1992)
2. Vekkot, S., Shukla, P.: A novel architecture for wavelet based image fusion. In: Proc. International Conference of Signal and Image Processing, Amsterdam, The Netherlands, vol. 57, pp. 372–377 (2009)
3. Vetterli, M., Herley, C.: Wavelets and filter banks: theory and design. IEEE transactions on signal processing 40(9), 2207–2232 (1992)
4. Yingjie, Z., Liling, G.: Region-based image fusion approach using iterative algorithm. In: Proc. Seventh IEEE/ACIS International Conference on Computer and Information Science (ICIS), Oregon, USA (2008)
5. Li, H., Manjunath, B.S., Mitra, S.K.: Multisensor image fusion using the wavelet transform. J. Graphical Models and Image Processing. 57(3), 235–245 (1995)
6. Graps, A.: An introduction to wavelets. IEEE Computational Science and Engineering 2(2) (1995)
7. Mallat, S.: A theory for multiresolution signal decomposition - the wavelet representation. IEEE Trans. on Pattern Analysis and Machine Intelligence. 11(7), 674–693 (1989)
8. Daubechies, I.: Ten lectures on wavelets. CBMS, SIAM 61, 198–202, 254–256 (1994)
9. Rockinger, O.: Various registered test images (2005), `http://www.imagefusion.org/`

Boosting Based Fuzzy-Rough Pattern Classifier

Prahlad Vadakkepat, Pramod Kumar P., Sivakumar Ganesan, and Loh Ai Poh

Department of Electrical and Computer Engineering, National University of Singapore, 4 Engineering Drive 3, Singapore - 117576
{prahlad,pramod}@nus.edu.sg, siva_ncst@yahoo.com, elelohap@nus.edu.sg
http://www.ece.nus.edu.sg

Abstract. A novel classification algorithm based on the rough set concepts of fuzzy lower and upper approximations is proposed. The algorithm transforms each quantitative value of a feature into fuzzy sets of linguistic terms using membership functions and calculates the fuzzy lower and upper approximations. The membership functions are generated from cluster points generated by the subtractive clustering technique. A certain rule set based on fuzzy lower approximation and a possible rule set based on fuzzy upper approximation are generated. A genetic algorithm, based on iterative rule learning in combination with a boosting technique, is used to generate the possible rules. The proposed classifier is tested with three well known datasets from the UCI machine learning repository, and compared with relevant classification methods.

1 Introduction

Inductive knowledge acquisition is a prime area of research in pattern recognition. Computational intelligence techniques are useful in such machine learning exercises. Fuzzy and rough sets are two computational intelligence tools used for decision making in uncertain situations. The proposed algorithm utilizes a fuzzy-rough set approach for the decision making in classification problems.

Fuzzy sets [1], a generalization of the classical sets, proposed by Zadeh in 1963, is a mathematical tool to model vagueness. The concept of fuzzy sets is important in pattern classification, due to its similarity with human reasoning. Traditionally, the rules in a fuzzy inference system are generated from expert knowledge. If no expert knowledge is available, the usual approach is to identify and train fuzzy membership functions in accordance with the clusters in the data. The subtractive clustering technique [2] is an attractive tool for determining data clusters. Subtractive clustering does not require the prior knowledge of the number of clusters.

Z. Pawlak introduced the rough set theory in the early eighties, as a tool to handle inconsistencies among data [3]. A rough set is a formal approximation of a vague concept by a pair of precise concepts, the lower and upper approximations.

Fuzzy and rough set theories are considered complementary in that they both deal with uncertainty: vagueness for fuzzy sets and indiscernibility for rough sets. These two theories can be combined to form rough-fuzzy sets or fuzzy-rough sets [4]. Combining the two theories provides the concepts of lower and

P. Vadakkepat et al. (Eds.): FIRA 2010, CCIS 103, pp. 306–313, 2010.

upper approximations of fuzzy sets by similarity relations, which is useful for addressing classification problems. In the present work the classification rules are generated utilizing the fuzzy-rough set concept.

2 Fuzzy-Rough Sets for Classification

The concept of equivalence classes that form the basis for rough set theory, can be extended to fuzzy set theory to form fuzzy equivalence classes [4]. The concept of crisp equivalence class can be extended by the inclusion of a fuzzy similarity relation S on the universe, which determines the extent by which two elements are similar in S. Using the fuzzy similarity relation, the fuzzy equivalence class $[x]_S$ for objects close to x can be defined as $\mu_{[x]_S}(y) = \mu_S(x, y)$. This definition degenerates to the normal definition of equivalence classes, when S is non-fuzzy. The family of normal fuzzy sets produced by a fuzzy partitioning of the universe of discourse, can play the role of fuzzy equivalence classes [4]. The fuzzy *P-lower* and *P-upper approximations* are defined as [4]:

$$\begin{aligned} \mu_{\underline{P}X}(F_i) &= \inf_x \max\{1 - \mu_{F_i}(x), \mu_X(x)\} \ \forall i, \\ \mu_{\overline{P}X}(F_i) &= \sup_x \min\{\mu_{F_i}(x), \mu_X(x)\} \ \forall i, \end{aligned} \tag{1}$$

where F_i denotes a fuzzy equivalence class belonging to U/P. The tuple $\langle \underline{P}X, \overline{P}X \rangle$ is a fuzzy-rough set.

The crisp positive region in traditional rough set theory is defined as the union of the lower approximations. By the extension principle, the membership of an object $x \in U$, belonging to the fuzzy positive region is defined by (2).

$$\mu_{POS_P(Q)}(x) = \sup_{X \in U/Q} \mu_{\underline{P}X}(x). \tag{2}$$

The present work proposes an algorithm that automatically generates low dimensionality fuzzy rules and corresponding membership functions for pattern classification, directly from the training dataset, based on fuzzy lower and upper approximations.

3 Boosting Based Fuzzy-Rough Classifier

The new classifier algorithm based on fuzzy lower and upper approximations is proposed in this section.

3.1 Stage 1: Membership Functions from Cluster Points

The fuzzy membership functions are created in this step using the feature cluster centers, and these are utilized for the fuzzy partitioning of the feature space. The subtractive clustering technique [2] is utilized to identify the feature cluster centers. The cluster centers are identified for each feature of every class, and the triangular membership functions are formed using the identified feature cluster centers. For example, if a set of cluster points $cp_1 = \{0.5, 0.7\}$ are obtained for an attribute A_1, which is within the range $[0.1, 0.9]$, then the triangular membership functions for A_1 are as shown in Fig. 1.

A set of membership functions $\{MF_{A_i}\}$ is obtained for each attribute per class from the cluster center points. This set of membership functions acts as the base for the crisp sets generated in Stage 2, and for the descriptive fuzzy if-then rules generated in Stage 3.

3.2 Stage 2: Generation of Certain Rules

The set of membership functions $\{MF_{A_i}\}$ obtained in Stage 1, serves as the starting point for defining a crisp set of *certain* regions $\{MF_L\}$. These crisp sets are formed at data clusters so as to maximize the fuzzy lower limit approximations as shown in Fig. 2.

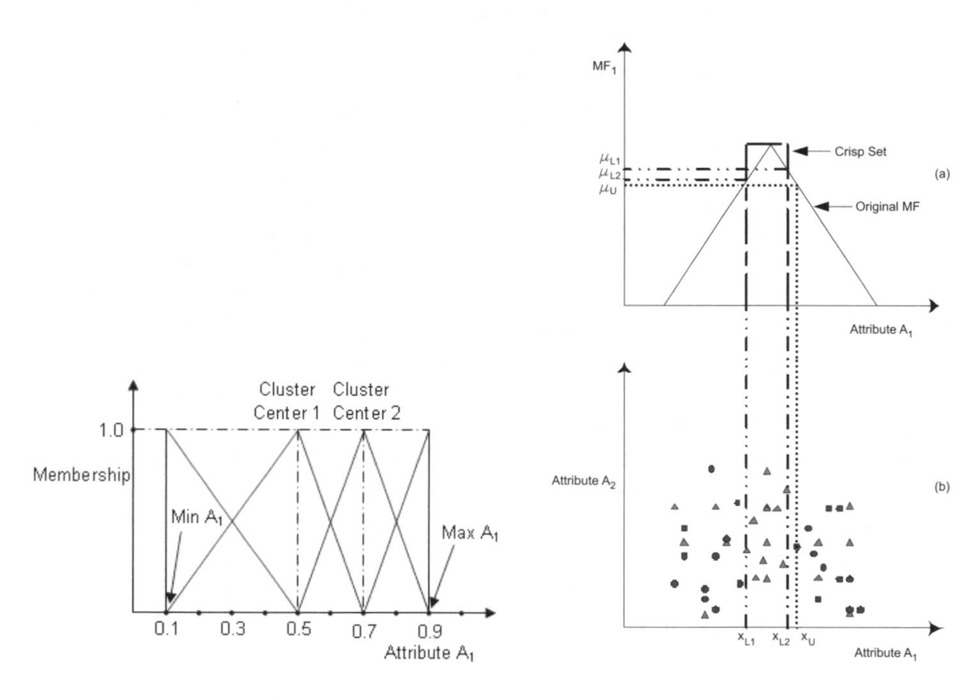

Fig. 1. Formation of membership functions from cluster center points

Fig. 2. Obtaining crisp sets from cluster center points

Fig. 2(b) plots a sample distribution of data for a two attribute classification case. Fig. 2(a) maps the data distribution on attribute A_1 to one of the triangular membership functions $\{MF_{A_1}\}$, obtained in Stage 1 from the cluster points. To obtain the fuzzy lower approximation, the triangular membership function is considered as a singleton rule. The consequent class c_{lj} for this rule is identified by determining the class C_{max} that dominates among the training samples for the attribute A_1 by (3).

$$C_{max}(x_{A_i}^k) = \text{argmax}_{C_m} \sum_{R_{lj}/c_{lj}=C_m} \mu_{R_{lj}}(x_{A_i}^k) \quad (3)$$

The maximum membership μ_u attained by samples belonging to other classes is as per (4).

$$\mu_u = \{ \max_{k|c^k \neq c_{lj}} (\mu_{A_i}(x_{A_i}^k)) \} \quad (4)$$

All the samples which have memberships greater than μ_u can be conclusively identified as from class c_{lj}. Attribute A_1 values for all the samples belonging to class c_{lj}, which have a membership larger than μ_u, are obtained. These attribute values are sorted and the smallest A_1 value, x_{L1} and the largest A_1 value, x_{L2} are identified. These values serve as the limits of a crisp set as indicated in Fig. 2. Only samples belonging to class c_{lj} attain a membership of 1, while samples from other classes attain a membership of 0. Each crisp set serves as a singleton attribute rule. The set of all such *certain* regions serves as the certain rule base and all samples which attain a membership of 1 for the certain rule base, form a fuzzy lower approximation for the partition induced by class c_{lj}.

3.3 Stage 3: Generation of Possible Rules

The training samples which are not classified by the certain rule set in Stage 2, serve as the training data set for Stage 3. In this stage, possible rules are generated based on fuzzy upper approximations derived from $\{MF_{A_i}\}$ through a boosting enhanced GA.

Boosting Enhanced Genetic Algorithm for Learning If - Then Rules. The genetic algorithm employed in the proposed method follows the Iterative Rule Learning (IRL) approach [5]. The architecture of the proposed approach is depicted in Fig. 3. The fuzzy rule chromosome population is created by the rule generator, from the set of membership functions $\{MF_{A_i}\}$ obtained in Stage 1. The genetic rule learner selects the best rule from the population based on fitness, after employing the *crossover* and *mutation* genetic operators. The generated rule is added to the rule base and the weights of the training samples are adjusted by the boosting algorithm for the current set of rules. These steps are iterated until the error rate converges, or the maximum number of iterations is reached.

Rule Generation and Encoding of Fuzzy If - Then Rules. Every chromosome in the population encodes a single fuzzy rule, with each of the attributes $[A_i, i = 1, \ldots, n]^T$ being represented by a membership function. The membership function for each of the attributes A_i, is chosen at random from the set $\{MF_{A_i}\}$ obtained in Stage 1. Each of the triangular membership functions in the set $\{MF_{A_i}\}$ is assigned an integer label. The membership function for each attribute is encoded in the chromosomes by its label. The rule is represented by

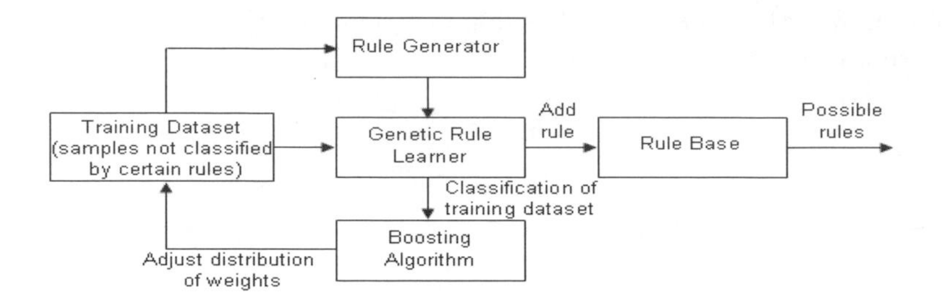

Fig. 3. Generation of possible rules

an integer valued vector $a_1, a_2, \ldots, a_n$ in the chromosome. The GA iterations are fast as the training set has become comparatively smaller after Stage 2.

A specific fuzzy rule may involve most of the attributes. However, general fuzzy rules which are shorter, usually involve only a few of the attributes. To allow for the generation of general rules in addition to specific rules, the chromosome also encodes an additional bit string $S = \{s_1, \ldots, s_n\}$. Each of the bits $(s_1, \ldots, s_n)$ indicate the presence or absence of the respective attribute $(A_1, \ldots, A_n)$ in the fuzzy rule antecedent.

If a fuzzy rule is activated, i.e if any of the attributes in the rule attain a membership value greater than zero, the degree of support for the rule is decided by applying a fuzzy operator *min* (5) to the fuzzy membership functions from the antecedent attributes.

$$\mu_{R_j}(x^k) = \mu_{R_j}(\{x_1^k, \ldots, x_n^k\}) = \min_{n=1}^{N} \mu_{A_{jn}}(x_n^k) \tag{5}$$

where,
R_j Label of the jth fuzzy if-then rule,
$A_{j1}, \ldots, A_{jn}$ Antecedent fuzzy sets corresponding to the attributes.

Each possible classification c_j accumulates the degree of activation of fuzzy rules R_j, with a matching consequent $c_j = C_m$. The sample x^k is classified according to the class label $C_{max}(x^k)$ by (6),

$$C_{max}(x^k) = \text{argmax}_{C_m} \sum_{R_j/c_j=C_m} \mu_{R_j}(x^k) \tag{6}$$

Fitness Criteria. The proposed fitness criteria is designed for optimizing four objectives based on the fuzzy-rough set perspective. Each objective is evaluated separately and finally aggregated into a single scalar fitness value in the range $[0, 1]$. The weights w_k assigned to individual samples by the boosting algorithm, based on the relative difficulty in classifying the sample, also forms a part in the fitness criteria.

The first objective is to ensure that the rule covers positive samples with large weights as compared to the negative samples. This is obtained by the weighted plausibility factor f_{j_1} as,

$$f_{j1} = \frac{\sum\limits_{k|c^k=c_j} w_k \mu_{R_j}(x_k)}{\sum\limits_{k} w_k \mu_{R_j}(x_k)} \tag{7}$$

At the later stages of iteration, the weights of the unclassified samples tend to be high. To ensure that the rule is generic rather than covering only large weighted samples in the later stages, a plausibility factor defined in (8) is utilized as another objective f_{j2}.

$$f_{j2} = \frac{\sum\limits_{k|c^k=c_j} \mu_{R_j}(x_k)}{\sum\limits_{k} \mu_{R_j}(x_k)} \tag{8}$$

The third measure f_{j3} maximizes the absolute individual membership attained by samples from class c_j.

$$f_{j3} = \frac{\sum\limits_{k|c^k=c_j} \mu_{R_j}(x_k)}{l = \left|\mu_{R_j}(x_k) > 0 \forall k|c^k = c_j\right|} \tag{9}$$

The final objective f_{j4} is to maximize the number of positive samples l_1 covered as compared to the negative ones l_2. This fitness value is normalized by the number of samples l_3 in the training set which belong to class c_j.

$$\begin{aligned} l_1 &= \left|\mu_{R_j}(x_k) > 0 \;\forall\; k|c^k = c_j\right| \\ l_2 &= \left|\mu_{R_j}(x_k) > 0 \;\forall\; k|c^k \neq c_j\right| \\ l_3 &= \left|k|c^k = c_j\right| \\ f_{j4} &= \begin{cases} 0 : l1 < l2 \\ \dfrac{l1 - l2}{l3} \end{cases} \end{aligned} \tag{10}$$

The aggregate fitness f_j of the rule is computed as,

$$f_j = \prod_{1 \leq i \leq 4} f_{j_i} \tag{11}$$

Iterative Boosting and Assignment of Weights. All the training samples are assigned with uniform weights $w^k = 1$ at the beginning of the iterations. After each iteration, the weights of all the classified / misclassified samples are adjusted as per (12). The weights of the unclassified samples remain unchanged.

$$w^k(t+1) = \begin{cases} 0 \;\text{ if classified and } w^k(t) - d < 0, \\ w^k(t) - d \;\text{ if classified by selected rule,} \\ w^k(t) + d \;\text{ if misclassified by selected rule} \end{cases} \tag{12}$$

Once the final rule set is generated, the classification of a new sample is done using a single winner rule. If more than one rule is activated, the winner is decided as the one which has the largest aggregate membership as in (5). This allows for the intuitive interpretation of the fuzzy rule base.

4 Performance Evaluation

The performance of the proposed classifier is evaluated on three datasets viz. Iris, Wine, and Glass from UCI machine learning repository [6]. Ten fold cross validation is done for all the datasets and the average accuracies are reported.

A crossover rate of 0.9 and a mutation rate of 0.1 are used for the GA. The GA is allowed to run for 50 generations before selecting the best rule. The rule generation is stopped when the error rate started converging (after 20-25 rules).

Table 1. Classification results of the ten fold cross test

Data set	**Avg. # certain rules**	**Avg. # possible rules**	**Avg. accuracy (%)**	**Best accuracy (%)**
Iris	10	6	96.7	98.4
Wine	11	10	98.3	99.5
Glass	3	17	61.7	63.4

The classification results of the ten fold cross tests are provided in Table 1. Table 2 provides a comparison of the performance of the proposed classifier with relevant classification algorithms in the literature. For the comparison, the samples are divided into training and testing sets in a similar way as that done in the compared work. The proposed algorithm provided equivalent or better accuracies for all the datasets considered.

Table 2. Performance comparison

Dataset	**Algorithm**	**Avg. Accuracy %**
Iris	Proposed classifier	**98.7**
	C4.5 [7]	93.7
	Naive Bayes	95.5
	Pittsburgh Approach [8]	98.5
Wine	Proposed classifier	**100**
	Ishibuchi without *CF* modification [9]	98.5
	Ishibuchi with *CF* modification [9]	100
	Corcoran [10]	99.5
Glass	Proposed classifier	**63.1**
	Ishibuchi [9]	64.4
	1R [11]	53.8
	C4 [12]	63.2

5 Conclusion

A classifier based on the concept of fuzzy equivalence relation is proposed. A certain rule set is generated based on the fuzzy lower approximation. A possible rule set based on fuzzy upper approximation is generated by a boosting enhanced genetic algorithm. Both certain and possible rule sets are evolved automatically from the training dataset. The performance of the proposed classifier is evaluated with some well known datasets and compared with relevant classification techniques. The classification results show the good discriminative power of the proposed algorithm.

References

1. Zadeh, L.A.: Fuzzy sets. Information and Control 8(3), 338–353 (1965)
2. Chiu, S.: Fuzzy model identification based on cluster estimation. Journal of Intelligent and Fuzzy Systems 2(3) (September 1994)
3. Pawlak, Z.: Rough sets. International Journal of Computer and Information Science 11, 341–356 (1982)
4. Dubois, D., Prade, H.: Putting rough sets and fuzzy sets together. In: Slowinski, R. (ed.) Intelligent Decision Support: Handbook of Applications and Advances in Rough Sets Theory, vol. 11, pp. 203–232. Kluwer Academic Publishers, Dordrecht (1992)
5. Gonzalez, A., Herrera, F.: Multi-stage genetic fuzzy systems based on the iterative rule learning approach. Mathware Soft Computing 4, 233–249 (1997)
6. Blake, C.L., Merz, C.J.: Uci machine learning repository (1998), `http://archive.ics.uci.edu/ml/`
7. Quinlan, J.R.: Neuro-Fuzzy and Soft Computing. Morgan Kaufmann, San Mateo (1993)
8. Shi, Y., Eberhart, R., Chen, Y.: Implementation of evolutionary fuzzy systems. IEEE Transactions on Fuzzy Systems 7(2), 109–119 (1999)
9. Ishibuchi, H., Nakashima, T., Murata, T.: Performance evaluation of fuzzy classifier systems for multi-dimensional pattern classification problems. IEEE Trans. on Systems, Man, and Cybernetics- Part B: Cybernetics 29(5), 601–618 (1999)
10. Corcoran, A.L., Sen, S.: Using real-valued genetic algorithms to evolve rule sets for classification. In: Proceedings of First IEEE International Conference on Evolutionary Computing, Orlando, FL, pp. 120–124 (June 1994)
11. Holte, R.C.: Very simple classification rules perform well on most commonly used dataset. Machine Learning 11(1), 63–90 (1993)
12. Quinlan, J.R.: Simplifying decision trees. International Journal of Man-Machine Studies 27, 221–234 (1987)

Face Detector Design for an Expression Robot

Ching-Chang Wong, Yu-Ting Yang, Hao-En Cheng, and Meng-Hung Tsai

Department of Electrical Engineering, Tamkang University
151, Ying-Chuan Rd. Tamsui, Taipei County, Taiwan 25137, R.O.C.
wong@ee.tku.edu.tw

Abstract. In this paper, the design and implementation of an expression robot with 10 degrees of freedom (DOF) is presented. The robot is capable to express some emotions to the human. A processer with an FPGA chip is utilized to control the robot. The robot has two eyebrows, two eyes, a mouth, and an audio and vision system, for the interactions with human. The robot is able to accomplish the movements of the different parts on its face to express the emotions. The paper also proposes a face detection system, which aids more natural human-robot interaction. The paper shows the use of FPGA in handling vision, face detection, and the expression of emotions.

Keywords: Interactive Robot, FPGA-based system, Face Detection.

1 Introduction

In the past, autonomous robots such as industrial robots, cleaning robots, and rescue robots are often designed to operate independently, with less interactions with human [1-2]. In this context, it is desirable for the new generation robots to have the capabilities for the natural communication with human. Social robots pose a dramatic and intriguing shift in the way a robot can be controlled. The new developments in the interactive technologies lead to the evolution of robots that can interact and cooperate with human as a partner. It makes the human-robot cooperation friendlier, and leads to the easier division of labor. The research in human-computer interaction (HCI) and human-machine interface (HMI) includes the recognition of facial expression, body posture, gesture, gaze direction, and voice [3-5].

In this paper, an expression robot with 10 degrees of freedom (DOF) is designed and implemented. The robot is capable to express some emotions to the human. In order to detect the environment, a vision system (one CMOS camera) is equipped on the robot. A processer with an FPGA chip is utilized to control the robot. Many of the functions are implemented on this FPGA chip, to make the hardware circuit simple. The robot can express more complex facial expressions by combining the movement of different parts of the face. The rest of the paper is organized as follows: The design of the mechanical structure of the robot is described in Section 2. The main processer design is explained in Section 3. In Section 4, face detector design is described. The

P. Vadakkepat et al. (Eds.): FIRA 2010, CCIS 103, pp. 314–321, 2010.

motion structure and behavior designs are explained in Sections 5 and 6 respectively. Finally, some concluding remarks are provided in Section 7.

2 Mechanical Structure Design

The expression robot project develops an anthropomorphic outlook called TKUER (TamKang University Expression Robot, see Fig.2.1(a)). The main design concept of the mechanical structure for the expression robot is detailed in this section. In order to present the correct expression, the movements of eyes, eyebrows and mouth are important. The robot has 10 DOF. One DOF direct the eyebrows, another six control the eyes and eyelids, and the remaining three controls its mouth movements. By exquisite motion design, TKUFR can show many different expressions. These mechanical structures are shown in Fig.2.1 (b)~(e).

The components of TKUFR are mainly fabricated using acrylics. Each actuator system for the DOF consists of a high torque motor and a gear. The eyes have six DOF. Four DOF are adopted for the rolling of the eyes. As the eye is important part of a face and eyelids can express vitality, we used the other two DOF to close and open the eyelids. The robot shows the weary expression by closing eyelids, and the intense vigor expression by opening it.

In the eyebrow design, the two eyebrows are moved using one motor. Eyebrows can express sentiments. The pleasure is expressed by raising the eyebrows, and the misery is expressed by lowering it. The mouth design is used to increase the depth of the expression of the robot. There are three DOF to control the mouth. The robot looks happier when the mouth rises. Table 2.1 shows the list of expressions with the corresponding movements.

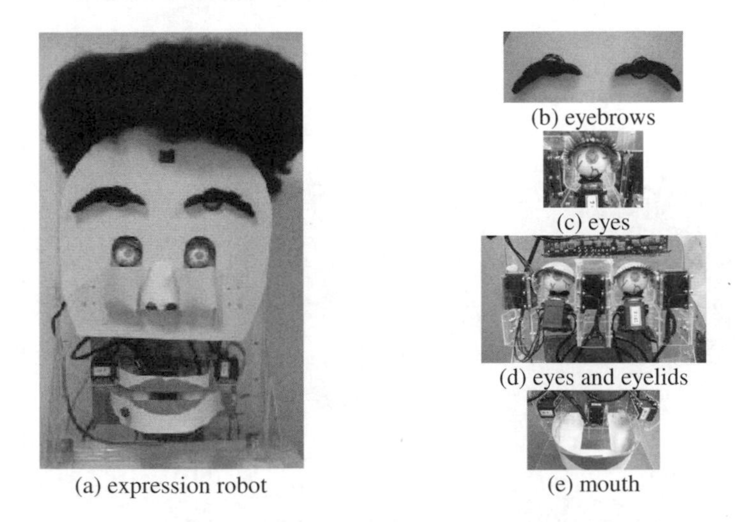
(a) expression robot
(b) eyebrows
(c) eyes
(d) eyes and eyelids
(e) mouth

Fig. 2.1. Photograph of the implemented robot

Table 2.1 The list of expressions with corresponding movements

Position	Motion	Expression
eye	up	x
	down	x
	right	x
	left	x
eyelib	open	vigor, excitation
	close	weary, calm
eyebrow	raise	pleasure
	lower	misery
mouth	raise one side of upper lip	abash
	raise both sides of upper lip	happy
	open lower lip	surprise

3 Processor Design

In order to visually perceive the person who interacts with it, TKUFR is equipped with one CMOS camera on the upper part of the face. In addition, an audio system is incorporated at the back of the expression robot. The electronic design of the expression robot is done using FPGA. The processor design of TKUFR is done by multi-core processor on a single SOPC system. In order to implement the vision system, audio system, motor controller, and to connect with the computer, the processor needs many peripherals, input/output ports, and plenty of logic cells. DE2-70 [6] is a multimedia development platform. It is equipped with 70,000 logic cells, audio/voice processor, memory, and ethernet. The Cyclone II 2C70 FPGA Chip on DE2 platform is used, and two soft-core processors are designed and placed on it. By the QuartusII [7] and SOPC Builder [8] programmer, the system can be implemented in the FPGA chip. The DE2-70 platform and the system block diagram are shown in Fig. 3.1.

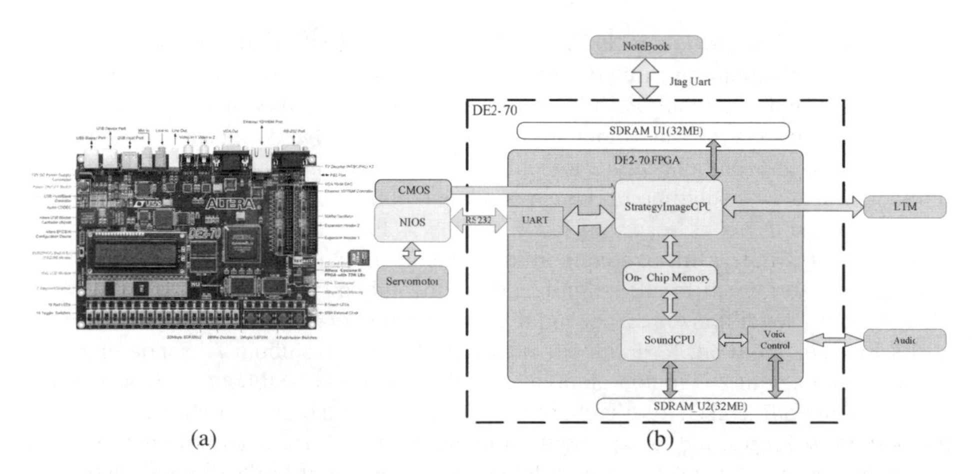

Fig. 3.1. (a) Multimedia development platform DE2-70. (b) System block diagram.

4 Face Detector Design

The visual functions flow of the system is depicted in Fig. 4.1. The proposed system utilizes dynamic gesture, that can be defined by a finite state automation model based on the temporal motion characteristics. The robot is designed to detect the face first. In order to increase the accuracy, mouth is also detected. Succinctly the algorithm in face feature design can be written as:

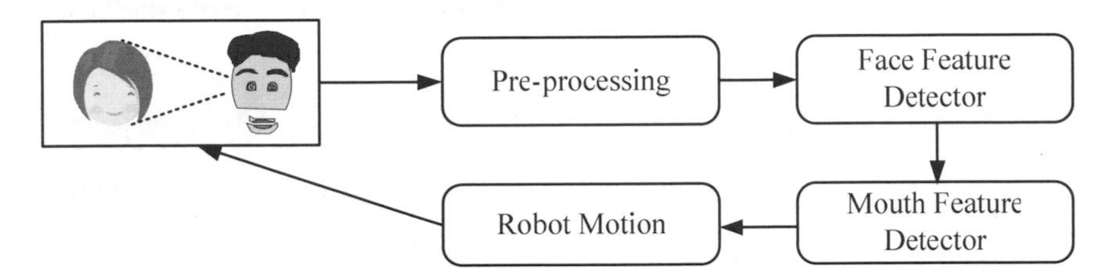

Fig. 4.1. Flow chart of the tasks

Color Space Transformation
Image data is captured by a CMOS camera placed on the upper body of the robot. The image sensor has 5 million pixels that can take 800x480 pixels image. Each pixel has three bytes which represents red, green, and blue, respectively. The color space is transformed from RGB to NCC (Normalized, Color, Coordinates) [9]. NCC space is normalization of the RGB space. NCC has lower variations with changes in illuminations conditions.

Threshold
Threshold process is a doorsill of pixels, if the pixel value is greater than the threshold value then the pixel will be classified as object. Typically, an object pixel is given a value of “1” while a background pixel is given a value of “0.” Finally, a binary image is created by coloring each pixel white or black, depending on a pixel's label [10]. The threshold process of the face is shown in Fig.4.2 (a). Mouth is darker than the rest of the face, as different threshold value is used in the mouth threshold process (Fig. 4.2 (e)).

Noise Filtering
Filtering of the noise is done by morphological opening and closing, to eliminate small noises without destroying the original image. By morphological opening, the image noise can be reduced. Fig.4.2 (b) and Fig.4.2 (f) show the images processed by morphological opening. By morphological closing, the small holes in the image can be removed. Fig.4.2 (c) and Fig.4.2 (g) show the images processed by morphological closing.

Sobel Edge detection
Detecting image edges helps the face detection. Sobel edge detection is shown in Fig.4.1(d) and Fig.4.1(h).

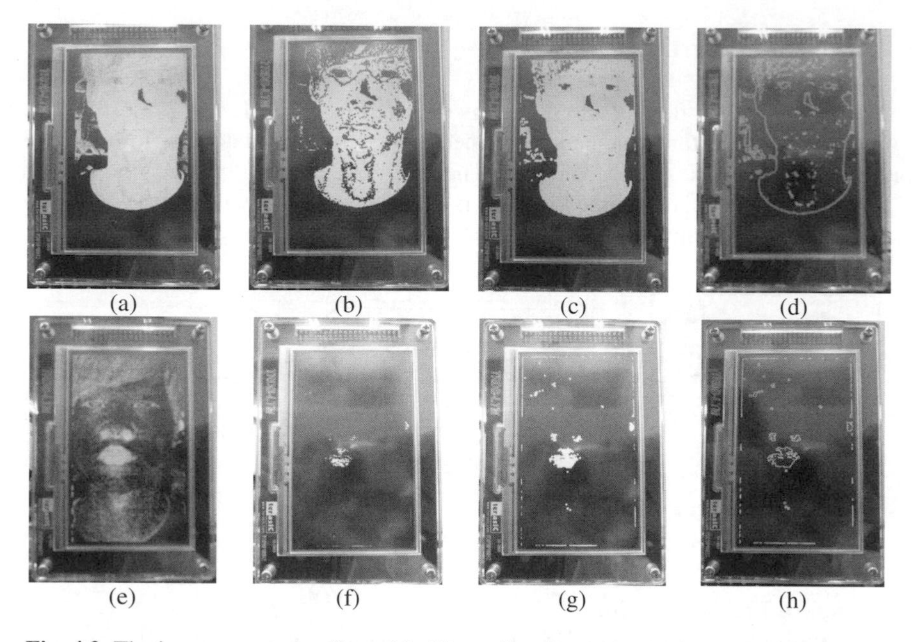

Fig. 4.2. The image processing flow. (a)~(d) are face features, (e)~(h) are mouth features.

Face Detector

The image edge should be analyzed before finding face. There is an easy way to dismember each piece of the image. The process flow of one piece are shown in Fig.4.3 (a)~(n). First, the first pixel is searched and pushed to stack (Fig.4.3 (a)). Then, the adjacent pixel is be pushed to stack (Fig.4.3 (b)). This is repeated till the last pixel, and then the search is finished. So the minimum and maximum of x-axis and y-axis are the end-points of the piece (Fig.4.4). And the finally recognized result is shown in Fig.4.4 (f).

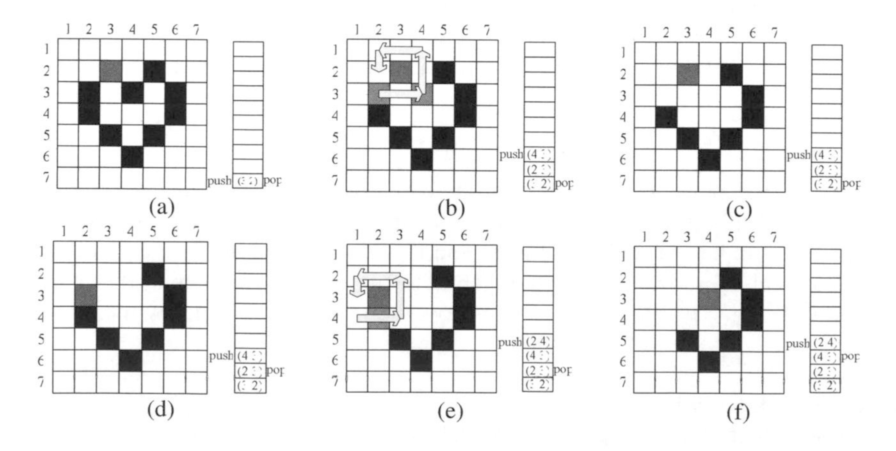

Fig. 4.3. Flow of dismembering of image

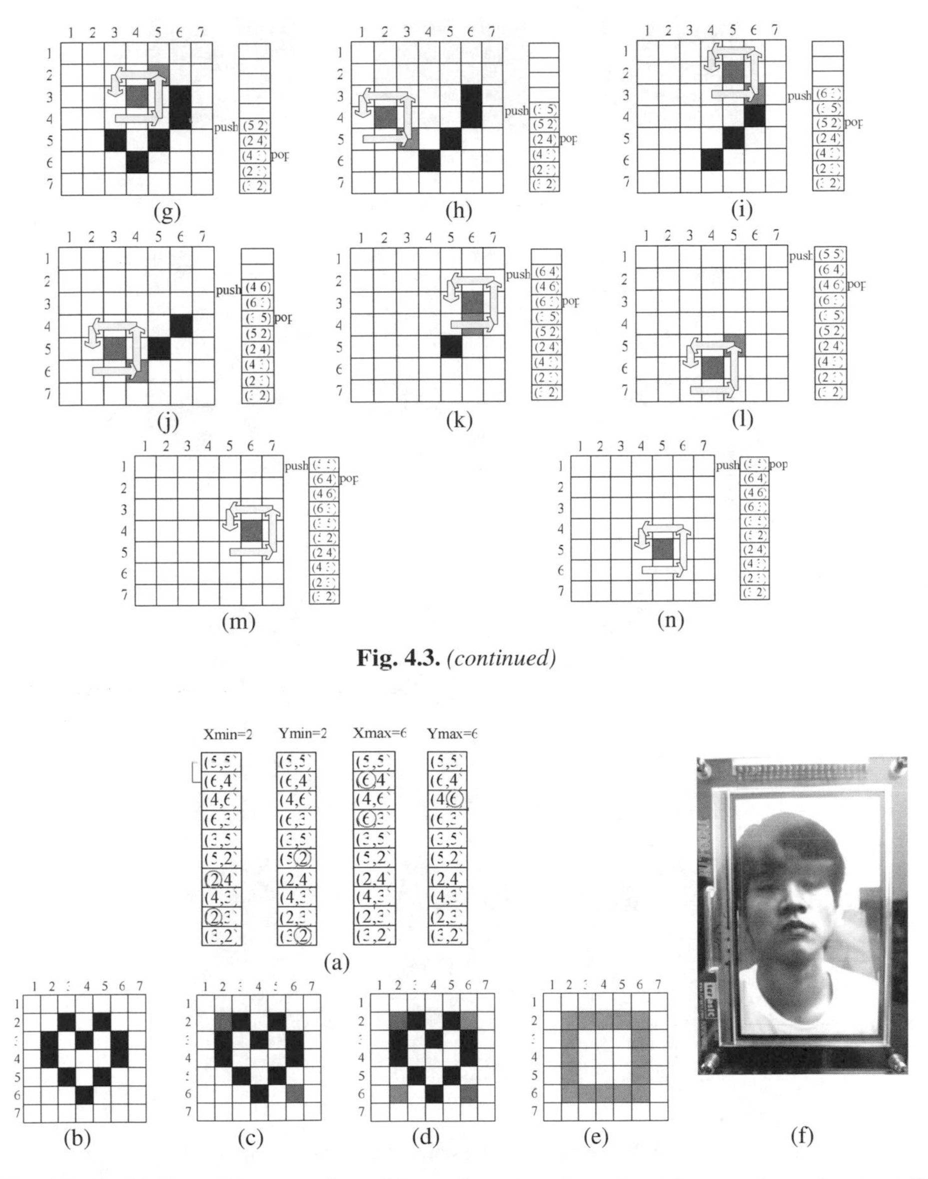

Fig. 4.3. *(continued)*

Fig. 4.4. (a)~(e) The minimum and maximum of x-axis and y-axis are became the end-points of the piece. (f) The finally recognized result

5 Motion Structure Design

The moving structure has two parts. First part is the mouth motion; another is eye and eyebrow motion. A mechanical structure with 10 degrees of freedom is proposed to implement the expression robot. The motions are shown in Fig.5.1 and Fig.5.2.

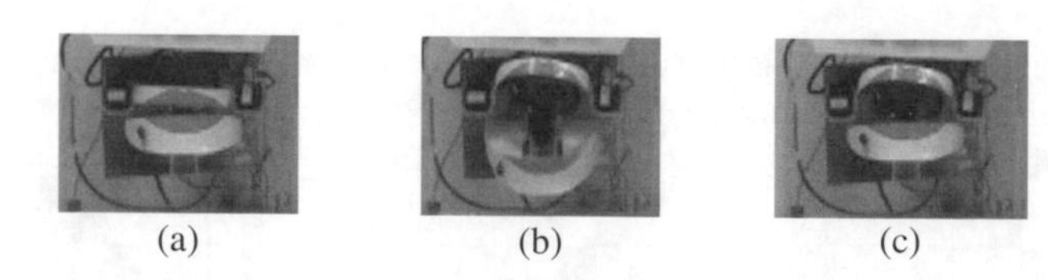

Fig. 5.1. Mouth motions

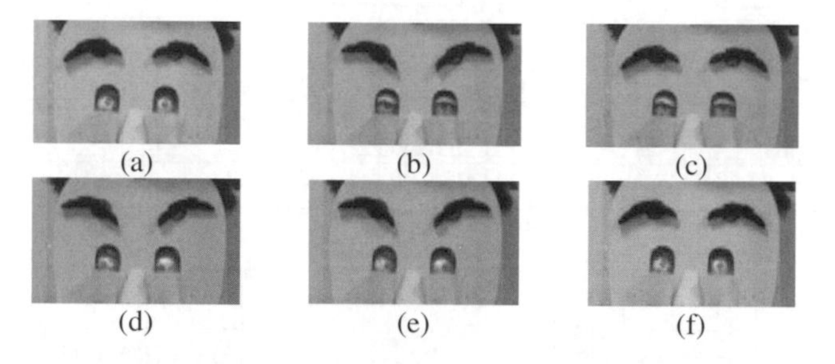

Fig. 5.2. Eye and eyebrow motions

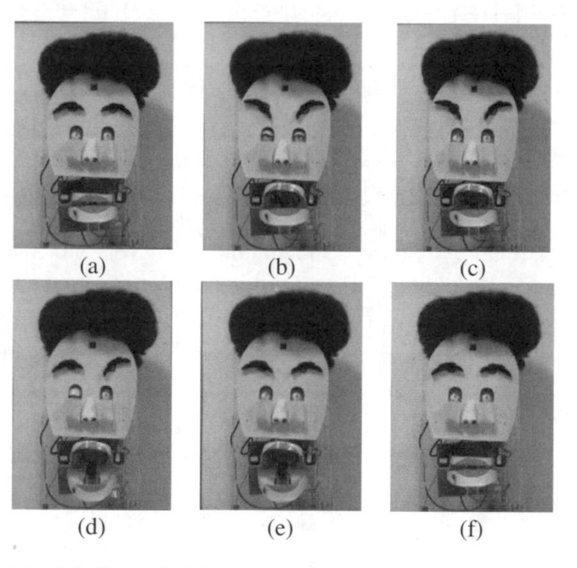

Fig. 5.3. Some facial expressions of the expression robot

Combinations of these two moving parts enable the expression robot to show more complex facial expressions (Fig.5.3).

6 Behavior Design

There are six parts in the behavior design: (1) face too close, (2) face too far, (3) face on top, (4) face on bottom, (5) face on right, and (6) face on left. Two experiment results of the robot in the "face on top" state and the "face too close" state are shown in Fig.5.1. We can see that expression robot can do different expressions.

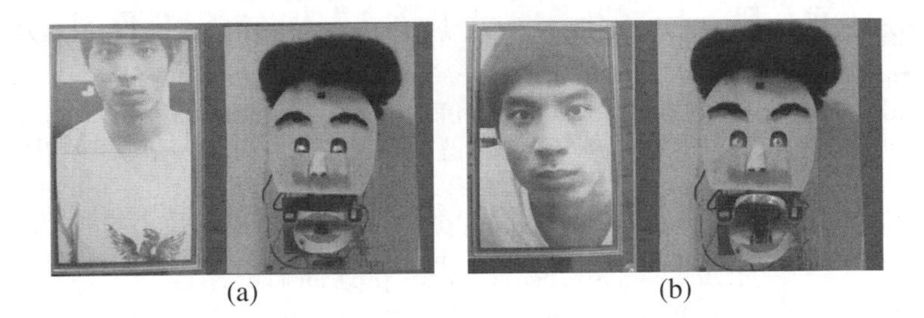

(a) (b)

Fig. 6.1. Behavior design of expression robot. (a) face on top, and (b) face too close.

7 Conclusion

A expression robot with 10 DOF is designed using a multi-core processor structure built on an FPGA chip. The implemented robot can do some basic face expressions. A face detector is used by the robot, to interact with the human. In order to increase the accuracy in face detection, the mouth is also detected. The paper discussed the image processing steps done. It utilized simulation to study the feasibility. In the future, some human face recognition system will be designed for the robot. By learning from the human facial expressions, the robot will be able to display better behaviors in different situations.

Acknowledgement

This research was supported in part by the National Science Council (NSC) of the Republic of China under contract NSC 98-2221-E-032-038.

References

1. Forlizzi, J., DiSalvo, C.: Service in the domestic environment: A study of the roomba vacuum in the home. In: Proc. of the 2006 ACM Conference on Human-Robot Interaction, pp. 258–265 (2006)
2. Pransky, J.: AIBO - the no. 1 selling service robot. Industrial Robot, 24–26 (2001)
3. Paviovic, V.I., Sharama, R., Huang, T.S.: Visual Interpretation of Hand Gestures for Human-Computer Interaction: A Review. IEEE Transactions on Pattern Analysis and Machine Intelligence 19(7) (1997)
4. Breazeal, C.: Emotion and sociable humanoid robots. International Journal of Human-Computer Studies 59(1-2), 119–155 (2003)
5. Iida, F., Tabata, M., Hara, F.: Generating Personality Character in a Face Robot through Interaction with Human. In: Proc. of 7th IEEE International Workshop on Robot and Human Communication, pp. 481–486 (1998)
6. ALTERA: Nios II Processor Reference Handbook (2009)
7. ALTERA: Introduction to the Quartus II Software (2009)
8. http://www.sopcco.com/
9. http://en.wikipedia.org/wiki/Wikipedia:WikiProject_Color/Normalized_Color_Coordinates
10. http://en.wikipedia.org/wiki/Thresholding_image_processing

Fuzzy-PI Force Control for Industrial Robotics

Nuno Mendes[1], Pedro Neto[1], J. Norberto Pires[1], and A. Paulo Moreira[2]

[1] Department of Mechanical Engineering (CEMUC), University of Coimbra, POLO II, 3030-788, Coimbra, Portugal
{nuno.mendes,pedro.neto,jnp}@dem.uc.pt

[2] Institute for Systems and Computer Engineering of Porto (INESC-Porto), University of Porto, Rua Dr. Roberto Frias, 4200-465, Porto, Portugal
amoreira@fe.up.pt

Abstract. Increasingly, robot programs are generated off-line, for example, through a virtual model of a robotic cell. However, when the virtual model does not reproduce exactly the real scenario or the calibration process is not performed correctly it is difficult to generate reliable robot programs. In order to circumvent this problem, it was introduced sensory feedback (force and torque sensing) in a robotic framework. By controlling the robot end-effector pose and specifying its relationship to the interaction/contact forces, robot programmers can ensure that the robot maneuvers correctly, damping possible impacts and also increasing the tolerance to positioning errors from the off-line programming process. In this paper Fuzzy-PI and PI reasoning was proposed as a force control technique. The effectiveness of the proposed approach was evaluated in a serie of 20 experiments that demonstrated that Fuzzy-PI controllers are more suitable to deal with this type of situations.

Keywords: Robotics, Fuzzy-PI, Force control.

1 Introduction

Increasingly, robot programs are generated off-line, without interrupt robot production. Through a virtual model of a robotic cell it is today possible to generate a robot program. These off-line programming systems present a satisfactory performance if the virtual model reproduces exactly the real scenario; however, in some situations this "ideal situation" is not always achievable. In certain circumstances robot programs are generated with errors, for example, when the virtual model has dimensional inaccuracies and in the presence of inaccuracies created in the calibration process (virtual-real environment). In these cases, robots have to autonomously acquire the information to support decision making.

In this paper it is proposed the introduction of sensory feedback (force and torque sensing) in a robotic framework. By controlling the end-effector pose and specifying its relationship to the interaction/contact forces, robot programmers can ensure that the robot maneuvers in an a real environment safely, damping possible impacts and increasing the tolerance to positioning errors from the off-line programming process. Also, when any contact occurs between the robot tool

P. Vadakkepat et al. (Eds.): FIRA 2010, CCIS 103, pp. 322–329, 2010.

and its surrounding environment the interaction forces are controlled properly, otherwise they may damage the working objects and robot tools. The proposed approach focuses on the use of two force control techniques, PI and Fuzzy-PI reasoning. The robot program was generated off-line from a CAD model [1]. The effectiveness of the proposed approach was evaluated in a series of experiments. The robot ability to track the desired path (extracted from CAD) and to adapt/adjust to the environment was analyzed. Finally, results are discussed and some considerations about future work directions are made.

1.1 Force Control for Robotic Systems – An Overview

Many robot tasks require contact with the surrounding environment of the robot. That interaction generates contact forces that should be controlled in a way to finish the task correctly. Those contact forces depend on the stiffness of the tool and working objects/surfaces and should be properly controlled. The option for a particular control technique depends on identifying if we have a passive force control or an active force control method [2]. In the first case, contact forces produce an undesirable effect on the task; they are not necessary for the process to be carried out. In the second case, the contact forces are necessary to finish the task correctly, i.e., the contact forces should be controlled, making them assume some particular value. Up to now, many kinds of robotic systems using force control strategies have been developed and successfully applied to various industrial processes such as polishing [3], deburring [4,5] and high-precision assembly [10]. A large number of force control techniques (Fuzzy, PI, PID, etc.) with varying complexity have been proposed thus far [8,9]. Fuzzy control was first introduced and implemented in the early 1970's [11] in an attempt to design controllers for systems that are structurally difficult to model due to naturally existing nonlinearities and other modeling complexities. Fuzzy logic control appears very useful when the processes are too complex for analysis by conventional quantitative techniques.

2 Proposed Approach

A force/torque (F/T) sensor was attached to the robot wrist. The real-time force and torque feedback data from the F/T sensor will be required to achieve displacement control of the robot end-effector. By analyzing the incoming data from the F/T sensor, the implemented control system decides which displacements should be applied to achieve satisfactory robot performance, making the system adaptive to the working environment as desired. The automatic end-effector adjustment is achieved by a closed loop position control of the robot end-effector provided by F/T sensor data and Fuzzy-PI reasoning. If the contact force exceeds an acceptable range, the force control system will maneuver the robotic arm until that force falls within an acceptable range. The experimental setup of the proposed robotic system is the following:

- An industrial robot Motoman HP6 equipped with the NX100 controller.
- A personal computer running Microsoft Windows Xp.
- A six degrees of freedom F/T sensor from JR3, equipped with a PCI receiver and processing board installed on the computer PCI bus.
- A local area network (LAN), Ethernet and TCP/IP based, used for robot-computer communication. The network is isolated from the laboratory traffic using a properly programmed high speed (100 Mbps) network switch.

The computer is running the software application that manages the cell (acquires data from the F/T sensor and sends displacement commands to the robot). The MotomanLib is a Data Link Library created in our laboratory to control and monitoring the robot remotely. Using an ActiveX component named JR3PCI [4], the F/T sensor measures forces and torques real-time with a sample rate of 8 KHz (Figure 1).

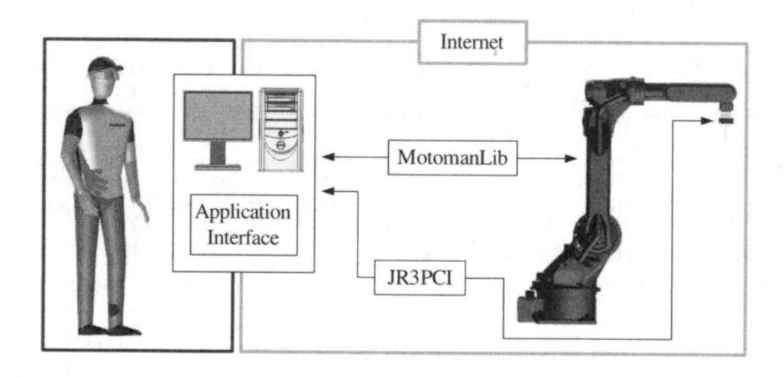

Fig. 1. Communications and system architecture

3 Force Control

Associating the proportional integrative control (PI) and Fuzzy logic we have a polyvalent controller with the ability to be adjusted to any object, regardless to its base or contact surface. Thus, a fuzzy logic controller type Mamdani based on the traditional PI controller was implemented [11]. The PI controller provides a good performance when applied in practical situations. The controller with derivative factor help to diminish the correction time, however, it is very sensitive to noise, precluding their use in our framework. Our preference for the controller type Mamdani is due to its easy implementation and the good results usually obtained, besides that do not need a rigorous mathematical model of the system. The fuzzy controller should respect the following stages:

- Definition of input and output variables.
- Fuzzification.
- Definition of a group of rules to model the application (knowledge base).
- Design of the computational unit that accesses the fuzzy rules.
- Defuzzification.

The force control system collects inputs from the F/T sensor and processes that inputs according to the rules specified in the fuzzy logic memberships. The outputs are presented as the displacement (mm) at which the robotic arm should be moved to obtain the desired contact range of forces and torques (Figure 2).

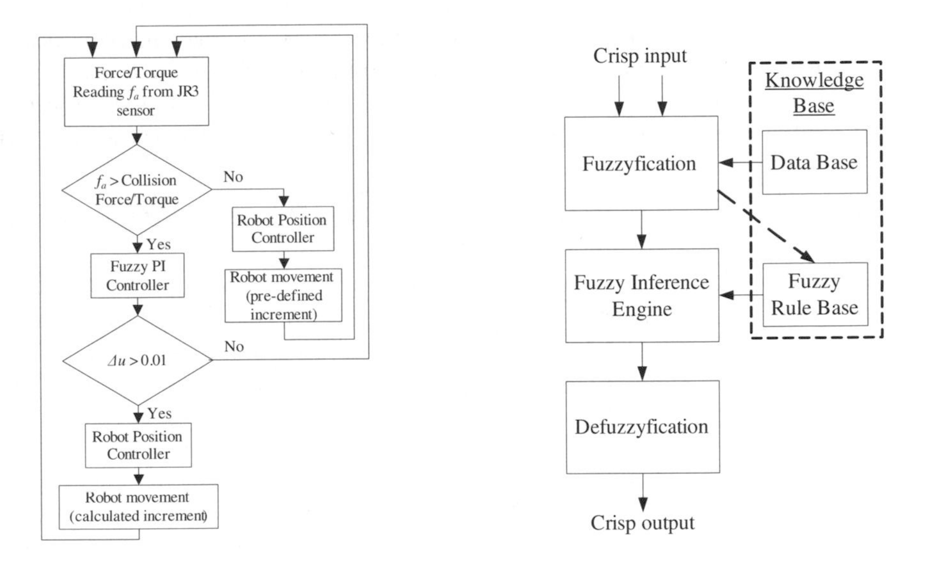

Fig. 2. Flowchart of the fuzzy-PI controller (left). Fuzzy-PI controller (right).

3.1 Fuzzy Control Architecture

The Fuzzy concept has been used to control the force limits of robotic arms [4]. It is ideal for controlling nonlinear systems and for modeling complex systems where an inexact model exists. It is also potentially very robust, maintaining good closed-loop system performance over a wide range of operating conditions. In this paper, the controller input variables are the force/torque error e (1) and the change of the error de (2), where f_a is the actual wrench and f_d is the desired wrench (set point). The controller output is the position/orientation accommodation for the robot (robot displacements).

$$e_k = f_{d_k} - f_{a_k} \tag{1}$$

$$de_k = e_k - e_{k-1} \tag{2}$$

3.2 Fuzzy-PI

From the conventional PI control algorithm:

$$u(t) = K_P \cdot e(t) + K_I \cdot \int e(t)dt \tag{3}$$

where u is the robot displacement and K_P and K_I are coefficients constants. Transforming (3):

$$u_k = u_{k-1} + \triangle u_k$$

$$\triangle u_k = K_P \cdot de_K + K_I \cdot e_K \tag{4}$$

If e and de are fuzzy variables, (4) becomes a fuzzy control algorithm. A practical implementation of our fuzzy-PI concept is simplified in Figure 3. Finally, the center of area method was selected for defuzzify the output fuzzy set inferred by the controller (5), where μ_i is the membership function which takes values in the interval [0, 1].

$$\triangle U = \frac{\sum_{i=1}^{n} \mu_i \cdot \triangle U_i}{\sum_{i=1}^{n} \mu_i} \tag{5}$$

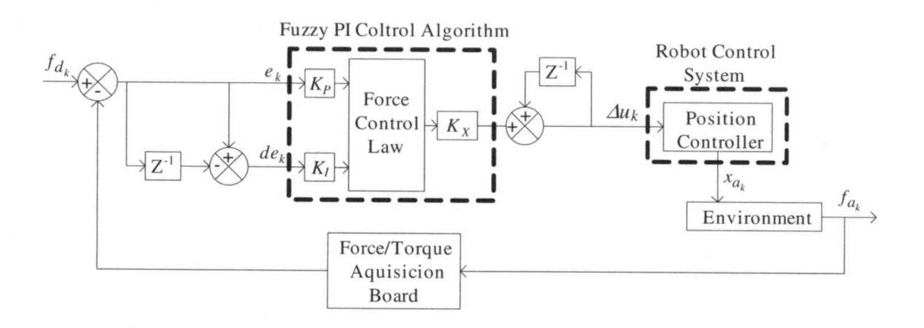

Fig. 3. Implemented fuzzy logic controller with PI function

3.3 Knowledge Base

The knowledge base of a fuzzy logic controller is composed of two components, namely, a data base and a fuzzy control rule base. Each control variable should be normalized into seven linguistic labels. The most common labels used are: positive large (PL), positive medium (PM), positive small (PS), zero (ZR), negative large (NL), negative medium (NM) and negative small (NS). The grade of each label is described by a fuzzy set. The function that relates the grade and the variable is called the membership function, Figure 4. The well-known PI-like fuzzy

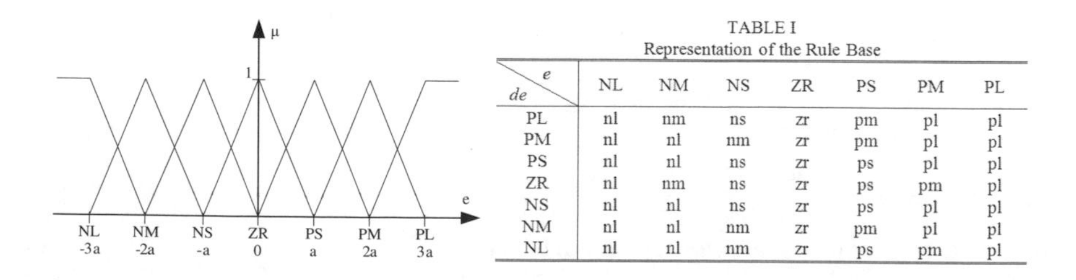

TABLE I
Representation of the Rule Base

de \ e	NL	NM	NS	ZR	PS	PM	PL
PL	nl	nm	ns	zr	pm	pl	pl
PM	nl	nl	nm	zr	pm	pl	pl
PS	nl	nl	ns	zr	ps	pl	pl
ZR	nl	nm	ns	zr	ps	pm	pl
NS	nl	nl	ns	zr	ps	pl	pl
NM	nl	nl	nm	zr	pm	pl	pl
NL	nl	nl	nm	zr	ps	pm	pl

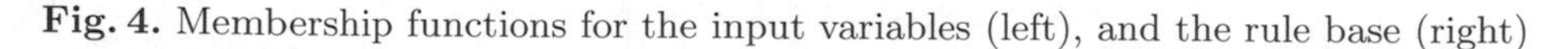

Fig. 4. Membership functions for the input variables (left), and the rule base (right)

rule base suggested by MacVicar-Whelan [12] is used in this paper (Figure 4), allowing fast working convergence without significant oscillations, and preventing overshoots and undershoots. Recently, the MacVicar-Whelan rule base has been improved [13].

3.4 Tuning Strategy

Fuzzy logic design is involved with two important stages, knowledge base design and tuning. The objective of tuning is to select the proper combination of all control parameters so that the resulting closed-loop response best meets the desired design criteria. In order to adapt the system to various contact conditions and scenarios the scaling factors should be tuned, namely K_P, K_I and K_X. Reference [7] proposes an adjustment where the scaling factors are dynamic and thus they have been adjusted along the task. The utilization of different tables of rules accordingly the task to be performed and the materials involved are presented in [6]. In our paper, the scaling factors are set to appropriate constant values, achieved by the method of trial and error.

4 Experiments, Results and Discussion

The effectiveness of the proposed approach was evaluated in a series of 20 experiments, more specifically making the robot follow a surface profile by maintaining a set force normal to the workpiece surface, Figure 5. This test case was chosen as many others with workpieces with similar stifness could have been chosen, even though they had different geometries. The initial robot program was generated off-line from a CAD drawing. The aim is that the force system absorbs the error from the calibration process, workpiece surface irregularities and from the positioning accuracy of the workpiece. When the robot starts the contact with the workpiece, the force control system assumes the robot control, adjusting the end-effector to the environment, in other words, adjusting the pre-programmed paths from CAD. The force control ensures that the contact forces and moments converge to a desired value.

The experiments showed force control results similar to that presented in Figure 5 for fuzzy-PI and PI. The figures report the contact forces between the end-effector and workpiece, and the robot displacement. Although both systems have the capability of disturbance rejection, in the PI control, when the robot starts to move down the contact force deviates from the set force. So, it was established that an approach based on fuzzy-PI reasoning produce better results (smaller offset and overshoot). It seems to be clear that force control improves significantly robot performance, making robots more flexible and with capacity to make decisions. However, Fuzzy-PI controllers provide adequate performance when they are tuned to a specific task, but if the environmental stiffness is unknown or varying significantly, performance is degraded.

As a future work, the speed of convergence to the set forces and torques should be increased and comparisons with other force controllers will be made.

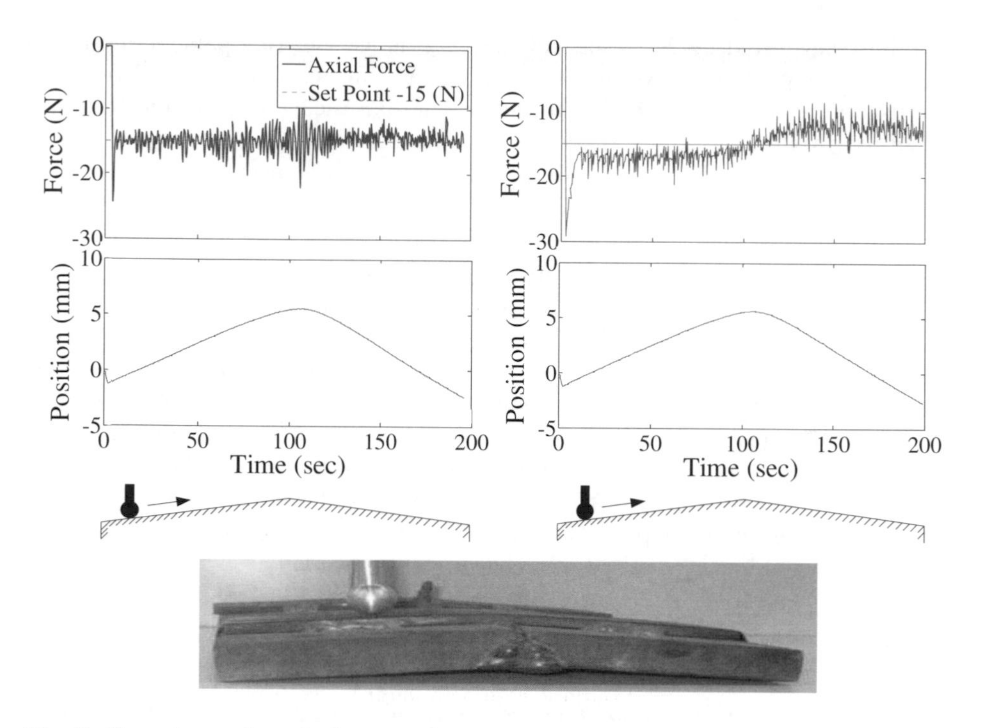

Fig. 5. Experimental results by using fuzzy-PI controller (left) and PI controller (right)

5 Conclusion

Owing to the implemented force control system the robot is able to respond to high degree of environment uncertainty. The effectiveness of the proposed approach was proved through the experiments, showing that force control improves significantly robot performance, making robots more human-like, flexible and with capacity to make decisions. The fuzzy-PI controller showed better results than the traditional PI controller (smaller offset and overshoot). The proposed robotic platform increases the "intelligence" of the robot in several ways, allowing the robot to deal with uncertainty and helping to reduce the set up time required to start robot operation.

References

1. Neto, P., Pires, J.N., Moreira, A.P.: CAD-Based Off-Line Robot Programming. In: 4th IEEE Int. Conference on Robotics, Automation and Mechatronics. IEEE Press, Singapore (2010)
2. Siciliano, B., Villani, L.: Robot Force Control. Kluwer Publishers, Boston (1999)
3. Nagata, F., Kusumoto, Y., Fujimoto, Y., Watanabe, K.: Robotic Sanding System for new Designed Furniture with Free-Formed Surface. Robotics and Comp. Int. Man. 23, 371–379 (2007)

4. Pires, J.N., Ramming, J., Rauch, S., Araujo, R.: Force/torque Sensing applied to Industrial Robotic Deburring. Sensor Review J. 22, 232–241 (2002)
5. Pires, J.N., Afonso, G., Estrela, N.: Force control experiments for industrial applications: a test case using an industrial deburring example. Assembly Automation 27, 148–156 (2007)
6. Pires, J.N., Godinho, T., Araujo, R.: Force control for industrial applications using a fuzzy PI controller. Sensor Review 24, 60–67 (2004)
7. Lin, S.T., Huang, A.K.: Hierarchical fuzzy force control for industrial robots. IEEE Trans. on Ind. Electronics 45, 646–653 (1998)
8. Li, H.X., Gatland, H.R.: A new methodology for designing a fuzzy logic controller. IEEE Trans. on Systems, Man, and Cybernetics 25, 505–512 (1995)
9. Tang, K.S., Man, K.F., Chen, G., Kwong, S.: An Optimal Fuzzy PID Controller. IEEE Trans. on Ind. Electronics 48, 757–765 (2001)
10. Chen, H., Wang, J., Zhang, G., Fuhlbrigge, T., Kock, S.: High-precision assembly automation based on robot compliance. The Int. J. of Advanced Man. Tec. 45, 999–1006 (2009)
11. Mamdani, E.H.: Application of fuzzy algorithms for control of simple dynamic plant. Proc. of IEEE 121, 1585–1588 (1974)
12. MacVicar-Whelan, P.J.: Fuzzy sets for man-machine interactions. Int. J. Man-Mach. Stud. 8, 687–697 (1977)
13. Eksin, I., Guzelkaya, M., Gurleyen, F.: A new methodology for deriving the rule-base of a fuzzy logic controller with a new internal structure. Engineering Applications of Artificial Intelligence 14, 617–628 (2001)

PSO-Optimized Fuzzy Logic Controller for a Single Wheel Robot

Abdullah Al-Mamun and Zhen Zhu

Electrical and Computer Engineering,
National University of Singapore
Singapore 117576
eleaam@nus.edu.sg, richard.zhuzhen@gmail.com

Abstract. A fuzzy logic controller (FLC) for steering control of a single wheel robot with fuzzy membership functions optimized using particle swarm optimization (PSO) is presented in this paper. The single wheel robot is statically unstable and has nonlinear dynamics when it is in motion. Design of a linear controller is anything but straightforward. At an earlier stage of development, the wheel was controlled manually using remote control console. That experience is transformed into a linguistic based control, e.g., FLC. However, the issue of selecting various functions and parameters for FLC still remains to be resolved. In this work, Particle Swarm Optimization (PSO) is used to optimize the membership functions of the FLC.

Keywords: Single wheel robot, Fuzzy logic controller, Particle Swarm Optimization (PSO).

1 Introduction

This paper presents an optimized fuzzy logic controller (FLC) for steering a gyroscopically stabilized single wheel robot, gyrobot [1]. The mechanism is designed and built according to the ideas used in the design of *gyrover* developed in Carnegie Mellon University [2]. It is a flying saucer shaped wheel shown in Fig. 1 (left) that moves forward/ backward by rolling on its edge. Tilting the wheel while on motion steers it to the side it is tilted, according to the principle of gyroscopic precession. We use the term *lean angle* to refer to gyrobot's angle compared to its vertical position.

A simplified schematic of the mechanism is shown in Fig. 1 (right). A two-axis gimbal assembly supporting a fast spinning flywheel is hung from a solid platform. The flywheel is spun by a dc motor (RE 25). The axle of the wheel passes through the platform, whereas, its outer shell is rigidly attached to the axle. The axle is driven by a motor (FAULHABER 3257...CR) making the wheel move forward/ backward. Another motor (HSR-5995TG) is used to tilt the spinning flywheel, mounted inside the two-axis gimbal. Reaction to the tilting of the spinning flywheel, which has large angular momentum, causes the entire wheel to lean sideways. This, in turn, makes the gyrobot steer. Proper control of lean angle is, therefore, crucial to steer the gyrobot. The term *tilt angle* refers to flywheel's position with respect to its equilibrium.

Dynamic behavior of gyrobot is described by nonlinear differential equation, given in the next section. The system is statically unstable. A number of researchers

P. Vadakkepat et al. (Eds.): FIRA 2010, CCIS 103, pp. 330–337, 2010.

published their works on similar statically unstable single wheel robot or unicycle-like mobile platform [3]-[5]. Our experiments with stabilizing the wheel manually using remote control shows that certain skills of the operator is necessary in manipulating remote control buttons. We chose FLC to realize an intelligent controller that can stabilize the wheel and steer it appropriately.

Complexity of designing fuzzy logic controller arises at the time of deciding its membership functions and the best parameters of these membership functions, number of rules to be used, and the best granularity resulting in the best solution for the given problem. Finding the FLC that optimizes certain pre-defined criteria is the issue in hand. Although FLC evolved as a popular approach for control design more than two decades ago, systematic design of FLC from the point of view of stability and performance optimization started fairly recently [7]. FLC can be optimized using neural network, Lyapunov method etc. Solution also comes in the form of bio-inspired optimization algorithms, e.g., Particle Swarm Optimization (PSO), Genetic Algorithm (GA), Ant Colony Optimization (ACO), etc [6]-[8]. These heuristic algorithms have the capability of solving nonlinear problems, well-constrained or even NP-hard problems. We use PSO to optimize the membership functions of FLC.

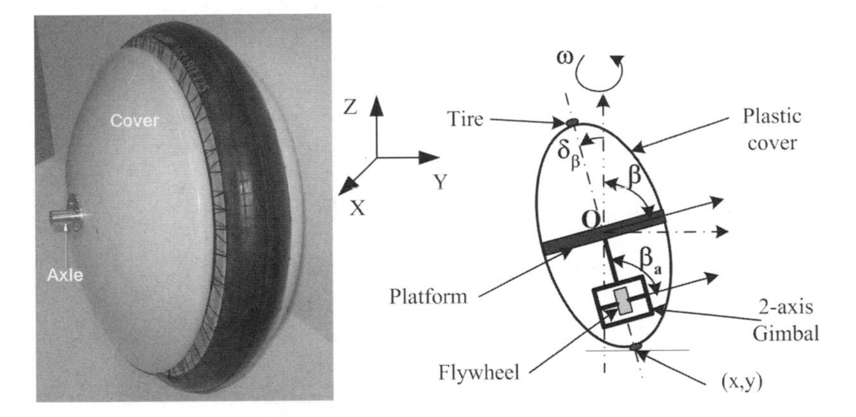

Fig. 1. View of the single wheel robot and a schematic drawing showing various components

2 Single Wheel Robot and Its Dynamic Model

The dynamic model and non-holonomic constraints of the wheel are derived using constrained Lagrangian method with matrix partition to eliminate the Lagrangian multipliers. In finding the model, the wheel is assumed to be a rigid, homogeneous disk that rolls without slipping on flat surfaces. The inner mechanism is modeled as a manipulator link hanging from the axle and with a spinning flywheel at the end. It is also assumed that the inner mechanism hangs just beneath the center of axle. These assumptions make the resulting model at the best an approximation of actual dynamics. The dynamic model is,

$$\overline{M}(q_0)\ddot{q}_0 = \overline{F}(q_0,\dot{q}_0) + \overline{B}u_0, \tag{1}$$

with $q_0 = [\alpha\ \beta\ \gamma\ \beta_a]^T$, where α, β, and γ represents the angle of precession, lean angle and rotational angle of the wheel, respectively, and β_a is the tilt angle of the spinning flywheel. Tilt motor torque and drive motor torque form the input vector u_0. This model is simplified further by decoupling tilt angle (β_a) to obtain,

$$M(q)\ddot{q} = F(q,\dot{q}) + Bu\,, \tag{2}$$

with $q = [\alpha\ \beta\ \gamma]^T$. In the decoupled model, the control input is $\dot{\beta}_a$.

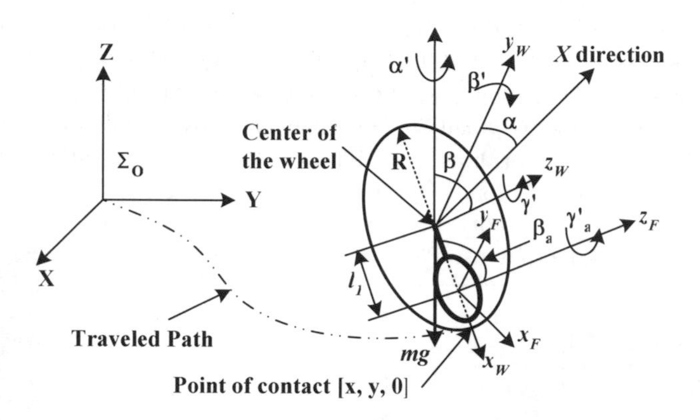

Fig. 2. Illustration of coordinates used to derive the dynamic model

Following parameters are associated with the mechanism of the gyrobot.

I_{xw}, I_{yw}, I_{zw}	:	moment of inertia of the wheel,
I_{xf}, I_{yf}, I_{zf}	:	moment of inertia of the flywheel,
m	:	mass of gyrobot,
R	:	radius of the gyrobot shell,
g	:	acceleration due to gravity, and
μ	:	friction coefficient.

The matrices used in the model of equation 2 are,

$$M = \begin{bmatrix} M_{11} & 0 & M_{13} \\ 0 & M_{22} & 0 \\ M_{13} & 0 & M_{33} \end{bmatrix},\ F = \begin{bmatrix} F_1 \\ F_2 \\ F_3 \end{bmatrix} \text{ and } B = \begin{bmatrix} 0 & B_{12} \\ 0 & 0 \\ 1 & 0 \end{bmatrix} \text{ with}$$

$$M_{11} = I_{xf}\left(1+\cos^2(\beta+\beta_a)\right) + I_{xw}\left(1+\cos^2\beta\right) + mR^2\cos^2\beta,$$

$$M_{13} = 2I_{xw}\cos\beta + mR^2\cos^2\beta,\ M_{22} = I_{xf} + I_{xw} + mR^2,\ M_{33} = 2I_{xw} + mR^2,$$

$$F_1 = \left(I_{xw} + mR^2\right)(\sin 2\beta)\dot{\alpha}\dot{\beta} + I_{xf}(\sin 2(\beta+\beta_a))\dot{\alpha}\dot{\beta} - \mu\dot{\alpha} - 2I_{xw}(\sin\beta)\dot{\beta}\dot{\gamma} + 2I_{xf}(\sin(\beta+\beta_a))\dot{\beta}\dot{\gamma}_a,$$

$$F_2 = -gmR\cos\beta - \left(I_{xw} + mR^2\right)(\cos\beta\sin\beta)\dot{\alpha}^2 - \left(2I_{xw} + mR^2\right)(\sin\beta)\dot{\alpha}\dot{\beta}$$
$$- I_{xf}\cos(\beta+\beta_a)\sin(\beta+\beta_a)\dot{\alpha}^2 - 2I_{xf}(\sin(\beta+\beta_a))\dot{\alpha}\dot{\gamma}_a,$$

$$F_3 = 2\left(I_{xw} + mR^2\right)(\sin\beta)\dot{\alpha}\dot{\beta},$$

$$B_{12} = I_{xf}\sin(2(\beta+\beta_a))\dot{\alpha} + 2I_{xf}(\sin(\beta+\beta_a))\dot{\gamma}_a.$$

The trajectory of the point of contact between the wheel and ground is,

$$\dot{x}_R = R\left(\dot{\gamma}\cos\alpha + \dot{\alpha}\cos\alpha\cos\beta - \dot{\beta}\sin\alpha\sin\beta\right), \tag{3}$$

$$\dot{y}_R = R\left(\dot{\gamma}\sin\alpha + \dot{\alpha}\sin\alpha\cos\beta + \dot{\beta}\cos\alpha\sin\beta\right). \tag{4}$$

Full derivation of this model is omitted here; interested readers may refer to [9]-[10].

Though variable speed of the flywheel can give better control of steering, we keep flywheel speed constant for the sake of simplicity at this stage. The driving speed of the gyrobot is also kept constant. So the only control input is the tilt velocity of flywheel. An inner control loop is used to control the tilt velocity whose response is faster than the response for changes in wheel lean angle (β).

3 Controller for the Single Wheel Robot

3.1 Fuzzy Logic Controller

Sensor (Crossbow CXTA01) for measuring lean angle is mounted on the platform and the sensor reading is sampled at intervals of 50 ms. Lean angle error $\delta_\beta(k) = \beta_{\text{ref}} - \beta(k)$ and error velocity $\Delta\delta_\beta(k) = \delta_\beta(k) - \delta_\beta(k\text{-}1)$ are two inputs to the fuzzy logic controller; k is sampling index. The output of the controller is the desired tilt rate ($\Delta\beta_a(k)$) for inner gimbal. Triangular membership function is used for input and output variables, each with five fuzzy sets: positive large (PL), positive small (PS), Zero (Z), negative small (NS) and negative large (NL). Lean error, error velocity and tilt velocity are bounded to ±30°, ±100°/sec, and ±100°/sec, respectively to ensure conformity with physical limitations of gyrobot.

For two inputs with five linguistic values each, the rule set contains 25 rules, summarized in Table 1.

Table 1. Fuzzy rule set

		$\Delta\delta_\beta(k)$				
		NL	NS	Z	PS	PL
$\delta_\beta(k)$	NL	PL	PL	PL	PL	PS
	NS	PL	PS	PS	PS	Z
	Z	PS	PS	Z	NS	NS
	PS	Z	NS	NS	NS	NL
	PL	NS	NL	NL	NL	NL

This design results in a fuzzy PD (Proportional plus Derivative) controller which is not suitable for eliminating steady state error and inclusion of integral action is necessary. It is possible to obtain fuzzy PI controller using error and change in error as inputs to the rule base. However, it is rather difficult to write rules for the integral action. Attention is also needed to address the issue of integrator windup. To avoid these problems, we configure the controller as an incremental controller. The controller is given by the following where u_{flc} is the output of the FLC, $u(k)$ is the tilt velocity command at the kth sampling, K_I is integral gain,

$$u(k) = K_I u(k-1) + u_{flc}\left(\delta_\beta(k), \Delta\delta_\beta(k)\right). \tag{5}$$

3.2 Tuning of FLC Using Particle Swarm Optimization (PSO)

PSO, originally developed by Kennedy and Eberhart [11], is a population based swarm algorithm. In this algorithm, a vector representing individual solution is treated as one particle. The population dynamics simulates bio-inspired behavior of these particles, such as that seen in "flock of birds" or "swarm of bees". It involves sharing of information and allows individuals to take profit from own discoveries and previous experience during the search for food. Many researchers studied the convergence properties of PSO [12]-[13].

To design an adaptive fuzzy controller, we have to choose the settings of FLC, i.e., number of MFs for each variable, number of rules, parameters associated with each membership function etc. Information required to construct the FLC is represented in a particle vector. Then PSO is used to find the optimum membership functions. We keep nature of membership function, i.e., triangular function, unchanged throughout optimization, but allow the centers of these MFs to change.

For all variables, two extreme MFs have their centers fixed at the extreme values of the corresponding variable. All intermediate MFs are flexible in nature and their peaks are adapted by PSO algorithm. For each MF, peaks of the adjacent MFs on either side form the left and right base support of it.

4 Results

The peaks of PL and NL for the variable δ_β are kept fixed at +0.5236 radian and -0.5236 radian, respectively (corresponding to ±30°). Similarly, the centers of NL and PL are kept fixed at -1.7453 rad/s and +1.7453 rad/s for $\Delta\delta_\beta$ and $\Delta\beta_a$. PSO algorithm minimizes $\frac{1}{N}\sum_{k=1}^{N}\left(\delta_\beta(k)\right)^2$ obtained from the response of gyrobot in following a given trajectory β_{ref} for 15 s. MFs for δ_β after 50 iterations of optimization are shown in Fig. 5. Membership functions for other variables are not shown due to space constraints. Results of the 50 iteration optimization are summarized in Table 2 that shows location of the center of triangular MF. Responses of the gyrobot to step change in β_{ref} for initial membership functions and PSO-optimized membership functions are shown in Fig 6.. Input to the tilt motor is shown in Fig. 7.

Table 2. Location of the center of triangular membership functions

		Initial values	Optimized values
δ_β	NegM	-0.2618	-0.1733
	Zero	0.0	+0.0214
	PosM	+0.2618	+0.2201
$\Delta\delta_\beta$	NegM	-0.8727	-0.8753
	Zero	0.0	+0.0431
	PosM	+0.8727	+1.1451
$\Delta\beta_a$	NegM	-0.8727	-1.2104
	Zero	0.0	-0.0777
	PosM	+0.8727	+0.7166

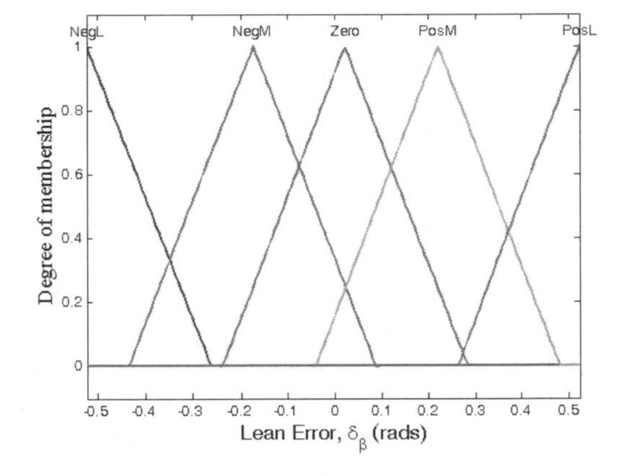

Fig. 5. Optimized membership function for lean angle error, δ_β

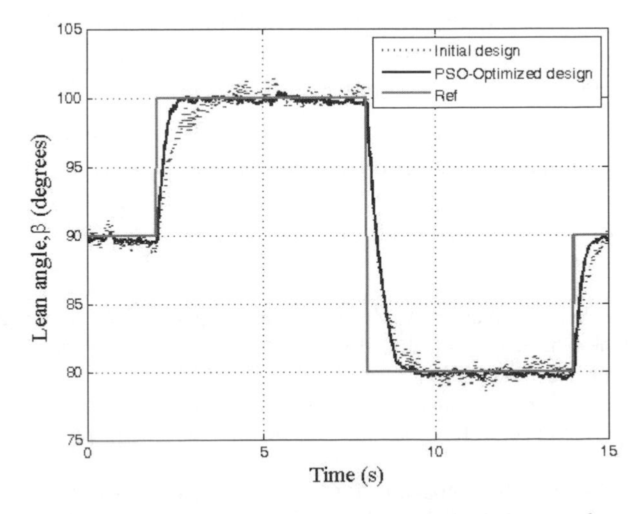

Fig. 6. Closed loop step response : gyrobot's lean angle

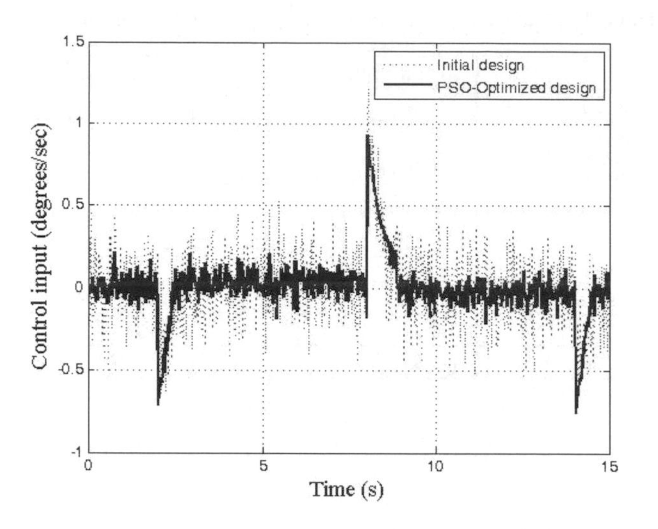

Fig. 7. Closed loop step response: control input or tilt velocity

5 Conclusion

Fuzzy logic controller with membership functions optimized by PSO for a single wheel robot is presented in this paper. The main difficulty in designing the FLC lies with selection of membership functions, choosing MF parameters and deciding the linguistic rules. Heuristic optimization methods can be employed to decide on these issues. In this paper, bio-inspired optimization algorithm PSO is used to determine the parameters of triangular membership functions of the FLC.

Five ranges for each of the linguistic variables are used with each range defined by a triangular membership function. Dimension of the search domain is reduced by fixing the centers of two extreme MFs of each variable and by pre-defining the base supports of MFs. We also predefine the membership functions. Removing these constraints may produce better solution which will be explored in the future.

Acknowledgments. This work is supported by the NUS research grant (R-263-000-244-112). We extend thanks to the funding authority.

References

1. Saleh, T., Yap, H.H., Al-Mamun, A., Zhen, Z., Vadakkepat, P.: Design of a gyroscopically stabilized single-wheeled robot. In: IEEE Conference on Robotics, Automation and Mechatronics, pp. 904–908 (2004)
2. Brown, H.B., Xu, Y.: A single wheel gyroscopically stabilized robot. In: IEEE International Conference on Robotics and Automation, pp. 3658–3663. IEEE Press, New York (1996)
3. Xu, Y., Au, K.W.: Stabilization and path following of a single wheel robot. IEEE/ASME Transactions on Mechatronics 9(2), 407–419 (2004)

4. Al-Mamun, A., Zhu, Z., Vadakkepat, P., Lee, T.H.: Tracking control of the GYROBOT - a gyroscopically stabilized single-wheeled robot. In: The 32nd Annual Conference of the IEEE Industrial Electronics Society, IECON 2005 (2005)
5. Guo, Z., Xu, J.X., Lee, T.H.: A gain-scheduling optimal fuzzy logic controller design for unicycle. In: The IEEE/ASME International Conference on Advanced Intelligent Mechatronics, pp. 1423–1428 (2009)
6. Martinez-Marroquin, R., Castillo, O., Soira, J.: Parameter tuning of membership functions of a Type-1 and type-2 fuzzy logic controller for an autonomous wheeled mobile robot using ant colony optimization. In: 2009 IEEE Int. Conf. on Systems, Man and Cybernetics, pp. 4770–4775 (2009)
7. Sharma, K.D., Chatterjee, A., Rakshit, A.: A hybrid approach for design of stable adaptive fuzzy controllers employing Lyapunov theory and particle swarm optimization. IEEE Transactions on Fuzzy Systems 17(2), 329–342 (2009)
8. Giordano, V., Naso, D., Turchiano, B.: Combining genetic algorithm and Lyapunov-based adaptation for online design of fuzzy controllers. IEEE Transactions on System, Man and Cybernetics, Part B 36(5), 1118–1127 (2006)
9. Nandy, G.C., Xu, Y.: Dynamic model of a gyroscopic wheel. In: IEEE Int. Conf. on Robotics and Automation, pp. 2683–2688 (1998)
10. Au, K.W., Xu, Y.: Decoupled dynamics and stabilization of single wheel robot. In: IEEE/RSJ Int. Conf. on Intelligent Robots and Systems, pp. 197–203 (1999)
11. Kennedy, J., Eberhart, R.C.: Particle swarm optimization. In: IEEE Int. Conf. on Neural Network, pp. 1942–1948 (1995)
12. Trelea, I.C.: The particle swarm optimization algorithm: convergence analysis and parameter selection. Information Processing Letters 85, 317–325 (2003)
13. Kadirkamanathan, V., Selvarajah, K., Fleming, P.J.: Stability analysis of the particle dynamics in particle swarm optimizer. IEEE Transactions on Evolutionary Computation 10(3), 245–255 (2006)

Neural Network Control of Nonlinear Time-Delay System with Unknown Dead-Zone and Its Application to a Robotic Servo System

Jing Na[1,*], Guido Herrmann[2], and Xuemei Ren[1]

[1] School of Automation, Beijing Institute of Technology, Beijing, 100081, P.R. China
[2] Department of Mechanical Engineering, University of Bristol, Bristol, BS8 1TR UK
{najing2120002,xmren}@bit.edu.cn, g.herrmann@bristol.ac.uk

Abstract. An adaptive controller is proposed for a class of nonlinear systems with unknown time-varying delays and a dead-zone input. Taking the dead-zone as a part of the system dynamics, the construction of the dead-zone inverse model is not needed and thus the characteristic parameters of the dead-zone are not necessarily known. Unknown time delays are handled by introducing improved Lyapunov-Krasovskii functions, where the requirements on the delayed functions/control coefficients are further relaxed without the singularity problem. A novel high-order neural network with only a scalar weight parameter is developed to approximate unknown nonlinearities. The closed-loop system is proved to be semi-globally uniformly ultimately bounded (SGUUB). Experiments on a robotic servo system are provided to verify the reliability of the presented method.

Keywords: Adaptive control, Dead-zone, Time-delay systems, Neural Networks, Servo systems.

1 Introduction

The existence of an unknown dead-zone in the control input may severely limit the system performances, which has created significant attention in recent years in this field, such as [1]-[6]. Similarly, the existence of time-delays in the system also renders the control design much more difficult and challenging. Neural-based local linearization controls are proposed for uncertain nonlinear systems with input delay [7]-[8]. To deal with delays in system states, Lyapunov-Krasovskii functions have been widely utilized [9]-[12]. Novel integral Lyapunov functions and discontinuous functions can avoid the control singularity [9]-[11].

Inspired by previous work, this paper focuses on the adaptive tracking control design for a class of strict-feedback nonlinear time-delay systems with an

* The work was supported by National Natural Science Foundation of China (No.60974046), Royal Society (Research Grant/Round 2007/R2), and a joint grant between the National Natural Science Foundation of China and Royal Society UK under grant No.61011130163/JP090823. Jing Na was also supported by Chinese Scholarship Council (No.2008603002).

P. Vadakkepat et al. (Eds.): FIRA 2010, CCIS 103, pp. 338–345, 2010.

unknown dead-zone input and time-varying delays. The control singularity problem and the unknown time-delays are all handled by introducing an improved Lyapunov-Krasovskii function. Compared with some existing results on the adaptive control for time-delay systems [9], [10], the slightly restrictive information on the bounds of the delayed functions or control coefficients is not required [13], [15]. The unknown dead-zone is taken as a part of the system dynamics, and then handled without using any characteristic parameter and inverse dead-zone model. Moreover, a novel class of neural networks, which have a simpler structure and only a scalar parameter independent of the number of hidden nodes, are developed to approximate unknown nonlinear functions.

From the viewpoint of robotic application, turntable systems have significant importance. They have been widely used in aerospace systems, industrial production (e.g. turntable-based robots supporting plasma spraying), even biological science and for the world's largest antenna measuring range [16]. These robotic servo systems are inevitably affected by dead-zone or backlash nonlinearities and common teleoperation introduce unavoidable transmission delays. Thus, the application of our control approach to a turntable system is topical and important.

2 Problem Statement and Preliminaries

Consider the nonlinear system with time-delays as

$$\begin{cases} \dot{x}_i(t) = f_i(\bar{x}_i) + g_i(\bar{x}_i)x_{i+1}(t) + h_i(t, \bar{x}_i(t - \tau_i(t))), \ 1 \le i \le n-1 \\ \dot{x}_n(t) = f_n(x) + g_n(x)u(t) + h_n(t, x(t - \tau_n(t))) \\ y(t) = x_1(t) \end{cases} \tag{1}$$

where $\bar{x}_i = [x_1, x_2 \cdots x_i]^T \in \mathrm{R}^i, i = 1, \cdots n$, $x = [x_1, x_2 \cdots x_n]^T \in \mathrm{R}^n$, $y(t) \in \mathrm{R}$ are state variables, and the controlled output, respectively; nonlinear functions $f_i(\cdot), g_i(\cdot), h_i(\cdot), i = 1, \cdots n$ are unknown but smooth; the functions $h_i(t, \bar{x}_i(t - \tau_i(t))), 1 \le i \le n$ may contain multiple delays, e.g.

$$h_i(t, \bar{x}_i(t - \tau_i(t))) = \sum\nolimits_{j=1}^{m_i} h_{ij}(t, \bar{x}_i(t - \tau_{ij}(t))),$$

to cover more general cases compared with [9]-[12], etc. The scalars $\tau_{ij}(t)$ are unknown and time-varying delays bounded by $\tau_{ij}(t) \le \tau_{im}$ and $\dot{\tau}_{ij}(t) \le \bar{\tau}_i < 1, i = 1, \cdots n, j = 1, \cdots m_i$. The control $u(t) \in \mathrm{R}$ is the output of an unknown dead-zone as

$$u(t) = D(v(t)) = \begin{cases} d_r(v(t) - b_r) & if \ v \ge b_r \\ 0 & if \ -b_l < v < b_r \\ d_l(v(t) + b_l) & if \ v \le -b_l \end{cases} \tag{2}$$

where d_r,d_l and b_l, b_r are the slopes and breakpoints of the dead-zone. The objective is to find a control $v(t)$ for system (1) with the dead-zone (2), such that $y(t)$ follows a reference $y_d(t)$.

Similar to [3], we can rewrite the dead-zone equation (2) as

$$u(t) = D(v(t)) = d(t)v(t) + \rho(t) \tag{3}$$

where $d(t) = \begin{cases} d_l & if\ v(t) \le 0 \\ d_r & if\ v(t) > 0 \end{cases}$, $\rho(t) = \begin{cases} -d_r b_r & if\ v \ge b_r \\ -d(t)v(t) & if -b_l < v < b_r \\ d_l b_l & if\ v \le -b_l \end{cases}$.

It is known that $\rho(t)$ is bounded by a positive constant $p = \max\{d_l, d_r\} \cdot \max\{b_r, b_l\}$ as $|\rho(t)| \le p$.

Assumption 1. The delayed functions $h_i(t, \bar{x}_i)$ fulfill $|h_{ij}(t, \bar{x}_i)| \le k_{ij}(\bar{x}_i)$, where $k_{ij}(\bar{x}_i) \ge 0$ are bounded, but not necessarily known functions on a large enough compact set C_i.

Assumption 2. The signs of functions $g_i(\cdot), i = 1, \cdots n$ are known, and there exist constants g_{i0} and g_{i1}, such that $0 < g_{i0} \le |g_i(\cdot)| \le g_{i1}$. Without loss of generality, it is assumed $0 < g_{i0} \le g_i(\cdot) \le g_{i1}$.

Assumption 3. Dead-zone parameters d_l,d_r,b_l, b_r are unknown but positive, and there exists a positive constant ℓ such that $\ell < d_l$ and $\ell < d_r$.

Remark 1. The bounded functions $k_{ij}(\bar{x}_i)$ or other restrictive assumptions on the delayed functions $h_{ij}(\cdot)$ in [9]-[13] are not utilized in this paper, and the constants g_{i0}, g_{i1} are only used for analytical purpose, which are assumed to be precisely known constants or smooth functions in [9], [11]. Moreover, the dead-zone model (2) studied here may be nonsymmetrical rather than the special symmetric case in [1]-[3], i.e. $b_r = b_l$, and the constants ℓ and p in Assumption 3 are not necessarily known.

3 Neural-Network Based Adaptive Control Design

High-order neural networks (HONN) [14] can approximate a nonlinear function $Q(Z)$ on a compact set Ω as

$$Q(Z) = W^{*T}\Phi(Z) + \varepsilon,\ \forall Z \in \Omega \subset \mathrm{R}^n \tag{4}$$

where the weights $W^* = [w_1^*, w_2^* \cdots w_L^*]^T \in \mathrm{R}^L$ and error $\varepsilon \in \mathrm{R}$ are bounded, $\Phi(Z) = [\Phi_1(Z), \cdots \Phi_L(Z)]^T \in \mathrm{R}^L$; is a basis vector with its element as $\Phi_k(Z) = \prod_{j \in J_k} [\sigma(Z_j)]^{d_k(j)}, k = 1, \ldots, L$, where J_k are collections of L not ordered subsets of $\{0, 1, \ldots, n\}$, and $d_k(j)$ are nonnegative integers. The function $\sigma(x)$ is a sigmoid function.

For controller design, define: $z_1 = x_1 - y_d$ and $z_i = x_i - \alpha_{i-1}, i = 2, \cdots, n$, where α_i is the virtual control for each sub-system of (1) specified as

$$\alpha_i = -k_i z_i - \frac{\hat{\theta}_i}{2} z_i \Phi_i^T(Z_i)\Phi_i(Z_i) - \hat{\varepsilon}_i \tanh(\frac{z_i}{\omega_i}), i = 1, \cdots, n-1 \tag{5}$$

where k_i, $\omega_i > 0$ are positive design parameters, $\Phi_i(Z_i)$ is a neural network basis function vector as in (4), to be defined via system nonlinearities in each step of the backstepping design procedure. The adaptive parameters $\hat{\theta}_i$ and $\hat{\varepsilon}_i$ in (5) are given by

$$\dot{\hat{\theta}}_i = \frac{\Gamma_i}{2}[z_i^2 \Phi_i^T(Z_i)\Phi_i(Z_i) - \sigma_i\hat{\theta}_i], \ i = 1, \cdots, n \tag{6}$$

$$\dot{\hat{\varepsilon}}_i = \Gamma_{ai}[z_i \tanh(\frac{z_i}{\omega_i}) - \sigma_{ai}\hat{\varepsilon}_i], \ i = 1, \cdots, n \tag{7}$$

with $\Gamma_i > 0$, $\Gamma_{ai} > 0$ and $\sigma_i > 0$, $\sigma_{ai} > 0$ being the design parameters. The choice of these parameters and in particular for the basis function vector, $\Phi_i(Z_i)$, will become evident in the following design procedure which is provided to deduce the required control v.

Step i $(1 \le i < n)$: Consider $z_i = x_i - \alpha_{i-1}$, we have

$$\dot{z}_i = \dot{x}_i - \dot{\alpha}_{i-1} = f_i(\bar{x}_i) + g_i(\bar{x}_i)(z_{i+1} + \alpha_i) + h_i(t, \bar{x}_i(t - \tau_i(t))) - \dot{\alpha}_{i-1} \tag{8}$$

The scalar $\dot{\alpha}_{i-1}$ is a function of $\bar{x}_{i-1}, y_d, \hat{\varepsilon}_1, \cdots, \hat{\varepsilon}_{i-1}, \hat{\theta}_1, \cdots, \hat{\theta}_{i-1}$, which can be represented as $\dot{\alpha}_{i-1} = \sum_{k=1}^{i-1} \frac{\partial \alpha_{i-1}}{\partial x_k}\dot{x}_k + \varphi_{i-1}$ with $\varphi_{i-1} = \frac{\partial \alpha_{i-1}}{\partial y_d}\dot{y}_d + \sum_{k=1}^{i-1} \frac{\partial \alpha_{i-1}}{\partial \hat{\theta}_k}\dot{\hat{\theta}}_k + \sum_{k=1}^{i-1} \frac{\partial \alpha_{i-1}}{\partial \hat{\varepsilon}_k}\dot{\hat{\varepsilon}}_k$, which is computable as explained in [9]-[11]. Then $\dot{\alpha}_{i-1}$can be considered as a function of $\bar{x}_i, \partial\alpha_{i-1}/\partial x_1, \cdots, \partial\alpha_{i-1}/\partial x_{i-1}, \phi_{i-1}$. Select the Lyapunov-Krasovskii function as:

$$\begin{aligned} V_i = {} & \frac{1}{2}z_i^2 + \frac{c_{i1}}{2}\sum_{j=1}^{m_i} \frac{e^{\varpi\tau_{im}}}{1-\bar{\tau}_i}\int_{t-\tau_{ij}(t)}^{t} e^{-\varpi(t-\varsigma)}k_{ij}^2(\bar{x}_i(\varsigma))d\varsigma + \frac{g_{i0}}{2\Gamma_i}\tilde{\theta}_i^2 \\ & + \frac{1}{2g_{i0}\Gamma_{ai}}(\varepsilon_i^* - g_{i0}\hat{\varepsilon}_i)^2 \end{aligned} \tag{9}$$

where $c_{i1} > 0$ is positive constant and g_{i0} is the lower bound of $g_i(x_i)$. The time derivative of V_i along (5) $\sim$ (7) and (8) is

$$\begin{aligned} \dot{V}_i \le {} & \frac{m_i}{2c_{i1}}z_i^2 + z_iQ_i(Z_i) + g_i(\bar{x}_i)z_iz_{i+1} + g_i(\bar{x}_i)z_i\alpha_i \\ & + \frac{c_{i1}}{2}\left(1 - 2\tanh^2(\frac{z_i}{\omega_i})\right)\sum_{j=1}^{m_i} \frac{e^{\varpi\tau_{im}}}{1-\bar{\tau}_i}k_{ij}^2(\bar{x}_i) - \varpi V_{di} - \frac{g_{i0}}{2}\tilde{\theta}_iz_i^2\Phi_i^T(Z_i)\Phi_i(Z_i) \\ & + \frac{\sigma_ig_{i0}}{2}\tilde{\theta}_i\hat{\theta}_i - (\varepsilon_i^* - g_{i0}\hat{\varepsilon}_i)z_i\tanh(\frac{z_i}{\omega_i}) + \sigma_{ai}(\varepsilon_i^* - g_{i0}\hat{\varepsilon}_i)\hat{\varepsilon}_i \end{aligned} \tag{10}$$

where $Q_i(Z_i) = f_i(\bar{x}_i) + \frac{c_{i1}}{z_i}\tanh^2(\frac{z_i}{\omega_i})\sum_{j=1}^{m_i}\frac{e^{\tau_{im}}}{1-\bar{\tau}_i}k_{ij}^2(\bar{x}_i) - \dot{\alpha}_{i-1}$ is an unknown but smooth function of $Z_i = [\bar{x}_i, z_i, \partial\alpha_{i-1}/\partial x_1, \cdots, \partial\alpha_{i-1}/\partial x_{i-1}, \phi_{i-1}] \in \mathrm{R}^{2i+1}$, which can be approximated by HONN (4) over the compact set Ω, so that $Q_i(Z_i) = W_i^{*T}\Phi(Z_i) + \varepsilon_i$. It follows

$$\dot{V}_i \le -\gamma_iV_i + \vartheta_i + c_{i2}g_{i1}^2z_{i+1}^2 + \frac{c_{i1}}{2}\left(1 - 2\tanh^2(\frac{z_i}{\omega_i})\right)\sum_{j=1}^{m_i}\frac{e^{\varpi\tau_{im}}k_{ij}^2(\bar{x}_i)}{1-\bar{\tau}_i} \tag{11}$$

where g_{i1} is the upper bounds of $g_i(\bar{x}_i)$, γ_i and ϑ_i are positive constants (if k_i is selected large enough), which are given as

$$\gamma_i = \min\{2(g_{i0}k_i - m_i/2c_{i1} - 1/4c_{i2}), \Gamma_i\sigma_i/2, \Gamma_{ai}\sigma_{ai}, \varpi\},$$

$$\vartheta_i = \sigma_i g_{i0}\theta_i^{*2}/4 + 1/2g_{i0} + \sigma_{ai}\varepsilon_i^{*2}/2g_{i0} + 0.2785\omega_i\varepsilon_i^*.$$

If z_{i+1} is bounded (This will be guaranteed in the next step.), then according to the extended Lyapunov Theorem [7], z_i, $\tilde{\theta}_i$, $\tilde{\varepsilon}_i$ are SGUUB for small enough ϑ_i and $c_{i2}g_{i1}^2(\bar{x}_i)z_{i+1}^2$, or large γ_i on a compact set C_i. This further implies the boundedness of $\hat{\theta}_i$, $\hat{\varepsilon}_i$, and α_i.

Step n: This is the last step, in which the control v is obtained. Consider $z_n = x_n - \alpha_{n-1}$, then

$$\dot{z}_n = \dot{x}_n - \dot{\alpha}_{n-1} = f_n(x) + g_n(x)\left(d(t)v(t) + \rho(t)\right) + h_n(t, x(t-\tau_n(t))) - \dot{\alpha}_{n-1} \quad (12)$$

where $\dot{\alpha}_{n-1}$ is a function of $x, y_d, \partial\alpha_{n-1}/\partial x_1, \cdots, \partial\alpha_{n-1}/\partial x_{n-1}, \varphi_{n-1}$, in which φ_{n-1} has a similar structure as in (8). Then the final control v is:

$$v = -k_n z_n - \frac{\hat{\theta}_n}{2} z_n \Phi_n^T(Z_n)\Phi_n(Z_n) - \hat{\varepsilon}_n \tanh(\frac{z_n}{\omega_n}) \quad (13)$$

where $\hat{\theta}_n$, $\hat{\varepsilon}_n$ are adaptive parameters specified in (6) and (7). The following Lyapunov function is selected:

$$\begin{aligned} V_n = &\frac{1}{2}z_n^2 + \frac{c_{n1}}{2}\sum_{j=1}^{m_n}\frac{e^{\varpi\tau_{nm}}}{1-\bar{\tau}_n}\int_{t-\tau_{nj}}^{t} e^{-\varpi(t-\varsigma)}k_{nj}^2(\bar{x}(\varsigma))d\varsigma + \frac{g_{n0}\ell}{2\Gamma_n}\tilde{\theta}_n^2 \\ &+ \frac{1}{2g_{n0}\ell\Gamma_{an}}(\varepsilon_n^* - g_{n0}\ell\hat{\varepsilon}_n)^2 \end{aligned} \quad (14)$$

with $c_{n1} > 0$ is a design parameter and $\hat{\varepsilon}_n$ is the estimation of the error bounds $\varepsilon_n^* = \varepsilon_n^* + g_{n1}p$. The time derivative of $V_{cn} + V_{dn}$ along (12) can be given as

$$\begin{aligned} \dot{V}_{cn} + \dot{V}_{dn} \leq &\frac{m_n}{2c_{n1}}z_n^2 + g_n(x)dz_n v + z_n Q_n(Z_n) + g_{n1}p\,|z_n| - \varpi V_{dn} \\ &+ \frac{c_{n1}}{2}\left(1 - 2\tanh^2(\frac{z_n}{\omega_n})\right)\sum_{j=1}^{m_n}\frac{e^{\varpi\tau_{nm}}}{1-\bar{\tau}_n}k_{nj}^2(x) \end{aligned} \quad (15)$$

where $Q_n(Z_n) = f_n(x) + \frac{c_{n1}}{z_n}\tanh^2(\frac{z_n}{\omega_n})\sum_{j=1}^{m_n}\frac{e^{\varpi\tau_{nm}}}{1-\bar{\tau}_n}k_{nj}^2(x) - \dot{\alpha}_{n-1}$ is an unknown but smooth function of $Z_n = [x, z_n, \partial\alpha_{n-1}/\partial x_1, \cdots, \partial\alpha_{n-1}/\partial x_{n-1}, \phi_{n-1}] \in \mathrm{R}^{2n+1}$ and estimated by a HONN (4) over the compact set Ω as $Q_n(Z_n) = W_n^{*T}\Phi(Z_n) + \varepsilon_n$. The time derivative of $V_{wn} + V_{an}$ along (6) and (7) is

$$\begin{aligned} \dot{V}_{wn} + \dot{V}_{an} \leq &-\frac{g_{n0}\ell}{2}\tilde{\theta}_n z_n^2\Phi_n^T(Z_n)\Phi_n(Z_n) + \frac{\sigma_n g_{n0}\ell}{2}\tilde{\theta}_n\hat{\theta}_n \\ &- (\varepsilon_n^* - g_{n0}\ell\hat{\varepsilon}_n)z_n\tanh(\frac{z_n}{\omega_n}) + \sigma_{an}(\varepsilon_n^* - g_{n0}\ell\hat{\varepsilon}_n)\hat{\varepsilon}_n \end{aligned} \quad (16)$$

Omitting some intermediate steps, it can be obtained that

$$\dot{V}_n \leq -\gamma_n V_n + \vartheta_n + \frac{c_{n1}}{2}\left(1 - 2\tanh^2(\frac{z_n}{\omega_n})\right)\sum_{j=1}^{m_n}\frac{e^{\varpi\tau_{nm}}}{1-\bar{\tau}_n}k_{nj}^2(x) \quad (17)$$

where γ_n and ϑ_n are positive constants (if k_n is set large enough) given as

$$\gamma_n = \min\{2(g_{n0}k_n\ell - m_n/2c_{n1}), \Gamma_n\sigma_n/2, \Gamma_{an}\sigma_{an}, \varpi\},$$

$$\vartheta_n = \sigma_n g_{n0}\ell\theta_n^{*2}/4 + 1/2g_{n0}\ell + \sigma_{an}\varepsilon_n^{*2}/2g_{n0}\ell + 0.2875\varepsilon_n^*\omega_n.$$

From (17) and according to the extended Lyapunov Theorem [7], z_n, $\tilde{\theta}_n$, $\tilde{\varepsilon}_n$ are SGUUB for small ϑ_n and c_{n1} or large γ_n over compact set C_n, which further implies the boundedness of $\hat{\theta}_n$, $\hat{\varepsilon}_n$. Moreover, the control v and the system state x_n are all bounded. From the above analysis, we know that z_n is proven to be bounded in the last Step n, and the extra term $g_{(n-1)1}^2 z_n^2$ in Step (n-1) is bounded. Thus the boundedness of z_{n-1} is proved according to the design in Step (n-1). Applying the fact that z_n is bounded for (n-1) times backward as shown in [14], it is readily concluded that the iterative design procedure can guarantee the boundedness of z_i,$\tilde{\theta}_i$, $\tilde{\varepsilon}_i$, and consequently the boundedness of $\hat{\theta}_i$, $\hat{\varepsilon}_i$, x_i and α_i.

Remark 2. The scalar $\theta_i^* = W_i^{*T}W_i^*$ is introduced as the weight of HONN (4), and only $\hat{\theta}_i$ is updated on-line such that the computation cost can be reduced significantly compared to other NN-based control schemes. The function $\frac{c_{i1}}{z_i}\tanh^2(\frac{z_i}{\omega_i})\sum_{j=1}^{m_i}\frac{e^{\varpi\tau_{im}}}{1-\bar{\tau}_i}k_{ij}^2(\bar{x}_i(t))$ is utilized as a part of $Q_i(Z_i)$ in each step, thus the possible singularity at $z_1 = 0$ can be avoided as $Q(Z_i)$ is well-defined anywhere. Consequently, the discontinuous function (e.g. [9], [11], [13]) or other restrictive assumptions on $h_i(\cdot)$ in [12] are avoided.

Remark 3. The construction of the inverse dead-zone model (See [1]-[2] and [6]) is not needed to achieve dead-zone identification. Moreover, the control gains with integration ([9], [11], [13]) can be replaced by constant gains k_i with the help of improved Lyapunov-Krasovskii functions V_{di} in (9) and (14).

4 Practical Results on a Robotic Servo System

To show the applicability of the proposed control, a robotic two-axis turntable servo system is employed. As shown in Fig. 1, the test-rig consists of a servo turntable, optical encoders, PWM amplifiers, Digital Signal Processing (DSP, TMS3202812) and PC. The turntable platform is driven by an AC motor (HC-UFS13) to track a given reference angle. This turntable servo system can be generally modeled as

$$M(q)\ddot{q} + C(q,\dot{q})\dot{q} + G(q) + D(q(t-\tau),\dot{q}(t-\tau)) = u \quad (18)$$

where $q, \dot{q} \in \mathrm{R}$ are position and velocity, respectively, $M(q)$ is the unknown positive inertia matrix, $C(q,\dot{q})\dot{q}$ is the unknown Coriolis and centrifugal torques, $G(q)$ is the unknown gravity, $D(q(t-\tau),\dot{q}(t-\tau))$ denotes the uncertainty and disturbances caused by the possible delays τ, and $u \in \mathrm{R}$ is the control input voltage perturbed by an unknown dead-zone. It is noted that system (18) can be represented as a special case of (1) and thus can be controlled via the proposed control.

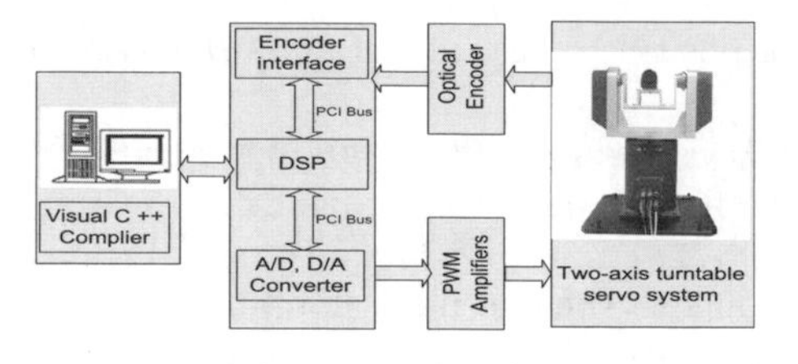

Fig. 1. Diagram of turntable servo test-rig

The proposed controller is implemented through a DSP via a C-program in CCS3.0. The sampling rate is 0.01s. The control parameters are specified as $k_1 = 200, k_2 = 0.4$, $\omega_1 = \omega_2 = 1$, $\Gamma_{a1} = \Gamma_{a2} = \Gamma_1 = \Gamma_2 = 0.0001$, $\sigma_1 = \sigma_2 = \sigma_{a1} = \sigma_{a2} = 0.01$, and HONN parameters are chosen as $L_1 = L_2 = 8$, $\sigma(x) = 2/(1+e^{-0.01x}) - 1$. The initial condition for experiment is set as $\hat{\theta}_1(0) = \hat{\theta}_2(0) = 0$, $\hat{\varepsilon}_1(0) = \hat{\varepsilon}_2(0) = 0$.

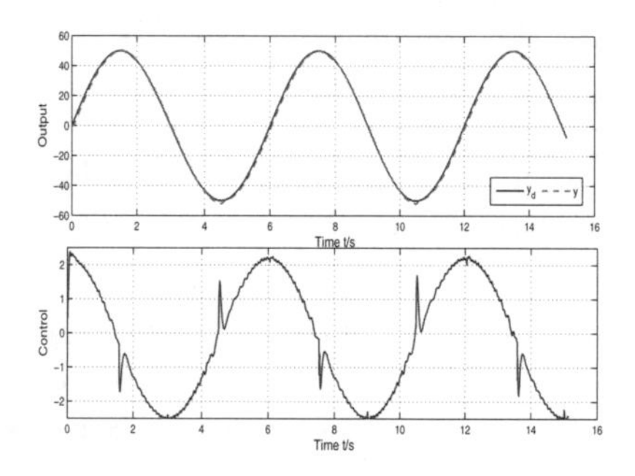

Fig. 2. Experiment result for $q_d(t) = 50\sin(0.5\pi t)$

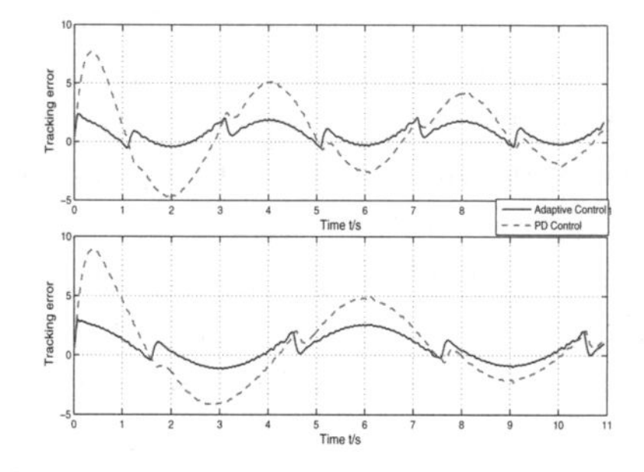

Fig. 3. Comparative tracking errors

Fig. 2 provides the corresponding experimental results for a reference signal $q_d(t) = 50\sin(0.5\pi t)$. In these figures, the top subplot shows the output tracking performance and the bottom one illustrates the practical control signal $v(t)$. It can be observed clearly that the presented control using the backstepping design has a satisfactory tracking performance. For comparison, a simple PD controller is also tested with $k_p = 0.4; k_d = 0.08$. It is shown in Fig.3 that the proposed control can provide a superior transient and steady-state performance due to the adaptive compensation of the unknown dynamics.

5 Conclusions

A NN-based control is studied for a class of nonlinear systems with unknown time-varying multiple delays and unknown dead-zone input. By introducing a novel Lyapunov-Krasovskii function without integral operation, the requirement on the upper bounds or other assumptions of the delayed functions can be relaxed. The possible control singularity problem and the discontinuous control are also avoided. A novel HONN with only one on-line updating parameter is developed for decoupled backstepping design. No characteristic parameter of the dead-zone and its inverse model are utilized, which renders the proposed control more practically suitable in the implementation. Practical experiments on a turntable servo test-rig show that the proposed control provides superior tracking performance than general PD control.

References

1. Tao, G., Kokotovic, P.V.: Adaptive Control of Systems with Actuator and Sensor Nonlinearities. Wiley, New York (1996)
2. Tao, G., Lewis, F.L.: Adaptive Control of Nonsmooth Dynamic Systems. Springer, London (2003)
3. Ibrir, S., Xie, W.F., Su, C.Y.: Adaptive tracking of nonlinear systems with non-symmetric dead-zone input. Automatica 43, 522–530 (2007)
4. Selmic, R.R., Lewis, F.L.: Deadzone compensation in motion control systems using neural networks. IEEE. Trans. Automat. Contr. 45, 602–613 (2000)
5. Zhang, T.P., Ge, S.S.: Adaptive dynamic surface control of nonlinear systems with unknown dead zone in pure feedback form. Automatica 44, 1895–1903 (2008)
6. Zhou, J.: Decentralized adaptive control for large-scale time-delay systems with dead-zone input. Automatica 44, 1790–1799 (2008)
7. Huang, J.Q., Lewis, F.L.: Neural-network predictive control for nonlinear dynamic systems with time delay. IEEE Trans. Neural Netw. 14, 377–389 (2003)
8. Na, J., Ren, X.M., Huang, H.: Time-delay positive feedback control for nonlinear time-delay systems with neural network compensation. ACTA Automat. Sin. 34, 1196–1203 (2008)
9. Ge, S.S., Hong, F., Lee, T.H.: Adaptive neural network control of nonlinear systems with unknown time delays. IEEE Trans. Automat. Contr. 48, 2004–2010 (2003)
10. Hong, F., Ge, S.S., Lee, T.H.: Practical adaptive neural control of nonlinear systems with unknown time delays. IEEE Trans. Syst., Man, Cybern. B, Cybern. 35, 849–854 (2005)
11. Ge, S.S., Hong, F., Lee, T.H.: Adaptive neural control of nonlinear time-delay systems with unknown virtual control coefficients. IEEE Trans. Syst., Man, Cybern. B, Cybern. 34, 499–516 (2004)
12. Wang, M., Chen, B., Shi, P.: Adaptive neural control for a class of perturbed strict-feedback nonlinear time-delay systems. IEEE Trans. Syst., Man, Cybern. B, Cybern. 38, 721–730 (2008)
13. Zhang, T.P., Ge, S.S.: Adaptive neural control of MIMO nonlinear state time-varying delay systems with unknown dead-zones and gain signs. Automatica 43, 1021–1033 (2007)
14. Kosmatopoulos, E.B., Polycarpou, M.M., Christodoulou, M.A., Ioannou, P.A.: High-order neural network structures for identification of dynamical systems. IEEE Trans. Neural Netw. 6, 422–431 (1995)
15. Ge, S.S., Tee, K.P.: Approximation-based control of nonlinear MIMO time-delay systems. Automatica 43, 31–43 (2007)
16. The World's Largest Antenna Measuring Range: Astrium and EurasSpace Deliver Compact Range to China, `http://classic.eads.net/1024/en/pressdb/archiv/2003/2003/en_20030321_euras_e.html`

The Computational Accuracy of Cover Time for Circular Sensing Range

Muhamad Azfar Ramli and Gerard Leng

Cooperative Systems Lab E1-03-06,
Department of Mechanical Engineering,
National University of Singapore,
1 Engineering Drive 2,
Singapore 117576

Abstract. Sensor coverage completion is usually tracked by use of a coverage map and discretizing the target region into smaller square cells or pixels. However, real sensors are typically fixed in range thus creating circular sensing regions. In this paper, we quantitatively discuss the effect of the size of the discretization of the domain on the accuracy of the covering time of a simple random walk algorithm. A simulation algorithm is proposed using coordinate pairs to track the covered regions in order to minimise memory usage. Finally we present how accurate estimates of mean and standard deviation of cover time can be obtained by comparing between simulation results with varying discretization ratio and the domain size.

Keywords: Random Coverage, Discretization Error.

1 Introduction

A sensor's 'footprint' is the region of the target domain within its detection range. Complete coverage of the target domain can be defined as the point in time when the combined union of all the footprints prior to that time is exactly equal to the target domain area. Specifically, we are interested in the first time that this happens and we call this the cover time. Coverage problems have important applications in unknown terrain exploration, search for multiple unknown targets and even simple applications such as floor cleaning and spray painting of car panels. A literature survey conducted reveals much work has been dedicated to finding efficient methods of covering the target region. Choset [2] has summarized much of the earlier work done and classifies them into various categories. In his paper, he notes that different decompositions may be taken as approximate or exact depending on the coverage algorithm and the sensor footprint used. Other approaches to coverage problems mostly focuses on designing efficient algorithms [1] [4]and finding optimal paths [7] and implementing multiple robots [6] for cooperative coverage such as to further decrease the cover time required.

The drawback of many of these algorithms is that most of them assume complete knowledge of the domain. This may not be a realistic situation in the case

P. Vadakkepat et al. (Eds.): FIRA 2010, CCIS 103, pp. 346–352, 2010.

of robotic search and exploration of unknown terrain. Random or heuristics approaches [3] [8] have been proposed earlier and are more practical for immediate usage for coverage operations. Integrating multiple sensors into such algorithms is also much easier as compared to deterministic algorithms. The main drawback of these methods is the obvious inefficiency of the method as compared to motion path planning and the fact that it is much harder to predict the performance of randomized methods.

It is decidedly difficult to develop accurate mathematical models predicting the performance of such robots. Wagner [8] derived performance bounds for such a random algorithm but does not give information on the distribution of the cover time. As such, we rely on computer simulations to try and predict the performance of realistic random covering algorithms. In this paper we compute the cover time using techniques based on raster graphics and bitmaps and discuss the accuracy of such a method. Our study will focus on circular sensor footprints even though the algorithm can accommodate other shapes.

2 Covering Algorithm

The proposed continuous space covering algorithm that we will analyse is similar to Wagner's [8] Probabilistic Covering (PC) algorithm on a continuous two dimensional domain for a ground robot with a proximity sensor of fixed range and a compass module. The target domain is assumed to be rectangular for simplicity. A pseudo-coverage algorithm is as follows: -

1) Sweep sensor in a disc around agent
2) Determine accessible positions in a 360° sweep (presence of obstacles)
3) Randomly select one of the unobstructed directions around sensor agent (uniformly distributed among all available positions)
4) Move in the direction selected in step 3 for distance equal to radius of sensor
5) Repeat Steps 1 - 4

Figure 1 shows an example of the movement of the sensor agent using the PC algorithm. Note the red region represents the area which is inaccessible due to the presence of the wall.

The main difference between our covering algorithm and Wagner's [8] PC algorithm is that the PC algorithm always takes a random angle between 0 to 360°. It adjusts to compensate for the boundaries by taking half the minimum distance to the nearest wall so as to ensure that the agent does not move into the wall. Our covering algorithm always moves the same distance for each time step however it avoids the walls simply by changing the range of the random angle that it moves towards in the new time step. The main advantage of this method is that if the agent were to move half a step towards the wall, there would clearly be no new area to cover. By avoiding all the angles that are pointed towards the walls, the agent has a higher probability of moving to a new uncovered space.

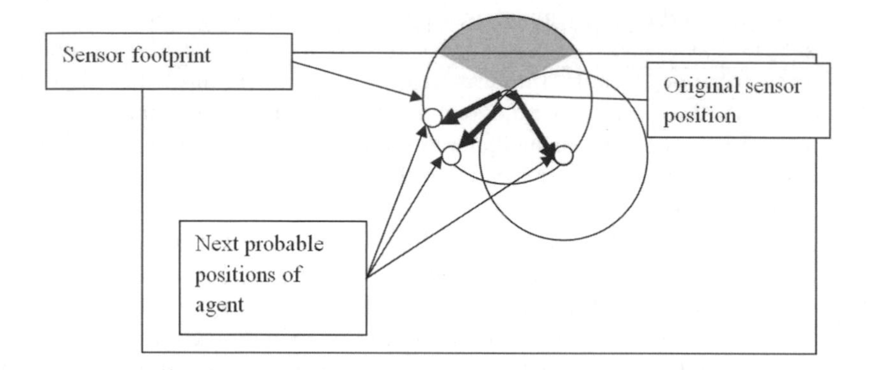

Fig. 1. Probabilistic Coverage Algorithm

3 Source of Discretization Error

A simple and direct method of tracking the coverage domain is by discretizing the domain into smaller squares ("pixels"). As the sensor moves across the domain, pixels that fall within the sensor footprint are deemed "covered" and the simulation proceeds until all the pixels are covered. However due to the fact that the coverage algorithm involved circular sensor footprints and that small pixel size are required for accurate representation, there arises a question on how small should the pixels be such as to give an accurate estimate of the cover time.

The error arises from the discrepancy between the actual sensor footprint and its representation on the pixel grid. In our algorithm, we use Bresenham's circle algorithm to discretize the sensor footprint [5]. We denote the grid discretization ratio as $\zeta = \frac{x}{d}$, where x is the length of each pixel and d is the diameter of the sensor footprint. The percentage discretization error is denoted as ϵ.

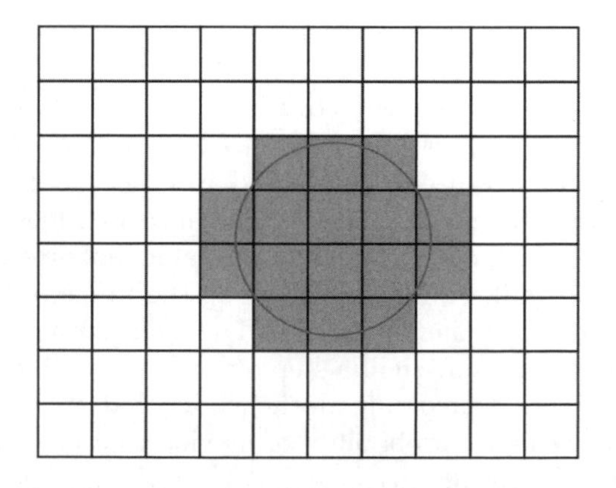

Fig. 2. Discretization error due to Circle drawing algorithm

Although it is possible to calculate the exact theoretical errors obtained from using the algorithm, we simply run the program for different values of and calculate the area obtained by counting out the number of pixels covered and subtracting the area of the circular footprint to find the error obtained. The results are shown in table 1 below: -

Table 1. ϵ vs ζ used for circle drawing

Actual footprint area with unit diameter $= \frac{\pi(1)^2}{4} \approx 0.785398$

ζ	No. of squares covered	Area covered $=\zeta^2\times$ no. of squares	$\epsilon(\%)$
0.1	89	0.89	13.32
0.05	341	0.8525	8.544
0.01	7997	0.7997	1.821
0.005	31689	0.7922	0.8692
0.001	786805	0.7868	0.1791

Note that that the variation of ϵ with ζ closely resembles a linear relationship.

4 The Coordinate Pairs Coverage Tracking Algorithm

We propose a simple method which retains the accuracy of directly tracking all squares within the domain while minimising both memory and computation required. This is by tracking the total area in the form of coordinate pairs. The idea arises from the simple fact that the process of area updating will only concern the boundaries of the area covered as well as the edges of the new sensor footprint to be updated in the memory. It is therefore unnecessary to store all the points; instead the area will be stored by a set of coordinate pairs. Each coordinate pair consists of two x coordinates and a y coordinate representing a strip of area covered in the form of (x1, x2; y). The total set of all the coordinate pairs stored therefore represent a region of covered area.

In Figure 3, the coordinates marked with O are the only ones that are required to be stored in memory in the form of the following coordinate pairs (4,6;2),(3,7;3),(3,7;4),(4,6;5). The process of combining the current set of coordinate pairs with a new update from the next time step is simple, each set of coordinate pairs are compared and coordinates that are found within the cancellation zones are simply deleted. This is represented by the coordinates marked with an x. The set is now updated to (4,6;2),(2,7;3),(1,7;4),(1,6;5),(2,5;6). The process will continue to update until the set becomes (1,9;1), (1,9;2),(1,9;3),(1,9;4), (1,9;5),(1,9;6),(1,9;7) which indicates that coverage has been completed.

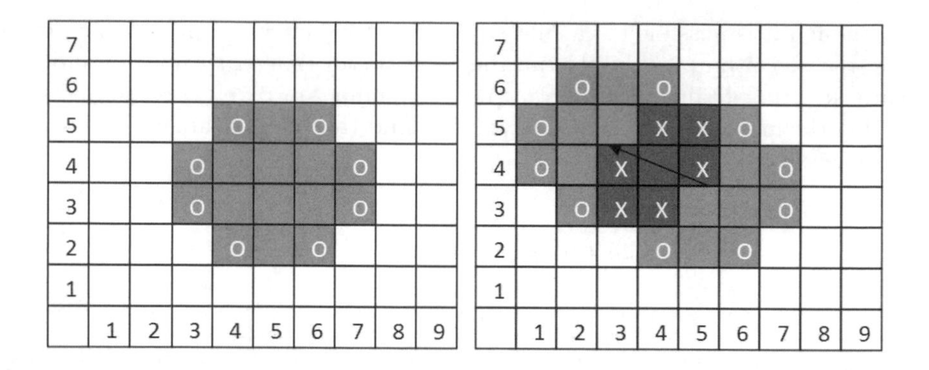

Fig. 3. The Coordinate Pairs Tracking Algorithm

5 Results and Discussion

60000 simulations were run on 8 parallel computers to ensure that the width of the 99% confidence interval for the estimation of the mean and variance is approximately within 1 time step. The same numbers of samples were then run for each set of values to ensure consistency. The first set of simulations was run in order to investigate the predicted error in the simulated cover time as the discretization ratio ζ is varied. A square domain with length of 4 x diameter of sensor footprint is used and the initial position of the sensor is randomly generated with a uniform distribution within the domain. The results are shown in Figure 5.

An estimate of the accurate mean and standard deviation of the cover time can be obtained simply by extrapolation of the graphs and finding the intercept on the vertical axis denoting the point where the theoretical value of $\zeta = 0$ which gives us a infinitely small grid size. Running a set of simulations with a coarse grid size takes a much shorter time to complete yet by combining several of the results in this way, the accuracy of the estimation can be maintained. The error % can therefore be calculated by subtracting the simulated cover time

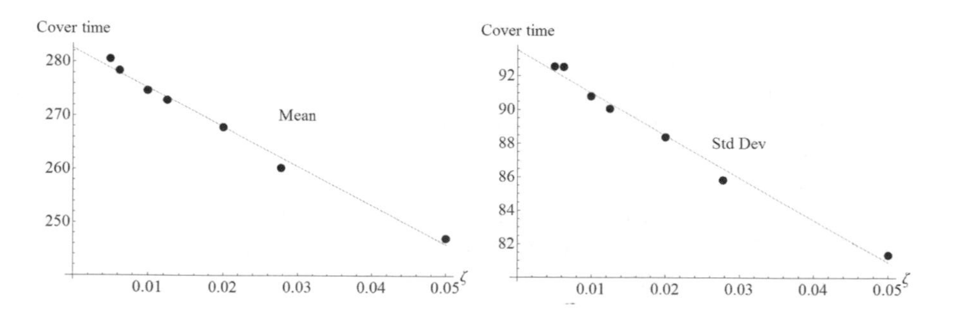

Fig. 4. Results of cover time vs ζ for domain size of 4 x 4 units of sensor diameter

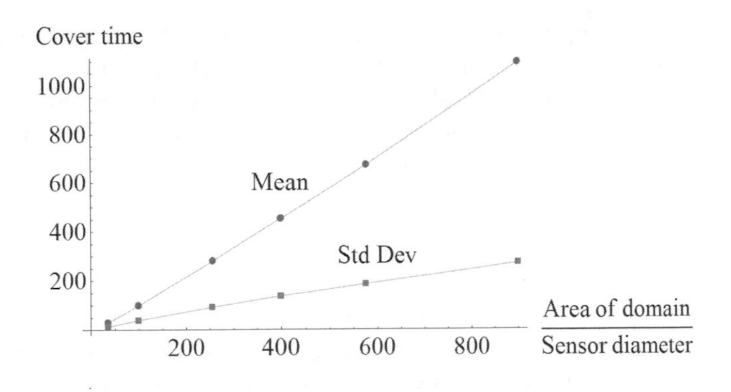

Fig. 5. Relationship between cover time and size of domain

from the estimated cover time and it is clear that the error increases linearly as the discretization error is increased.

Further simulations were conducted to investigate the variation of the mean and standard deviation of the cover time with respect to the room size. The simulation results shown in Figure 5 here were run with a constant discretization ratio of 0.01. It is clear that both the mean and the standard deviation are linearly related to the domain size. As an increase in domain size causes a significantly higher amount of computation and running time, this knowledge allows us to estimate the cover times for various sizes of domain given the knowledge of cover time properties of the smaller domain size.

6 Conclusion and Further Work

We have verified that the error caused by discretization of grid in simulating a random coverage process is linearly related to the discretization ratio ζ and show how this knowledge allows us to obtain a highly accurate result without needing to simulate using a fine grid discretization.

We have also verified from simulations that the mean and standard deviation of cover time for our given probabilistic coverage algorithm is linearly related to the relative size of the given square domain.

Note that the results shown in Section 5 are valid only for a circular sensor footprint. However, the coverage tracking algorithm described in Section 4 can easily be adapted to other shapes of realistic sensor footprints such as a circular sector. However the accuracy of the shapes when subjected to similar pixelation will vary and may no longer obey a linear relationship such as the one shown in Figure 4. By using the same technique of using varying discretization ratios and plotting the relationship and extrapolating the obtained graphs, we may still be able to eliminate any errors and obtain accurate results for cover time for various shapes of sensor footprints.

Further work includes finding the probability distribution of the cover time by using the results already obtained in the simulations. By varying the shape of the domain, as well as implementing multiple covering sensors into the algorithm, we can investigate also other effects on the cover time which can be useful for real applications. Varying methods of covering algorithms may also be used to find different more efficient coverage methods whilst maintain as little reliance on sensors and communication as possible.

References

1. Choi, Y.H., Lee, T.K., Baek, S.H., Oh, S.Y.: Online Complete Coverage Path Planning for Mobile Robots Based on Linked Spiral Paths Using Constrained Inverse Distance Transform. In: Proceedings of the 2009 IEEE/RSJ International Conference on Intelligent Robots and Systems, St. Louis, USA, pp. 5788–5793 (2009)
2. Choset, H.: Coverage for Robotics - a Survey of Recent Results. Annals of Mathematics and Artificial Intelligence 31, 433–464 (2001)
3. Gage, D.W.: Randomized Search Strategies with Imperfect Sensors. In: Proceedings of SPIE Mobile Robots VIII, vol. 2058, Boston, pp. 270–279 (1993)
4. Hazon, N., Kamika, G.A.: Redundancy, Efficiency and Robustness in Multi-Robot Coverage. In: Proceedings of the 2005 IEEE International Conference on Robotics and Automation, Barcelona, Spain, pp. 735–741 (2005)
5. Hearn, D., Pauline Baker, M.: Computer Graphics C Version. Prentice Hall, New Jersey (1997)
6. Rekleitis, I., Lee-Shue, V., New, A.P., Choset, H.: Limited Communication, Multi-Robot Team Based Coverage. In: Proceedings of the 2004 IEEE International Conference on Robotics and Automation, New Orleans, Louisiana, pp. 3462–3468 (2004)
7. Sarmiento, A., Murrieta-Cid, R., Hutchinson, S.: A Multi-robot Strategy for Rapidly Searching a Polygonal Environment. In: Lemaître, C., Reyes, C.A., González, J.A. (eds.) IBERAMIA 2004. LNCS (LNAI), vol. 3315, pp. 484–493. Springer, Heidelberg (2004)
8. Wagner, I.A., Lindenbaum, M., Bruckstein, A.M.: MAC vs PC: Determinism and Randomness as Complementary Approaches to Robotic Exploration of Continuous Unknown Domains. The International Journal of Robotics Research 19(1), 12–31 (2000)

Frontier Based Multiple Goal Search in Unknown Environments

V.R. Jisha* and D. Ghose

Guidance, Control and Decision Systems Laboratory, Department of Aerospace Engineering, Indian Institute of Science, Bangalore
{jisha,dghose}@aero.iisc.ernet.in
http://www.guidance.aero.iisc.ernet.in

Abstract. We present a frontier based algorithm for searching multiple goals in a fully unknown environment, with only information about the regions where the goals are most likely to be located. Our algorithm chooses an "active goal" from the "active goal list" generated by running a Traveling Salesman Problem (TSP) routine with the given centroid locations of the goal regions. We use the concept of "goal switching" which helps not only in reaching more number of goals in given time, but also prevents unnecessary search around the goals that are not accessible (surrounded by walls). The simulation study shows that our algorithm outperforms Multi-Heuristic LRTA* (MHLRTA*) which is a significant representative of multiple goal search approaches in an unknown environment, especially in environments with wall like obstacles.

Keywords: Frontier cell, Occupancy grid, Multiple goal search.

1 Introduction

For search and rescue missions in hazardous environments or in a situation where nuclear leakage or forest fire occurs a robot has to search for victims of the disaster or sources of nuclear spill or fire, in an unknown environment. So, in many realistic situations like these, the search space may contain multiple goals. In such situations, the exact locations of the disaster will not be known *a priori*. However, an approximate idea about the area in which the disaster might be located is likely to be known from other information sources. This is a practical scenario and is different from when there is no information about the location of the target points. The robot or the search and rescue agent starts from an initial location and reaches the locations of interest through completely unknown terrain. Thus, the problem is to find a route from a start position passing through all the goals, through an unknown intervening area which is cluttered with unknown obstacles. The robot can only sense the surrounding area within the range of its sensors.

---

* V.R. Jisha is a Lecturer in the Department of Electrical Engineering, College of Engineering, Trivandrum, Kerala, 695 016, India and is presently pursuing doctoral studies at the Guidance, Control, and Decision Systems Laboratory, Department of Aerospace Engineering, Indian Institute of Science, Bangalore, 560012, India.

P. Vadakkepat et al. (Eds.): FIRA 2010, CCIS 103, pp. 353–360, 2010.

Davidov and Markovitch [1] present a new framework for multiple goal heuristic search. In this framework the task is to collect as many goals as possible within the allocated resources. They show that the traditional distance-estimation heuristic is not sufficiently effective for multiple-goal search. Then the sum and progress heuristics are introduced, which take advantage of an explicitly given goal set to estimate the direction to larger and closer groups of goals. Unlike this, our algorithm tries to reach maximal number of goals in a completely unknown environment with minimal distance traveled.

The work described by Jarvis and Marzouqiin [2] is a probabilistic approach in which a single mobile robotic agent might discover targets in an obstacle strewn environment in a strategic manner which minimizes the average time taken to find a target. Here also the environment is initially known, and visibility regions could be calculated a priori. The search path is then planned by finding the optimal order in which different convex regions are searched.

Real-time search methods are an efficient tool for agents with limited sensing capabilities that are interacting with an initially unknown environment. Real time search methods in [3] and [4] are closest to our method in terms of the framework used. These methods restrict the planning activity to the area around the robot, executes those local actions, and then repeats the planning phase from the new location. The MHLRTA* algorithm in [4] associate to each state of the search space a vector of heuristic estimates, where each element is the heuristic estimate of the path cost from a state to the goal and it updates each heuristic estimate separately.

In this paper we describe a frontier based algorithm for successfully reaching all the goals, given the probable locations of the goal region (GR), where the goals are most likely to be located. Till date the concept of frontiers have been successfully implemented for autonomous exploration in unknown environments [5], [6]. Frontiers are regions on the boundary between explored space and unexplored space. An efficient algorithm for multiple goal search must not only reach all the goals (if paths exist), but it should be accomplished with minimal traveling distance. Our algorithm focusses on a proper choice of the frontier cell which ultimately reduces the number of moves to reach the goal, that is, it chooses a frontier cell having maximum "goal seeking index". As this algorithm is frontier based, search in unknown environment ensures avoiding repeated coverage, that is, robot will avoid going back again to an already explored region. In previous papers we have presented our work for single goal search [7], [8].

The remaining part of the paper is organized as follows: Section 2 gives the description of the problem of multiple goal search in unknown environments. Section 3 outlines the algorithm for multiple goal seeking strategy. Section 4 shows the efficacy of the approaches through simulations in environments with random obstacles and wall like obstacles. Finally, Section 5 concludes this paper.

2 Problem Definition

A robot, equipped with sensing and localization capabilities, starts searching for goals in an unknown environment. The only information available to the robot is the goal regions (GR), scattered in the environment where goals are likely to be present. But the exact location of goals are not known *a priori*. The objective is to find a path through all the goals from the start position. Once the robot reaches a GR, exploration of that region is done, till the goal is found and then moves towards the next GR. If at the start of the mission, the information about the location of GRs are not available, the robot will start its search in exploration mode and will exhaustively search the entire search space and reach all the goals if a solution exists. The discretization of the terrain, motion etc. are the same as in[7],[8].

3 Frontier Based Multiple Goal Search via Goal Switching

Here we assume that the probable locations of GRs are available to the robot at the beginning of the mission. As multiple goals are to be searched, the order in which the goals are to be searched is important and the order is decided by the "active goal list". The "active goal list" is generated by running a Genetic Algorithm (GA) based Traveling Salesman Problem (TSP) routine for a fixed number of iterations with the known locations of the centroids of the Goal Regions. This will be an optimistic approach, as the robot is not aware of the obstacles present in the environment beyond the sensing range. Even though GA based TSP routine will not yield the optimal solution, this will be good enough in such a scenario where the environment is completely unknown. As the algorithm has to run in real time choosing the exact solution will be time consuming and will not be feasible. The first goal in the an"active goal list" is selected as an "active goal".

The algorithm tries to find a series of intermediate target points for the robot from starting position till the active goal is reached. By performing a sensor scan at these points the robot can have a view of its surroundings and update the cells within the sensor range, with value Po_{xy} calculated [9], [10] from sensor measurements. These group of cells are referred to as explored cells and the collection of explored cells is known to the robot at that moment. From this set of explored cells, frontier cells which are on the boundary between explored and unexplored cells are identified. A frontier cell is an already explored cell which is an immediate neighbor of an unexplored cell. The algorithm chooses the best fit frontier cell, that is, the cell with maximum "goal seeking index", among these and a route is found from the current location to the frontier cell at each iteration. Once the robot reaches the "active goal", that goal is deleted from the "active goal list" and the next goal from the "active goal list" is chosen as active goal. The details of computing the cost of reaching a frontier cell from the current robot position, "goal seeking index" calculations and finding the optimal path from the current position of the robot to all the frontier cells at each step are explained in detail in [8].

If there are no obstacles on its way, the robot will make a move toward GR, that is, as time passes the distance between the present robot position and

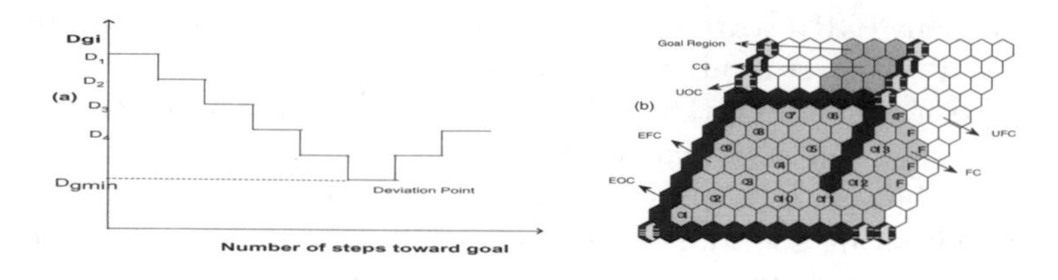

Fig. 1. (a) Distance D_{gi} between robot and CG of goal region as the robot moves toward the goal (only the GR corresponding to active goal shown) and deviation point is the move at which obstacle cells are encountered, preventing motion directly toward the goal (b) D_{gi} decreases from position 1 to 6 and shows an increasing trend from there onwards (**EOC**-Explored obstacle cell, **UOC**-Unexplored obstacle cell, **EFC**-Explored free cell, **UFC**-Unexplored free cell, **FC**-Frontier cell, **GR**-Goal Region)

centroid (CG) of goal region decreases, until it reaches the GR. We refer to the distance between robot position and CG of G_i (i^th goal) as D_{gi}. But as it encounters an obstacle, D_{gi} will start increasing as shown in Fig.1(a). The minimum distance achieved in this process is referred to as D_{gmin}. This is illustrated in Fig. 1(b). The robot starts from position 1 and the distance decreases till it reaches position 6. From there onwards D_{gi} keeps increasing because the robot encountered a series of obstacle cells like a wall at that point. The distance D_{gi} at position 6 is referred to as D_{gmin}. If obstacles are encountered in such a way that current robot position is deviated away after D_{gmin} is reached "goal switching" is done. If after reaching D_{gmin}, the robot is deviating away (D_{gi} is increasing) from the "active goal" for more than a fixed number of steps and while deviating away (because of wall like obstacles), if the robot identifies another goal nearer to it than the present active goal, then TSP routine is run again. A new "active goal list" is created and a new "active goal"is selected. This is referred to as "goal switching".

As the environment is completely unknown, "goal switching" helps in reaching a maximum number of goals in a given time. In free environments and in environments with random obstacles, "goal switching" may not happen. But when wall like obstacles are encountered which will prevent the robot motion directly toward GR, "goal switching" will direct the robot to search for a new"active goal". This is an added advantage especially when inaccessible goals are present (goal surrounded by walls) in the environment. Instead of searching around these inaccessible goals, the robot will try to reach other goals because of "goal switching". Hence more number of goals can be reached for given distance traveled.

We use the following notation to describe the algorithm. GR_1, $GR_2 \dots GR_n$ is the set of goal regions, C_{gi} denotes the centroid of the ith GR, N_g denotes the number of goals to be searched, N_f denotes the number of frontier cells at the current step. Given the coordinates of the centroids, running a TSP routine with fixed starting location yields active goal list G_1, $G_2 \dots G_n$. A_g denotes the active goal, the goal which is to be searched from current position, D_{gi} denotes the distance from the current robot position to the CG of G_i.

Alg. 1. Frontier based Multiple Goal Search via Goal Switching

Run TSP routine with starting position as the first location

While $N_g \neq 0$ repeat the following steps:

1. Select the active goal from the yielded Active Goal List
2. Identify the frontier cells from the current position. If $N_f = 0$, go to step 9
3. Evaluate goal seeking index of each candidate cell
4. Make a move towards the cell having maximum goal seeking index
5. If goal is reached, remove the goal from active goal list. Else check whether present $D_{gi} <$previous D_{gi}. If yes go step 2 else to step 7
6. $N_g = N_g - 1$. Go to step 1
7. If D_{gi} is greater than D_{gmin} and any $D_{gj} < D_{gi}$, then Run TSP routine again else go to step 2
8. Go to step 1
9. Reachable frontier cells exhausted, remaining goals cannot be reached

As the environment is completely unknown, the algorithm cannot claim optimality. The total distance traveled to reach all the goals can be reduced considerably by this algorithm over the time taken by an exploratory search. Our algorithm efficiently chooses the best candidate frontier cell, in an attempt to find a route through all the goals in minimal number of moves.

Completeness of the algorithm. In a bounded search space in which there exists a path from every free cell to all the goal cells, the algorithm will find a solution. Assume the converse, that there exists a path through all the goal states, but that the algorithm will never find it. This may happen only when the robot is always moving in a cyclic path. This will never happen, as the robot's target points are frontier cells. As the frontier cells are on the boundary between explored space and unexplored space, the robot will try to move toward the unexplored space, rather than moving toward already explored region. So if a path exists, then the frontier cells will be accessible and the algorithm will be able to find the path.

4 Simulation Results

The proposed algorithm has been tested through a series of simulation experiments upon environment maps and compared to the multiple goal search algorithm proposed in [4] MHLRTA* which is a significant representative of multiple goal search approaches in an unknown environment.

Fig. 2 shows four different simulation environments (the set of figures to the left show the final paths of proposed algorithm and the set of figures to the right show the final paths of MHLRTA* algorithm in same environments). The one in Fig. 2(a) is a free environment and the one in Fig. 2c is an environment with

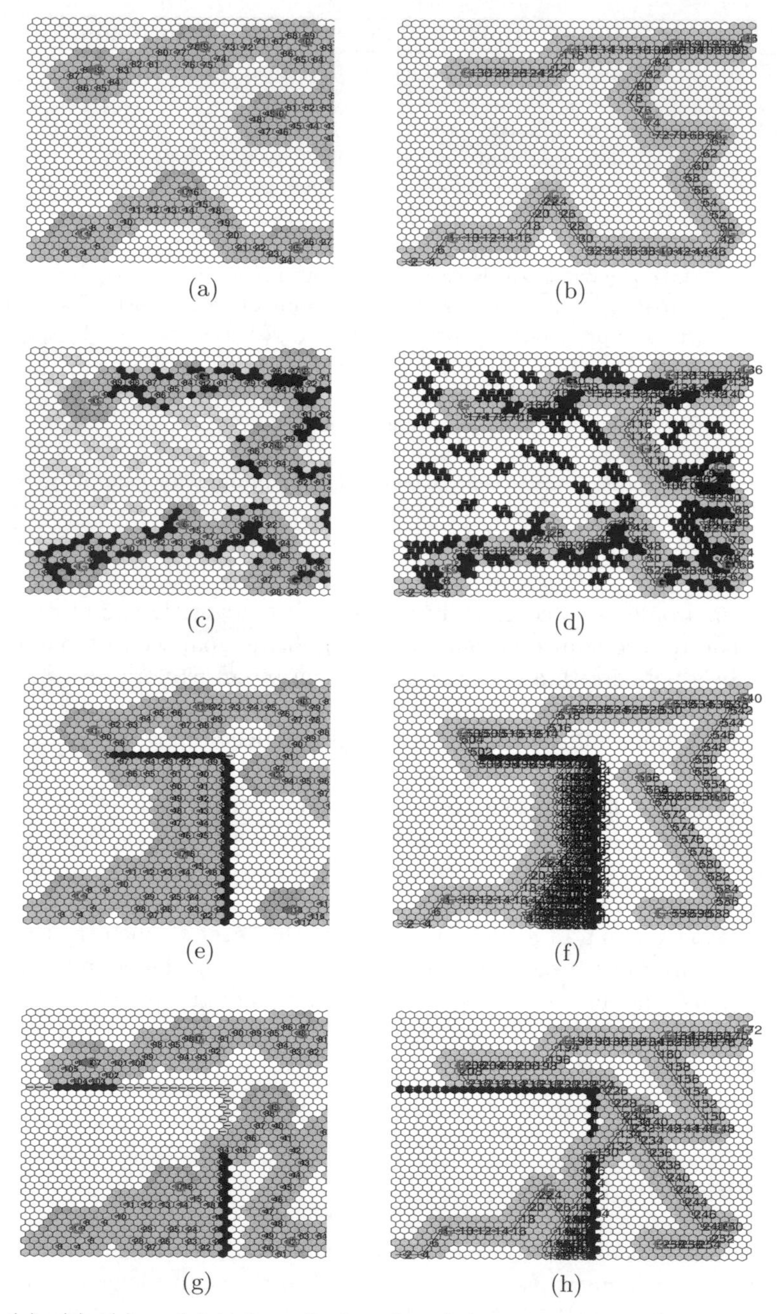

Fig. 2. (a), (c), (e) and (g) show final paths of the proposed algorithm in free env, env with random obstacles, first env with wall like obstacles and second env with wall like obstacles. (b), (d), (f) and (h) show final paths of the MHLRTA* in corresponding environments. Explored obstacle cells are indicated with black, unexplored obstacle cells are hatched, explored free cells with yellow, unexplored free cells with white and goal region with green.

Table 1. Performance comparison in terms of the number of steps to reach the goal, between proposed algorithm and MHLRTA* in different environments

Free Environment		Random Obstacles		Wall like Obstacles	
Prop. Algorithm	MHLRTA*	Prop. Algorithm	MHLRTA*	Prop. Algorithm	MHLRTA*
94	122	170	176	188	345
106	148	195	300	210	594
164	163	145	145	204	258
94	95	210	325	230	580
86	95	125	143	250	450

random obstacles having 22% occupancy. All maps represent 40×40 cell environment. All these maps are internally represented, as occupancy grids. Each cell is assumed to have size greater than the robots size so that it can move to any its unoccupied neighbors. The start position of the robot is same in all the experiments. But the obstacles are placed randomly, in the environment with random obstacles and are kept like a wall, in the environment with wall like obstacles. The robot knows its own and all goal regions coordinates. Table 1 shows the performance comparison between proposed algorithm and MHLRTA*, in terms of the number of steps to reach the goal. For comparison in free environment, five different goal configurations are chosen. For comparison in environments with random obstacles as well as in environments with wall like obstacles, five different obstacle configurations are chosen with the same set of goal configurations.

The robot is moving toward each active goal with the free space assumption and is showing near optimal behavior in free environments and in the environments with random obstacles. In environments with wall like obstacles also, it is capable of reaching all the goals if a path exists. But as the environment is completely unknown, it is not possible to find an optimal path. As the robot's target point at each step is a frontier cell, it is efficiently reaching all the goals avoiding repeated visits to already visited cell, which will reduce the total number of steps to reach the goals in a completely unknown environment. The performance of MHLRTA*[4] in the same set of environments is shown in Fig. 2(b),(d),(f),(h). MHLRTA* associate to each state x of the search space a vector of heuristic estimates $(h0, h1, \ldots h^{|G|-1})$ where each element h_i is the heuristic estimate of the path cost from x to goal g_i. The initial heuristic estimates are defined as $h_0^i(x) = d(x, g)$. It updates heuristic vector of the state x as, $h_x \leftarrow \min_{y \in Suc(x)}[k(x, y) + hi_(y)], 0 \leq i < G$ where $Suc(x)$ denotes the set of successors. The agent chooses the successor state y as $y \in \text{argmin}[k(x, z) + \min_{0 \leq i < |G|}[h^i(z)]$. Our algorithm outperforms MHLRTA* in terms of the total number of steps to reach the goals (Table1), mainly in environments with wall like obstacles, because of the proper choice of frontier cells when a wall is encountered. This is because frontier based approach will not choose target points in the already explored region and will be able to come out faster from local minima present in environments with wall like obstacles. But in MHLRTA* the heuristics update of the corresponding states are done till it comes out from local minima.

5 Conclusions

The basic idea behind frontier based multiple goal search algorithm is to find a series of intermediate target points for the robot from the starting position through all the goals scattered in a completely unknown environment. The algorithm is capable of performing goal seeking as well as exploration, so can be used even when exact locations of goals are not available. The performance of the algorithm in free environments as well as environment with random obstacles will be nearly optimal, as the order in which goals are to be searched is decided by the TSP routine, which other real time heuristic search algorithms used in completely unknown environment cannot achieve. The total distance traveled to reach all the goals in environments with wall like obstacles is considerably less compared to one of the state of the art real time heuristic search algorithm MHLRTA*, because it is capable of exiting from local minima (trap situations) present in these environments.

References

1. Davidov, D., Markovitch, S.: Multiple-goal heuristic search. Journal of Artificial Intelligence Research 26, 417–451 (2006)
2. Jarvis, R., Marzouqi, M.: Probabilistic target search strategy. In: Studies in Computational Intelligence. Autonomous Robots and Agents, vol. 76, pp. 17–23. Springer, Heidelberg (2007)
3. Korf, R.E.: Real time heuristic search. Artificial Intelligence 42(2), 189–211 (1990)
4. Kerrache, S., Drias, H.: Extending real-time heuristic search part I: Dynamically-changing goal sets. Multiagent and Grid systems An International Journal 2, 289–298 (2006)
5. Yamauchi, B.: Frontier based approach for autonoumous exploration. In: Proc. of the IEEE Int. Symposium on Computational Intelligence in Robotics and Automation, pp. 146–151 (1997)
6. Burgard, W., Moors, M., Stachniss, C., Schneider, F.E.: Coordinated multi-mobot exploration. IEEE Trans. on Robotics 21(3), 376–386 (2005)
7. Jisha, V.R., Ghose, D.: Goal directed route planning for robots in unknown environments. In: Proc. of the Int. Conference and Exhibition on Aerospace Engineering, Bangalore, India, May 19-22, pp. 1309–1318 (2009)
8. Jisha, V.R., Ghose, D.: Goal Seeking for robots in unknown environments. To Appear in the Proc. of IEEE/RSJ Int. Conference on Intelligent Robots and Systems (IROS 2010), Taipei, Taiwan (2010)
9. Moravec, H., Elfes, A.: High resolution maps from wide angle sonar. In: Proc. of IEEE Int. Conference on Robotics and Automation, vol. 2, pp. 116–121 (1985)
10. Murphy, R.: Introduction to AI Robotics, pp. 376–434 (2005)

Multiple Sensor Based Autonomous Monitoring and Control for Energy Efficiency

Liyanage C. De Silva[1,2], Titty Dewana[1], M. Iskandar Petra[1], and G. Amal Punchihewa[2]

[1] Faculty of Science, University of Brunei Darussalam, Brunei Darussalam
liyanagecd@yahoo.co.nz, merce5964@yahoo.com
[2] School of Engineering and Advanced Technology (SEAT), Massey University, Palmerston North, New Zealand
L.desilva@massey.ac.nz, g.a.punchihewa@massey.ac.nz

Abstract. Recently the importance of autonomous monitoring has become an essential part of the society. Autonomous monitoring and control in a home environment can lead to energy savings. Also autonomous monitoring can serve as an alerting mechanism when something is out of the norm so that the controlling mechanism can rectify the problem. In this paper we present a system we developed using multiple sensors for autonomous monitoring and control with relatively low cost components. The sensors we integrated together in to one central operation unit can monitor room temperature, humidity, human motion activity etc. We present a prototype implementation of an energy efficient home environment using our proposed centralized monitoring platform. Using a sample scenario we show that the energy consumption in a smart home is about 15kWh while that of a normal home is about 35kWh.

Keywords: Multiple Sensors, Energy Efficiency, Autonomous Monitoring, Smart Homes, Sensor Integration, Environment Monitoring.

1 Introduction

Use of multiple sensors for autonomous monitoring and control is applicable in a wide area including the environment monitoring, health monitoring, commerce and industry. In fact many sensors are used in our home environment such as inside refrigerators to control inner temperature, in the water pump assembly to control the water flow etc. However one should note that the number of sensors used in a motor car is much more than one might see in their own home. This leads us to the thinking of a Smart Home with multiple sensors to monitor and control the home environment.

In autonomous monitoring and control of a home environment one may employ many sensors to continuously monitor the activities in the home. There are various techniques available which can be used in smart homes such as temperature and humidity measurement and light intensity measurement etc.

Due to the ever increasing energy price and scarcity of the resources the cost of energy bills plays an important role in our household expenditure. Hence we need to find a way of constantly monitor the usage and reduce the waste. Energy is used to operate most of the appliances such as cooling, heating, lighting, and many more.

P. Vadakkepat et al. (Eds.): FIRA 2010, CCIS 103, pp. 361–368, 2010.

There are many ways to reduce the energy consumption they use. One of the best ways is to use sensor technology which helps to monitor and control the energy usage as well as make the user aware of the usage using display technologies or graphical user interfaces (which will help the user to observe the energy usage and control) which in turn persuades them to change the pattern of their energy usage.

2 Related Research

Autonomous monitoring and control has been used in many research works for monitoring environmental conditions using smart sensors. The research work presented by De Silva et al. 1 is a technique based on smart monitoring, control and communication to improve energy efficiency and provide eldercare. In their research, they also looked into the use of multiple of different sensors to monitor activities in a smart home and the result of the monitoring was used to control the home environment to reduce energy usage and provide eldercare. The application of the smart monitoring to a prototype system of audio based sensor was also presented in the paper.

Chao Chen et al. 2 proposed a monitoring system to monitor human movements at a home environment using wearable wireless sensors. This project designed and developed a wearable wireless sensor system called a waist-mounted tri-axial accelerometer unit to obtain data concerning physical activities of a person in need of medical care. The unit was used to record human movements. And the Sampled data are transmitted using an IEEE 802.15.4 wireless transceiver to a data logger unit and passed to a PC for analysis.

In the paper by Cameron Lach et al. 3 the authors proposed a graphical user interface to monitor and control home appliances to reduce energy usage. They have presented data from their prototype implementation.

Diane J. Cook et al. 4 proposed a technique of monitoring environmental conditions in a home called MavHome (Managing An Intelligent Versatile Home). The aim of the MavHome project was to create a home that acts as an intelligent agent. In their paper, they introduced architecture for an intelligent home and demonstrated the effectiveness of some monitoring and controlling algorithms on observed smart home data.

In the papers 5 and 6 the authors describe human activity detection techniques in home environments. This information can be used for home automation to control the level of lighting used at a given time.

In the literature review we looked at many research work that are being carried out to provide better insight and awareness towards the methods of increasing the quality of the daily life at the same time addressing the energy efficiency. In this paper we propose a centralized multiple sensor based platform for environment monitoring to reduce the energy usage in a typical home.

3 Autonomous Monitoring for Energy Efficiency

In a given home there are many appliances. But all of them run independently and at present no effort is being used to centralize the monitoring of their operation. Appliances such as air conditioning systems, heating systems require a large amount of energy to run but they tend to operate irrespective of the activities of the householders. This makes the homes one of the environments that waste precious energy. Hence the design of smart home environments using sensor network has a great potential to reduce the energy consumption by actively monitoring the household activities and controlling the related appliances.

If you have an active sensor network at home they can monitor and automatically control the appliances. For example the motion sensors can sense human presence and it can allow keeping the appliances for example television, air conditioner running as desired. Then when the sensor does not detect movement for long time within the detection area, it will start to power down and reduce energy wastage. Also motion sensors can help in saving energy by switching off the lights if no one passes by the sensor or detection zone.

Another advantage of a centralized system is that it is a way to keep occupants aware of their energy efficiency. Laptop screen or mobile phone could set up in a home to track electricity use, water use and other metrics and fed into an online interface or text messages can be sent via mobile phones. The occupants will get updates on their home's energy usage, children's and elderly conditions at any time. Hence if you have a sensor driven smart home people are constantly made aware of the energy they're using and are more conscious about their affect on the environment.

4 Motion and Temperature Sensor Based Air-Conditioner Control

Among the home appliances air conditioners are the equipments those contribute to high energy consumptions. In our platform the air conditioner system was designed and programmed in such a way that it can maintain temperatures given to each room in a house and ablc to improve the energy efficiency by monitoring the availability of occupants in each room by using distance measuring sensors (PING sensors 7) in addition to temperature sensors. The controlling steps of the air conditioner are given in the flow chart in the Figure 1.

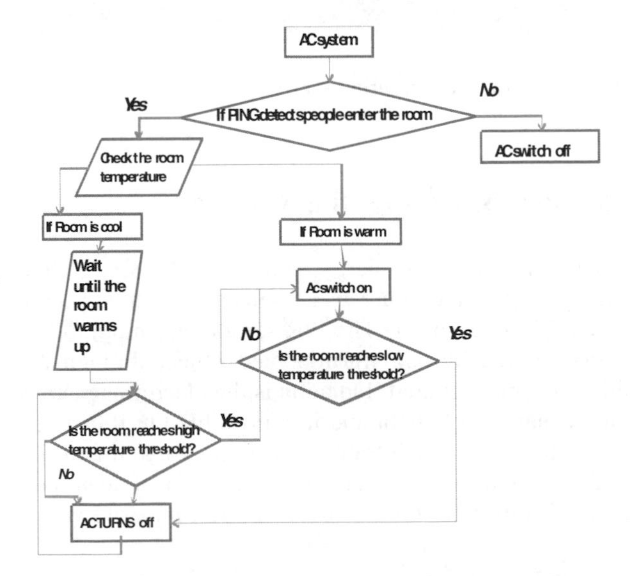

Fig. 1. Motion sensor and temperature sensor based air-conditioner control

Air conditioning is the process of maintaining comfortable conditions inside a closed space. In country like Brunei Darussalam, the surrounding temperature is higher and hence air-conditioners act as a cooler to keep the room cool. Figures 2 and 3 show our prototype implementation of a distance measuring sensor and temperature sensor based air-conditioner control arrangement.

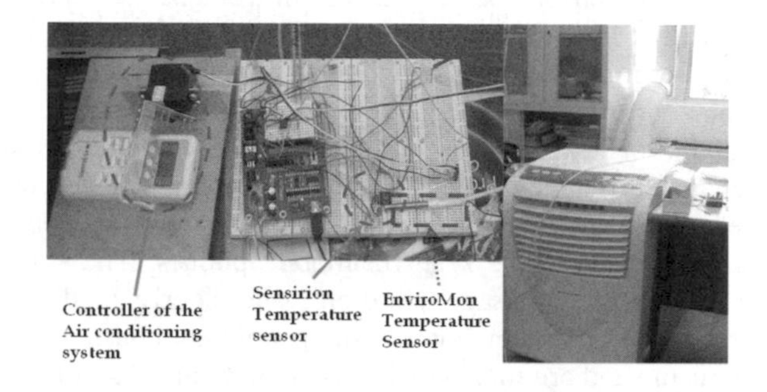

Fig. 2. The temperature sensing and controlling circuits of an Air-conditioner operated by a pair of distance measuring sensors

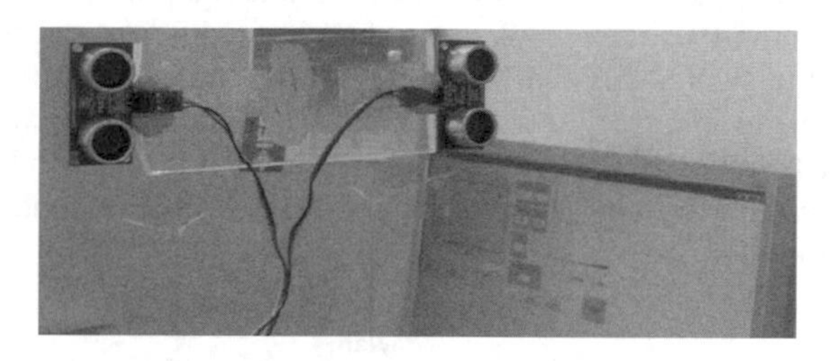

Fig. 3. Two PING type distance measuring sensors to detect enter and exit of a person to the Smart Home

5 Motion Sensors Based Lighting Control

Although lighting consumes less energy, similar to the air-conditioning control we can control the lighting in the house using sensors. For this purpose we have used Passive Infra-Red (PIR) Sensors 7. The PIR sensors are motion detectors that senses infrared radiation emitted by any warm-bodied object through a Fresnel lens and infrared-sensitive element. Infrared Radiation is the electromagnetic spectrum that has wavelength longer than visible light and it is invisible but it can be detected. Human body and animals that generate infrared radiation is maximum at a wavelength of 9.4μm. Infrared radiation will not pass through types of material such as window glass and plastic. However it will penetrate through material with some attenuation that is opaque to visible light e.g Germanium and Silicon. We used PIR sensors to switch on lights when there are off movement in the room and switch on the lights when someone enters.

6 Normal and Smart Home Scenarios for Energy Consumption

After setting up various sensors we then try to calculate the energy consumption in a normal home and smart home. We have assumed the following two scenarios for the comparison of the energy consumption in a normal home and a smart home.

Table 1. Scenario in a Normal Home Environment

Time	**Scenarios in a Normal Home Environment with two persons**
7 am to 5pm	Lights at foyer/garden, Living room , Room 1 and Room 2 are switched off
5 pm to 6 am	Lights at foyer/garden are switched on (Night time)
5.30 pm to 9 pm	Lights in the Living room are switched on. AC system is also switched on. During this hours person A and B may move back and forth from living room and their rooms .Lights and the AC are not switched off.
9 pm to7 am	Person A and B are in their own rooms and lights in living rooms are switched off. The AC is also switched off.

Table 2. Scenario in a Smart Home Environment

Time	**Scenarios in a Smart Home Environment with two persons**
7 am to 5pm	Lights at foyer/garden, Living room, Room 1 and Room 2 are switched off.
5pm to 6am	Lights at foyer/garden are switched on when dusk is approaching and go off at Dawn time, (around 5.30 am) using photo sensors. • At 5.30 am, lights in rooms 1 and room 2 also switched off when approaching Dawn time. • If outside the room is bright then the light inside the room will automatically match bright to the outdoor lighting. These are controlled by photo sensors.
5.30 pm to 9 pm	Lights in the Living room are switched on within this hours when person A in the room. AC system is also switched on. • However, at 6.30 pm to 7.30 pm, person A leaves the living room and Lights are automatically switched off. Person B also not in his room (room2), thus no energy from lighting system is being used. These are controlled by motion sensors.
9 pm onwards	Lights in the living room are switched off when person A and B are not in the room. AC system is also switched off. These are controlled by motion sensors.
10.30 pm	Person B went out of his room 1 for a while, lights in the room are automatically switched off. When Person B back to his room, then lights are on again. These are controlled by motion sensors.

After assuming the above two scenarios we have obtained the results shown in the Figure 4 below.

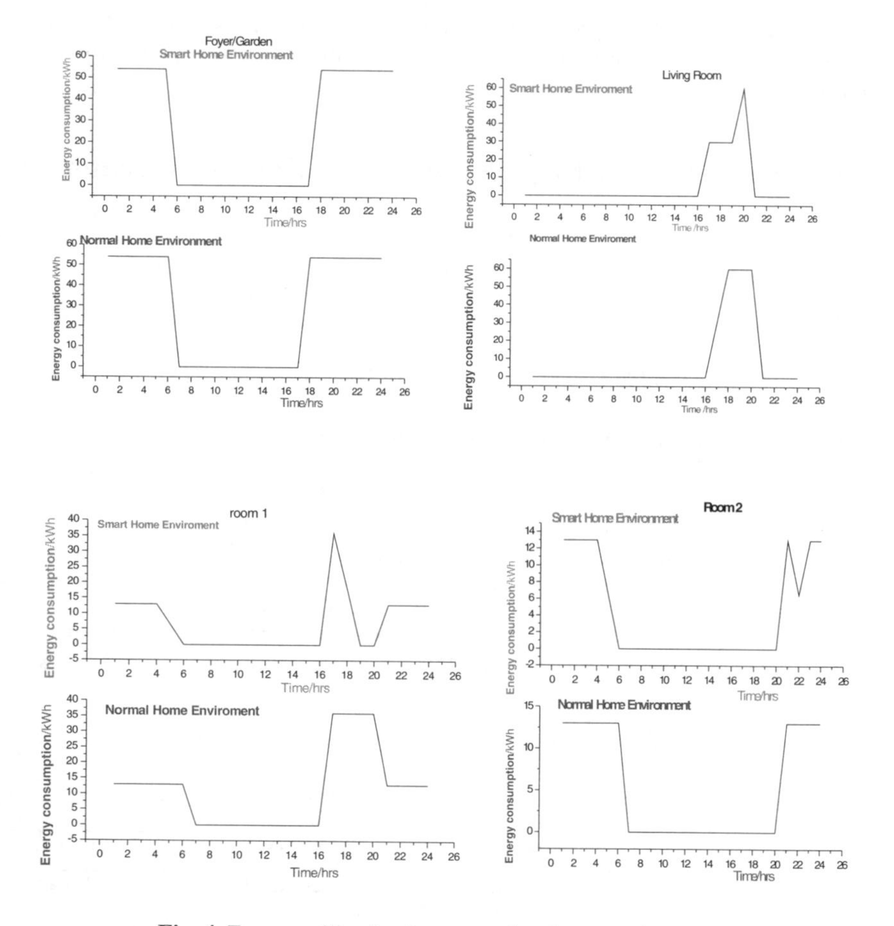

Fig. 4. Energy utilization in a normal and a smart home

Then we have calculated the total energy consumed in the normal home and in the smart home. They were 35kWh and 15 kWh per day respectively. This shows that about 20kWh energy savings per day using smart home technologies. Note that in our calculation we did not take some appliances such as rice cookers, microwaves into account as the energy consumption of these appliances are same in the smart home and the normal home.

7 Graphical User Interface

We have developed a GUI (Graphical User Interface) which can monitor and display the total energy consumption in the smart home (Figure 5). This an improved version of the GUI presented in 8.

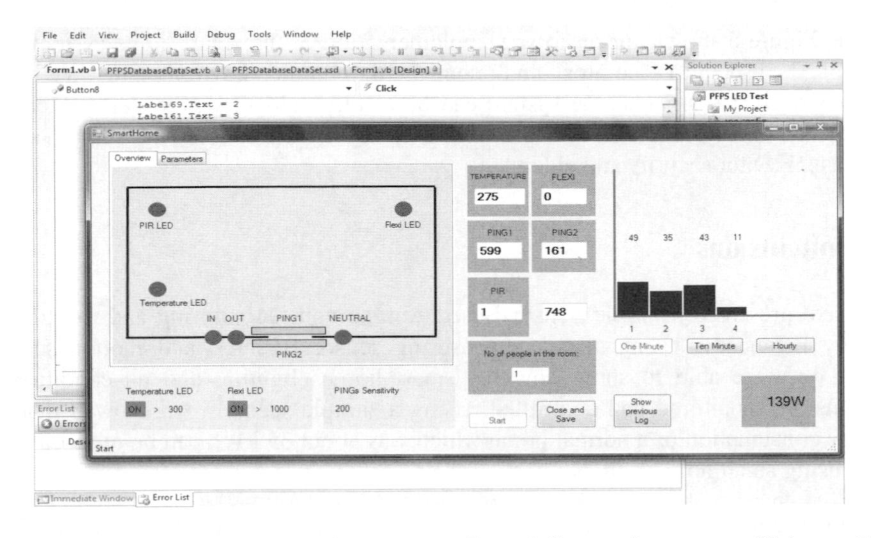

Fig. 5. GUI of the Smart Home Autonomous Control System for energy efficiency. This diagram shows the increase and decrease in energy level as the integrated sensors controlling the System.

The smart sensors can also be used to keep occupants aware of their total energy usage. Apart from the status of various sensors this interface provides an on screen visual display of current energy usage of the home. Currently this interface is running on the local machine only but in the future it can be coupled to the internet for remote monitoring of the house. This will be an ideal technique for a large industrial plant monitoring application.

Using this system one could track the family electricity usage (room by room) so that occupants of the house will be able to get a better feedback of their energy usage. In the future we can extend this to show the water etc. coupled with an online interface. When the usage of energy exceeds the normal usage then the system will alert the householder. The householder will get updates on their home's energy efficiency at anytime and anywhere. This way, people are constantly made aware of the energy they're using and are more conscious about their role in the environment and the energy conservation.

Fig. 6. The Prototype Smart Home System with Multiple Sensors

The Figure 6 shows our prototype implementation of the smart home system with multiple sensors to monitor and control the home environment. We have demonstrated that smart home system can be a fully autonomous system which can support the energy efficiency in residential and industrial premises with some features such as RFID for security and eldercare.

8 Conclusions

We have presented Multiple Sensor Based Autonomous Monitoring and Control for Energy Efficiency. Using distance measuring sensors (PING) and motion sensors (PIR) we were able to show how the smart homes lighting and air-conditioning systems are monitored and controlled. Using a sample scenario we showed that the energy consumption of a normal home which was about 35 kWh can be reduced to 15 kWh using smart home technologies.

Acknowledgements. Authors would like to thank Mohammad Auzi Bin Ahadani, Mohammad Solehin Bin Hj Masri and Arulanantha Pushparaja Rejikanth for their contributions towards the implementation of the prototype system.

References

1. De Silva, L.C., Iskandar Petra, M., Amal Punchihewa, G.: Ambient Intelligence in a Smart Home for Energy Efficiency and Eldercare, vol. 44, p. 187. Springer, Heidelberg (2009) ISBN 978-3-642-03985-0
2. Chen, C., Pomalaza-Ráez, C.: Monitoring Human Movements at Home Using Wearable Wireless Sensors (2010), http://www.ismict2009.org/pdf/1D06.pdf (retrieved on May 20, 2010)
3. Lach, C., Punchihewa, A., De Silva, L.C., Mercer, K.: Smart Home System Operating Remotely Via 802.11b/g Wireless Technology. Published in the Proceedings of the 4th International Conference Computational Intelligence and Robotics and Autonomous Systems (CIRAS 2007), Held in Palmerston North, New Zealand, November 28-30 (2007)
4. Cook, D.J., Youngblood, M., Heierman III, E.O., Gopalratnam, K., Rao, S., Litvin, A., Khawaja, F.: MavHome: An Agent-Based Smart Home. In: First IEEE International Conference on Pervasive Computing and Comm (PerCom 2003), p. 521 (2003) 0-7695-1893-1
5. Henry, T.C.C., Ruwan Janapriya, E.G., De Silva, L.C.: An Automatic System for Multiple Human Tracking and Action Recognition in Office Environment. Published in the Proceedings of IEEE International Conference on Acoustics, Speech and Signal Processing (ICASSP 2003), Hong Kong, April 6-10, IMSP-L2.6 (2003)
6. Chi, N.P., De Silva, L.C.: Head Gesture Recognition. Published in Proceedings of IEEE-International Conference on Image Processing (ICIP 2001), Thessaloniki, Greece, vol. II, pp. 266–269 (October 2001)
7. Parallax web page, http://www.parallax.com/ (retrieved on May 20, 2010)
8. De Silva, L.C.: Audiovisual Sensing of Human Movements for Home-care and Security in a Smart Environment. Published in the International Journal on Smart Sensing and Intelligent Systems 1(1), 220–245 (2008)

Author Index

Printing: Mercedes-Druck, Berlin
Binding: Stein + Lehmann, Berlin